New Frontiers in Signal Processing

New Frontiers in Signal Processing

Edited by Amelia Turner

CLANRYE
INTERNATIONAL
www.clanryeinternational.com

Clanrye International,
750 Third Avenue, 9th Floor,
New York, NY 10017, USA

ISBN: 978-1-64726-127-6

Cataloging-in-Publication Data

New frontiers in signal processing / edited by Amelia Turner.
p. cm.
Includes bibliographical references and index.
ISBN 978-1-64726-127-6
1. Signal processing. 2. Signal theory (Telecommunication). 3. Information measurement. I. Turner, Amelia.
TK5102.9 .N49 2022
621.382 2--dc23

For information on all Clanrye International publications
visit our website at www.clanryeinternational.com

CLANRYE INTERNATIONAL

Contents

Preface

Signal processing is the field that is concerned with the analysis, modification and synthesis of signals that represent information about the behavior or characteristics of some phenomenon. It is a subfield of information engineering, electrical engineering, and mathematics. Signal processing techniques serve to improve signal transmission fidelity, subjective quality, storage efficiency, and detect or emphasize signal characteristics that are of particular interest. Signal processing can be of different kinds based on its application, such as analog signal processing, discrete-time signal processing, continuous-time signal processing, digital signal processing, etc. Such techniques are useful in image and video processing, wireless communication, process control and audio signal processing, besides several other important applications. Different approaches, evaluations, methodologies and advanced studies on signal processing have been included in this book. There has been rapid progress in this field and its applications are finding their way across multiple industries. With state-of-the-art inputs by acclaimed experts of this field, this book targets students and professionals.

This book has been the outcome of endless efforts put in by authors and researchers on various issues and topics within the field. The book is a comprehensive collection of significant researches that are addressed in a variety of chapters. It will surely enhance the knowledge of the field among readers across the globe.

It gives us an immense pleasure to thank our researchers and authors for their efforts to submit their piece of writing before the deadlines. Finally in the end, I would like to thank my family and colleagues who have been a great source of inspiration and support.

Editor

1

Comparative Study between the Discrete-Frequency Kalman Filtering and the Discrete-Time Kalman Filtering with Application in Noise Reduction in Speech Signals

Leandro Aureliano da Silva, [1] **Gilberto Arantes Carrijo**, [1] **Eduardo Silva Vasconcelos**, [1] **Roberto Duarte Campos**, [2] **Cleiton Silvano Goulart**, [2] **and Rodrigo Pinto Lemos** [3]

[1] *Department of Electrical Engineering, Universidade Federal de Uberlândia, Av. João Naves de Ávila, 2160 Bloco 3N, Campus Santa Mônica, Uberlândia, MG, Brazil*
[2] *Department of Electrical Engineering, Faculdade de Talentos Humanos, R. Manoel Gonçalves de Rezende, 230 São Cristóvão, Uberaba, MG, Brazil*
[3] *Department of Electrical Engineering, Universidade Federal de Goiás, Av. Esperança, s/n. Campus Universitário, Goiânia, GO, Brazil*

Correspondence should be addressed to Leandro Aureliano da Silva; leandro_aureliano@yahoo.com.br

Academic Editor: Shunyi Zhao

This article aims to carry out a comparative study between discrete-time and discrete-frequency Kalman filters. In order to assess the performance of both methods for speech reconstruction, we measured the output segmental signal-to-noise ratio and the Itakura-Saito distance provided by each algorithm over 25 different voice signals. The results show that although the two algorithms performed very similarly regarding noise reduction, the discrete-time Kalman filter produced smaller spectral distortion on the estimated signals when compared with the discrete-frequency Kalman filter.

1. Introduction

Even with the advent of the Internet, voice transmission is still one of the most important ways of communication. The quality and intelligibility of speech signals play a key role in the ease and precision during information exchange. Practically in almost all voice transmission applications, the quality can be affected by factors such as ambient noise, losses due to digital link encoding, and interference from other conversations or even from other signal sources [1].

In order to overcome their harmful effects, digital speech processing techniques can be employed to reduce or even eliminate them. In recent years, some techniques and methods such as spectral subtraction, Kalman filtering, psychoacoustics, and wavelet transforms gained more prominence, especially in noise reduction, so that many research efforts have been made for improving them.

In [2, 3], the authors enhance speech quality by removing the musical noise introduced by spectral subtraction. In [1], the authors combined spectral subtraction and wavelets on a prefiltering approach for noise reduction in speech signals and used the result as an initial guess for a Kalman filter. When compared to Kalman filtering using only wavelets or spectral subtraction alone to produce the initial guess, their method showed the least spectral distortion and a similar segmental output signal-to-noise ratio.

Since wavelet-based denoising is highly dependent on thresholding the approximation and detail coefficients, recent research in this area focuses on new thresholds [4, 5].

Shao and Chang [6] concatenated the Kalman filter to a bank of wavelet filters with a perceptual weighting filter. They used a technique of masking the psychoacoustic model to derive the weighting filter. According to the authors, that work brought two contributions. The first one was the wavelet-based auditory model with a perceptual wavelet filter bank that maps the frequency response of the human auditory system through subband decomposition. The second was the Kalman filter using a voice state space model in the wavelet domain, whose computational cost was reduced when compared to the discrete-time Kalman filter. They were able to reduce the noise in different environments with low signal degradation.

Dhivya and Justin [7] proposed a noise reduction based technique that applies spectral subtraction to the wavelet approximation coefficients and soft thresholding to the detail coefficients. They used five wavelet filters and compared them according to their output signal-to-noise ratios. Besides the output SNR, they also considered the correlation coefficient and the perceptual evolution of speech quality (PESQ) criteria.

However, although these algorithms show significant advances in noise removal, most of them do not evaluate spectral distortion nor do they attempt to minimize it. So, since the method in [1] provided low spectral distortion, this article proposes a comparative study between discrete-time and discrete-frequency Kalman filters simply using the noisy signal as initial estimate. According to Fujimoto and Ariki [8], the main difference between the two approaches is that the operation of the Kalman filter is more computationally efficient in the frequency domain than in the time domain.

On the other hand, transforming the set of Kalman filter equations to/from the frequency domain produces a significant distortion in the estimated signal. Then, we used prefiltering based on spectral subtraction to reduce this distortion. In order to assess the performance of the proposed algorithms, we measured both the segmental signal-to-noise ratio of the outputs and the Itakura-Saito distance.

This article is structured as follows: Sections 2 and 3 describe the discrete-time and discrete-frequency Kalman filtering algorithms, respectively. Section 4 brings the experimental results and finally, in Section 5, the conclusions are presented.

2. Discrete-Time Kalman Filtering (DTKF)

In the 1960s, Rudolf Emil Kalman published the paper "A New Approach to Linear Filtering and Prediction Problems", describing a recursive solution to the discrete-time linear filtering problem [1]. Since then, due to the major advances of digital computing, Kalman filtering has become a very important technique in several areas such as navigation, monitoring processes, economics, and signal reconstruction from noisy samples.

In this article, the Kalman filtering development follows the heuristics described by Vaseghi [9]. Thus, the speech signal is modeled as an autoregressive process of order P, AR(P), according to

$$x(n) = \sum_{k=1}^{P} a_p(k)\, x(n-k) + w(n) \tag{1}$$

where $a_p(k)$ are the linear prediction coefficients of order P, $w(n)$ is the prediction error associated with the excitation of the source-filter model of speech production, and $x(n)$ is the nth sample of the speech signal.

It can be observed that, in the acquisition process of audio and speech signals, most of the signals are captured in the presence of some type of additive noise. Consequently, we can model the noisy signal as shown in

$$y(n) = x(n) + v(n) \tag{2}$$

where $y(n)$ is the noisy speech signal and $v(n)$ is a white Gaussian additive noise.

From (1) and (2), we can set up a state space model described by (3) and (4), respectively [9]:

$$\mathbf{x}(n) = \mathbf{A}(n-1)\,\mathbf{x}(n-1) + \mathbf{w}(n) \tag{3}$$

$$\mathbf{y}(n) = \mathbf{H}(n)\,\mathbf{x}(n) + \mathbf{v}(n) \tag{4}$$

where $\mathbf{x}(n)$ is the $P\times 1$ state vector at time n; $\mathbf{A}(n-1)$ is the state transition matrix with dimensions $P \times P$ that relates current time n with past time $(n-1)$; $\mathbf{w}(n)$ is the $P\times 1$ input vector of the state equation and it is modeled as a white noise; $\mathbf{y}(n)$ is the $M \times 1$ observation vector; $\mathbf{H}(n)$ is the channel distortion matrix of dimensions $M \times P$; and $\mathbf{v}(n)$ is an $M \times 1$ additive white noise vector [9].

According to Vaseghi [9], $\mathbf{w}(n)$ and $\mathbf{v}(n)$ are assumed to be independent white noise processes so that

$$E\left[\mathbf{v}(n)\,\mathbf{v}^{\mathrm{T}}(k)\right] = \mathbf{R}(n) = \begin{cases} r(n)\,\mathbf{I}_{M\times M} & k = n \\ 0 & k \neq n \end{cases} \tag{5}$$

$$E\left[\mathbf{w}(n)\,\mathbf{w}^{\mathrm{T}}(k)\right] = \mathbf{Q}(n) = \begin{cases} q(n)\,\mathbf{I}_{P\times P} & k = n \\ 0 & k \neq n \end{cases} \tag{6}$$

where $\mathbf{R}(n)$ and $\mathbf{Q}(n)$ are diagonal covariance matrices, respectively, related to the additive noise and the prediction error.

The Kalman filtering estimates a process by using a kind of feedback control: first, the filter estimates the state of the process at a given time, then the feedback is obtained in the form of a new measurement.

Brown and Hwang [10] and Vaseghi [9] divided the Kalman filtering equations into two groups. The first ones are the time-update equations (prediction) and the second are the measurement-update equations (correction). Equation (7) describes the time-update:

$$\hat{\mathbf{x}}\left(\frac{n}{n-1}\right) = \mathbf{A}(n-1)\,\hat{\mathbf{x}}\left(\frac{n-1}{n-1}\right) \tag{7}$$

while measurement-update equations are shown in (8) and (9), respectively.

$$\begin{aligned} \mathbf{K}(n) &= \mathbf{P}\left(\frac{n}{n-1}\right)\mathbf{H}^{\mathrm{T}}(n) \\ &\quad\times \left[\mathbf{H}(n)\,\mathbf{P}\left(\frac{n}{n-1}\right)\mathbf{H}^{\mathrm{T}}(n) + \mathbf{R}(n)\right]^{-1} \end{aligned} \tag{8}$$

$$\begin{aligned} \hat{\mathbf{x}}\left(\frac{n}{n}\right) &= \hat{\mathbf{x}}\left(\frac{n}{n-1}\right) \\ &\quad + \mathbf{K}(n)\left[\mathbf{y}(n) - \mathbf{H}(n)\,\hat{\mathbf{x}}\left(\frac{n}{n-1}\right)\right] \end{aligned} \tag{9}$$

$$\mathbf{P}\left(\frac{n}{n}\right) = \left[\mathbf{I} - \mathbf{K}(n)\,\mathbf{H}(n)\right]\mathbf{P}\left(\frac{n}{n-1}\right) \tag{10}$$

where $\mathbf{P}(n/n)$ is the error covariance matrix at time n; $\mathbf{K}(n)$ is the Kalman gain matrix, responsible for minimizing $\mathbf{P}(n/n)$; and $\hat{\mathbf{x}}(n/n)$ is the state estimate at time n, according to the previous observations of $\mathbf{y}(n)$.

3. Discrete-Frequency Kalman Filtering (DFKF)

Fujimoto and Ariki [8] introduced the discrete-frequency Kalman filtering (DFKF) in 2000 to provide more computationally efficient algorithm. This is accomplished by transforming the Kalman filter equations to be iterated in the frequency domain and then inverse transforming the estimated spectrum back to the time domain to find the estimated signal. In order to do so, they divide the frequency domain into multiple frames in such a way that the lth frame $X(k,l)$ is the complex spectrum of the noiseless signal $x(n,l)$ and $y(n,l)$ is the white Gaussian noise. Thus, the noise-corrupted signal $y(n,l)$ is given by the following equation [8]:

$$y(n,l) = x(n,l) + v(n,l) \quad (11)$$

Since $x(n,l)$ can be replaced by the Inverse Discrete Fourier Transform (IDFT) of $X(k,l)$, we have

$$y(n,l) = \sum_{k=0}^{N-1} X(k,l)\exp\left(j2\pi\frac{kn}{N}\right) + v(n,l) \quad (12)$$

In matrix notation, (12) can be represented as shown in

$$\mathbf{y}(n,l) = \underbrace{\begin{pmatrix} 1 \\ \exp\left(j2\pi\frac{n}{N}\right) \\ \vdots \\ \exp\left(j2\pi\frac{(N-1)n}{N}\right) \end{pmatrix}^T}_{\mathbf{F}_n} \underbrace{\begin{pmatrix} X(0,l) \\ X(1,l) \\ \vdots \\ X(N-1,l) \end{pmatrix}}_{\mathbf{X}_l} + \mathbf{v}(n,l) \quad (13)$$

that can be simply written as

$$\mathbf{y}(n,l) = \mathbf{F}_n\mathbf{X}_l + \mathbf{v}(n,l) \quad (14)$$

where n represents time within lth frame, N is the number of samples in the frame, and $\mathbf{F}_n$ is the $N \times 1$ vector containing the basis of the Discrete Fourier Transform (DFT). $\mathbf{X}_l$ is the complex spectrum vector of the lth frame. Since time n has no meaning for $\mathbf{X}_l$, there is no state transition matrix in the Kalman equations for the frequency domain, so that the computational effort of the DFKF is significantly reduced.

Analogous to the DTKF, the DFKF can be represented by the following equations:

$$\mathbf{K}_{(n,l)} = \mathbf{P}_{(n-1,l)}\mathbf{F}_n^H\left[\mathbf{F}_n\mathbf{P}_{(n-1,l)}\mathbf{F}_n^H + \mathbf{R}_{(n,l)}\right]^{-1} \quad (15)$$

$$\widehat{\mathbf{X}}_{(n,l)} = \widehat{\mathbf{X}}_{(n-1,l)} + \mathbf{K}_{(n,l)}\left[\mathbf{y}(n,l) - \mathbf{F}_n\widehat{\mathbf{X}}_{(n-1,l)}\right] \quad (16)$$

$$\mathbf{P}_{(n,l)} = \mathbf{P}_{(n-1,l)} - \left[\mathbf{K}_{(n,l)}\mathbf{F}_n\mathbf{P}_{(n-1,l)}\right] \quad (17)$$

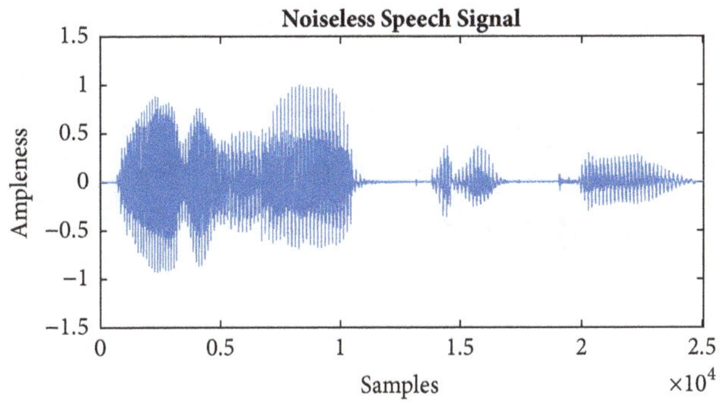

FIGURE 1: Noiseless signal used for comparison with the estimated signal.

where $(\cdot)^H$ means the complex conjugate transpose of a matrix.

In order to obtain the estimated signal of the Kalman filter in the time domain, we must apply the Inverse Discrete Fourier Transform (IDFT) on (16).

4. Results

In order to compare the performances of the studied techniques, we used 25 different recorded speech signals sampled at 22050 Hz and coded with 16 bits per sample. Each signal was windowed by a Hamming window of size 512 with 50% overlap. All tests were performed using Matlab R2013B on a Core i7 processor computer with 8 GB RAM.

The quality of the estimated speech signal in the output of each filter was evaluated using the segmental signal-to-noise ratio (SNRseg). We have chosen the SNRseg because it can be calculated over short segments of the speech signal, in order to balance the weights assigned to each segment of higher or lower signal strength. SNRseg is given by [11]

$$\text{SNRseg} = \frac{10}{M}\sum_{j=0}^{M-1}\log_{10}\left[\sum_{n=mj-N+1}^{mj}\frac{x^2(n)}{[x(n)-\hat{x}(n)]^2}\right] \quad (18)$$

where mj are the limits of each one of the M frames of length N. To carry out the tests, the signals were contaminated by additive white noise and the input segmental signal-to-noise ratio (SNRI) was adjusted to 3 dB.

As reported by Rabiner and Schafer [12], a suitable way to measure spectrum variations is the Itakura-Saito distance. Such measure can be calculated as

$$d(\mathbf{b},\mathbf{a}) = \log\left[\frac{\mathbf{b}R\mathbf{b}^T}{\mathbf{a}R\mathbf{a}^T}\right] \quad (19)$$

where $\mathbf{a}$ and $\mathbf{b}$ are the linear prediction coefficients (LPC) vectors of the original and estimated signals, respectively, and $\mathbf{R}$ is the autocorrelation matrix of the original signal. The closer to each other the spectra of the original and estimated signals, the smaller $d(\mathbf{b},\mathbf{a})$. Thus, an Itakura-Saito distance equal to zero indicates that the spectra are the equal [12].

The DTKF algorithm was employed in the first test, which used the utterance elétrica (electrical in Portuguese). The results are shown in Figures 1, 2, and 3, respectively.

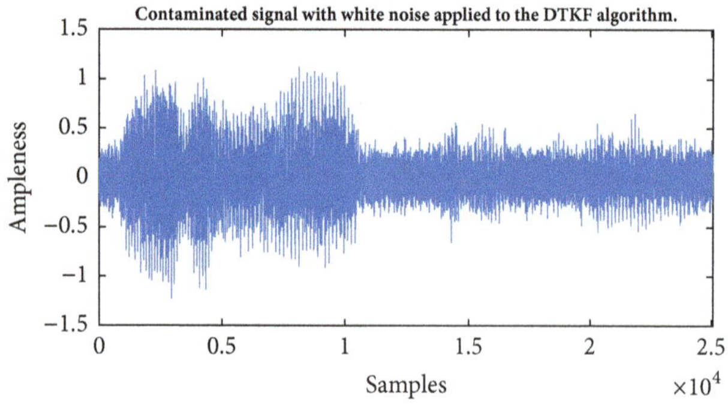

FIGURE 2: Contaminated signal with white noise applied to the DTKF algorithm.

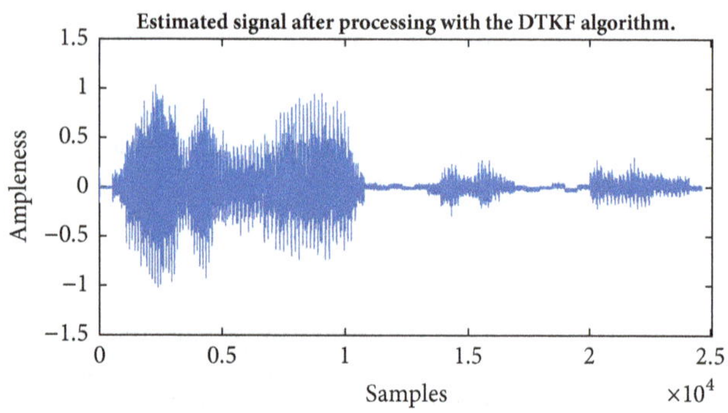

FIGURE 3: Estimated signal after processing with the DTKF algorithm.

Figures 2 and 3 evidence the noise reduction, especially during the silence parts of the signal. The SNRO in this case was 10 dB and the Itakura-Saito distance was 0.3250.

The second test preserved the same parameters of the first test except for the use of DFKF. The results are shown in Figures 4 and 5, respectively. The SNRO was 8 dB and the comparison of Figures 4 and 5 shows a considerable reduction in the noise. However, the Itakura-Saito distance was 0.3782, which indicates a larger distortion in the filtering.

Therefore, the DTKF algorithm produced smaller spectral distortion than the DFKF but provided a larger SNRO.

The results of the tests for the 25 words are presented in Figures 6 and 7. Figure 6 shows that the SNRO in targeted tests was almost always the same for DTKF and DFKF, with an average of 9 dB.

Figure 7 shows that the DTKF algorithm produced smaller signal distortion for all tests. Thus, we can affirm that the DTKF is more suitable than the DFKF for speech processing.

Tests were also performed after prefiltering the noisy signals. The prefiltering was based on spectral subtraction like in [1]. All results showed that the DTKF produced smaller spectral distortion than DFKF. The spectral distortions for the 25 words are shown in Figure 8 for an SNRI of 3 dB.

The comparison of Figures 7 and 8 indicates that prefiltering allowed only a tiny improvement in the reduction of spectral distortion provided by the DTKF algorithm.

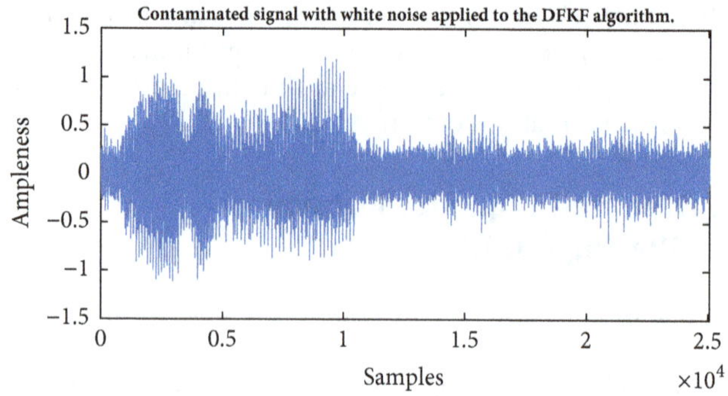

FIGURE 4: Contaminated signal with white noise applied to the algorithm DFKF.

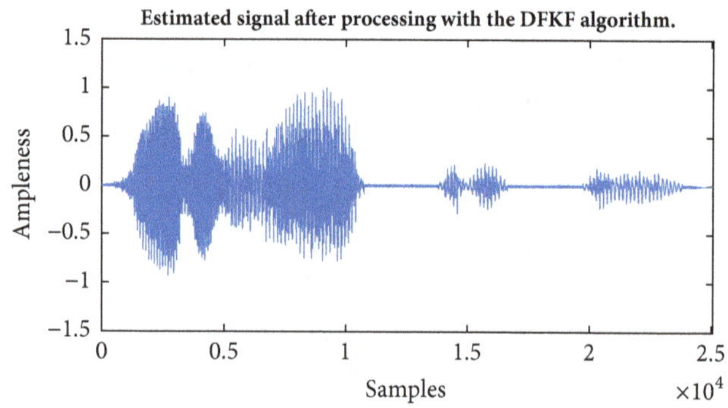

FIGURE 5: Estimated signal after processing with the DFKF algorithm.

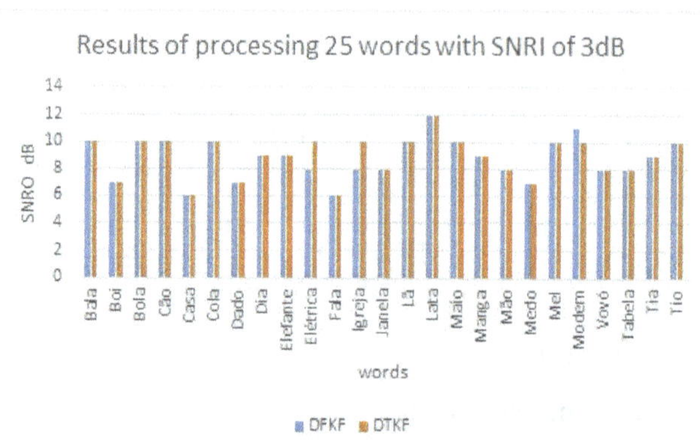

FIGURE 6: Comparison for segmental signal-to-noise ratio output (SNRO) with 25 words contaminated by white noise with signal-to-noise ratio input (SNRI) of 3 dB.

5. Conclusions

This paper presented a comparative study between discrete-time and discrete-frequency Kalman filtering algorithms. Tests were carried out with 25 different words using Itakura-Saito distance to measure the spectral distortion and the segmental signal-to-noise ratio to evaluate the noise reduction.

Although the two algorithms performed very similarly regarding noise reduction, discrete-time Kalman filtering produced smaller spectral distortion on the estimated signals for all targeted tests. This shows that discrete-time Kalman

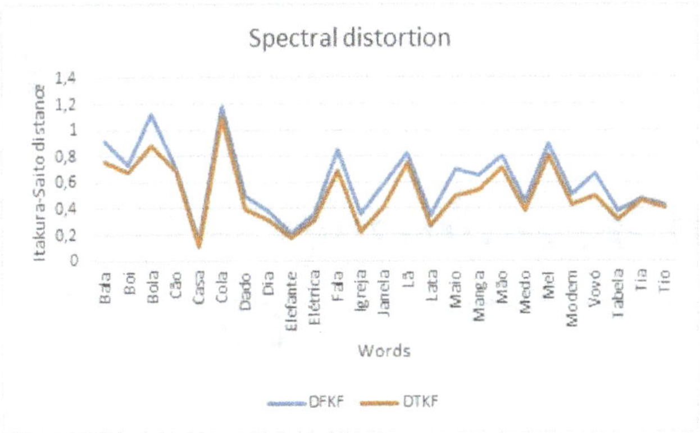

FIGURE 7: Comparison for spectral distortion for 25 words contaminated by white noise with signal-to-noise ratio input (SNRI) of 3 dB.

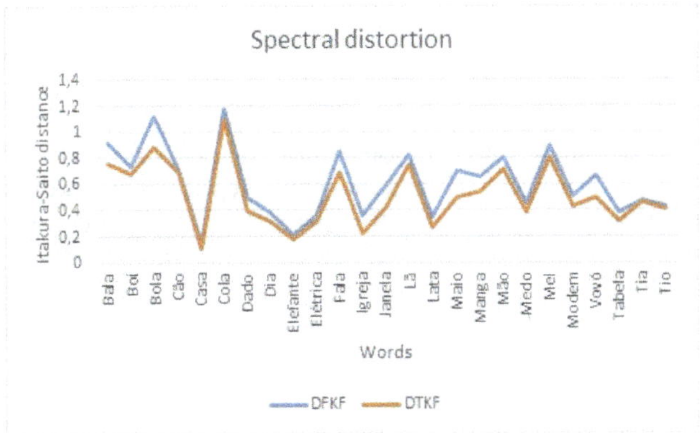

FIGURE 8: Comparison for spectral distortion for 25 words contaminated by white noise with signal-to-noise ratio input (SNRI) of 3 dB, using spectral subtraction with prefiltering of the contaminated signal.

filtering is more suitable than discrete-frequency Kalman filtering for the reconstruction of speech signals corrupted by additive white noise.

Conflicts of Interest

The authors declare that they have no conflicts of interest.

References

[1] L. A. da Silva and M. B. Joaquim, "Redução de Ruído em Sinais de Voz Usando, Filtros de Kalman de Tempo e Freqüência Discretos, Combinados com Subtração Espectral de Potência e/ou Wavelets," in *Proceedings of the XXV Simpósio Brasileiro de Telecomunicações-SBrT*, pp. 3–6, 2007.

[2] C. Lu, K. Tseng, and C. Chen, "Reduction of Musical Residual Noise Using Hybrid Median Filter," in *Proceedings of the 2012 Spring Congress on Engineering and Technology (S-CET)*, pp. 1-4, Xi'an, China, May 2012.

[3] R. Miyazaki, H. Saruwatari, T. Inoue, Y. Takahashi, K. Shikano, and K. Kondo, "Musical-noise-free speech enhancement based on optimized iterative spectral subtraction," *IEEE Transactions on Audio, Speech and Language Processing*, vol. 20, no. 7, pp. 2080–2094, 2012.

[4] R. Li, C. Bao, B. Xia, and M. Jia, "Speech enhancement using the combination of adaptive wavelet threshold and spectral subtraction based on wavelet packet decomposition," in *Proceedings of the 2012 11th International Conference on Signal Processing, ICSP 2012*, pp. 481–484, chn, October 2012.

[5] R. Aggarwal, J. Karan Singh, V. Kumar Gupta, S. Rathore, M. Tiwari, and A. Khare, "Noise Reduction of Speech Signal using Wavelet Transform with Modified Universal Threshold," *International Journal of Computer Applications*, vol. 20, no. 5, pp. 14–19, 2011.

[6] Y. Shao and C.-H. Chang, "A Kalman filter based on wavelet filter-bank and psychoacoustic modeling for speech enhancement," in *Proceedings of the ISCAS 2006: 2006 IEEE International Symposium on Circuits and Systems*, pp. 121–124, grc, May 2006.

[7] R. Dhivya and J. Justin, "A novel speech enhancement technique," *International Journal of Research in Engineering and Technology*, vol. 3, no. 19, pp. 98–102, 2014.

[8] M. Fujimoto and Y. Ariki, "Noisy speech recognition using noise reduction method based on Kalman filter," in *Proceedings of the 25th IEEE International Conference on Acoustics, Speech, and Signal Processing, ICASSP 2000*, vol. 3, pp. 1727–1730, June 2000.

[9] S. V. Vaseghi, *Advanced Digital Signal Processing and Noise Reduction*, John Wiley & Sons, 2008.

[10] R. G. Brown and P. Y. C. Hwang, *Introduction to random signals and applied Kalman filtering: with MATLAB exercises and solutions*, Wiley, New York, NY, USA, 1997.

[11] J. R. Deller, J. G. Proakis, and J. H. Hansen, *Discrete Time Processing of Speech Signals*, Prentice Hall PTR, 1993.

[12] L. R. Rabiner and R. W. Schafer, *Digital Processing of Speech Signals*, Prentice Hall, 1978.

2

Design of Fixed Wideband Beamformer through Improved Maximum Energy Approach

Rui Du, Yangyu Fan, and Jianshu Wang

School of Electronics and Information, Northwestern Polytechnical University, Xi'an, Shaanxi 710072, China

Correspondence should be addressed to Rui Du; rdu@mail.nwpu.edu.cn

Academic Editor: George Tsoulos

A maximum energy approach is investigated in this paper to design fixed wideband beamformer. This approach has been improved by integrating response variation (RV) into the target function to maintain the frequency invariant property of wideband beamformer over the whole passband. Two methods for designing null to suppress interference signal also have been proposed to make the wideband beamformer robust in complicated environment. Comparisons among other methods are provided to illustrate the effectiveness and enhancement of performance of the new approaches.

1. Introduction

The term beamforming is derived from early spatial filters that were designed for beams in order to receive a signal radiated from a specific location and attenuate others from other locations [1, 2]. Beamforming is widely adopted by signal processing of radar, sonar, communication, seismology, geophysics, and so on [3–7].

Wideband beamforming has attracted increasing attentions in recent years because of the advantages of wideband signal such as larger channel capacity. Wideband beamforming can be achieved mainly through two approaches: adaptive wideband beamforming and fixed wideband beamformer. Fixed wideband beamformer directly constrains the response of beamformer, and the beam pattern is constant once set. Three approaches are mainly adopted to design the fixed wideband beamformer: iterative optimization, least square approach, and eigenfilter approach [2]. Eigenfilter approach is efficient because no matrix inversion (least square approach) or iteration (iterative optimization) is required during the optimization.

Eigenfilter is a filter whose coefficients are the elements of an eigenvector. It was first proposed to design digital filters and then extended to the design of beamformer [8]. In [9], fixed wideband beamformer was achieved by maximizing the ratio of mainlobe's energy to sidelobe's energy through eigenfilter approach. It can also be achieved by minimizing total least squares error between reference response and responses over wide passband, both in near field and in far field [9, 10].

All the methods above have not discussed the design of null for interference, which is of great importance for beamformer to get a good result in complicated environment with inferences. The width of mainlobe varies over the passband. It will produce distortion if the signal is not coming from the main direction. Response variation (RV) was introduced in [11–14] to maintain a good frequency invariant property in wideband beamforming.

An improved maximum energy approach is proposed to design fixed wideband beamformer. In this approach, RV is integrated into the target function with a trade-off coefficient, which makes it possible to get a good frequency invariant property and constrain the beamformer's response directly at reference frequency simultaneously. Two methods for null designing are discussed: linear constraint approach and null's expanded energy approach. Both methods are transformed into a standard maximum energy problem and can be solved through a generalized vector approach.

The outline of the paper is as follows.

In Section 1, a brief introduction about the wideband beamforming is made.

In the next section, a TDL (Tapped Delay Line) structure and beam patterns are described. The standard maximum energy problem and generalized vector approach are also discussed in this section.

In Section 3, we will describe the method we proposed with improved target function and simplified constraint. Two methods for designing null are shown as well, both of which can be casted into a standard maximum energy problem.

The methods are tested in Section 4. The results demonstrate the effectiveness of the proposed methods compared with other methods.

2. Wideband Beamforming through Maximum Energy Approach

2.1. Wideband Beamforming Model with TDL Structure. A TDL structure is shown in Figure 1, where there is a linear array that consists of M sensors with a J order FIR connecting to each of them.

The symbol "△" in the structure represents a TDL unit which produces a delay of τ, which has been set to sampling time T_s for simplicity.

The output of this TDL structure can be expressed as follows:

$$y(t) = \mathbf{w}^H \mathbf{x}(t), \tag{1}$$

where

$$\begin{aligned}\mathbf{w} &= [w_{0,0}, \dots, w_{M-1,J-1}]^T \\ \mathbf{x}(t) &= [x_0(t), \dots, x_{M-1}(t), x_0(t-T_s), \dots, \\ &\quad x_{M-1}(t-(J-1)T_s)]^T,\end{aligned} \tag{2}$$

where $\mathbf{w}$ is $MJ \times 1$ weight vector and $\mathbf{x}(t)$ is the received signal through TDL structure.

We denote the received signal at the first sensor by $x_0(t)$. It can be represented as an inverse Fourier transformation within the bandwidth $[\omega_l, \omega_h]$:

$$x_0(t) = \frac{1}{2\pi}\int_{\omega_l}^{\omega_h} X_0(\omega) e^{j\omega t} d\omega. \tag{3}$$

Thus, the signal at mth and ith unit of the TDL structure can be represented as

$$x_{m,i}(t) = \frac{1}{2\pi}\int_{\omega_l}^{\omega_h} X_0(\omega) e^{j\omega(t-(\tau_m+iT_s))} d\omega. \tag{4}$$

Hence, (1) can be derived in a new form:

$$\begin{aligned} y(t) &= \frac{1}{2\pi}\int_{\omega_l}^{\omega_h} X_0(\omega) e^{j\omega t} \sum_{m=0}^{M-1}\sum_{i=0}^{J-1} e^{-j\omega(\tau_m+iT_s)} w_{m,i}^{*} d\omega \\ &= \frac{1}{2\pi}\int_{\omega_l}^{\omega_h} X_0(\omega) e^{j\omega t} P(\theta,\omega) d\omega, \end{aligned} \tag{5}$$

where $P(\theta,\omega)$ is the response of the beamformer and is expressed as

$$P(\omega,\theta) = \sum_{m=0}^{M-1}\sum_{i=0}^{J-1} e^{-j\omega(\tau_m+iT_s)} w_{m,i}^{*} = \mathbf{w}^H \mathbf{a}(\omega,\theta), \tag{6}$$

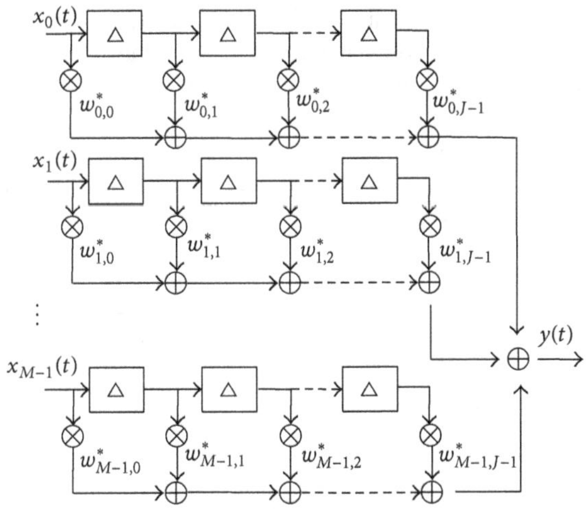

FIGURE 1: Structure of TDL.

where $\mathbf{a}(\omega,\theta)$ is the steering vector of the TDL structure; that is,

$$\begin{aligned}\mathbf{a}(\omega,\theta) &= \left[e^{-j\omega\tau_0}, \dots, e^{-j\omega\tau_{M-1}}, e^{-j\omega(\tau_0+T_s)}, \dots, \right. \\ &\quad \left. e^{-j\omega(\tau_{M-1}+(J-1)T_s)}\right]^T,\end{aligned} \tag{7}$$

where ω is the frequency and θ is the arriving direction of the signal. $\tau_0, \tau_1, \dots, \tau_{M-1}$ is the time delay in the array, which can be computed through $\tau_m = md \times \sin(\theta)/c$, where $m = 0, \dots, M-1$ and c is the speed of light.

For fixed wideband beamformer, $P(\omega,\theta)$ is only related to the weight vector and steering vector of the array.

2.2. Maximum Energy Approach. Maximum energy approach can provide a closed-form solution to the fixed wideband beamformer and it is computationally efficient because no matrix inversion is involved.

Given a Hermitian matrix $\mathbf{R}$ and a positive definite matrix $\mathbf{B}$, generalized Rayleigh ratio is described as

$$\frac{\mathbf{w}^H\mathbf{R}\mathbf{w}}{\mathbf{w}^H\mathbf{B}\mathbf{w}}. \tag{8}$$

Equation (8) reaches its maximum when $\mathbf{w}$ is the generalized eigenvector corresponding to the maximum generalized eigenvalue of matrix pair $\mathbf{R}$ and $\mathbf{B}$, $\mathbf{R}\mathbf{w} = \lambda_{\max}\mathbf{B}\mathbf{w}$, and reaches its minimum when $\mathbf{w}$ is the generalized eigenvector corresponding to the minimum generalized eigenvalue of matrix pair $\mathbf{R}$ and $\mathbf{B}$, $\mathbf{R}\mathbf{w} = \lambda_{\min}\mathbf{B}\mathbf{w}$.

According to the property of (8), the design of fixed wideband beamformer can be implemented by maximizing the Rayleigh ratio of beamformer's energy on angle range of interest to the energy over the whole direction area on the passband. For simplification, the angle range of interest is referred to as mainlobe in the paper, and the angle range that is not of interest is referred to as sidelobe. Hence,

$$\max_{\mathbf{w}} \frac{\int_{\Omega_{\text{PB}}}\int_{\Theta_{\text{ML}}} \left|\mathbf{w}^H\mathbf{a}(\omega,\theta)\right|^2 d\theta\, d\omega}{\int_{\Omega_{\text{PB}}}\int_{\Theta} \left|\mathbf{w}^H\mathbf{a}(\omega,\theta)\right|^2 d\theta\, d\omega}. \tag{9}$$

In (9), the numerator is the energy of the beamformer on mainlobe Θ_{ML} on passband Ω_{PB} and the denominator represents the beamformer's energy over the whole angle region Θ on Ω_{PB}.

A variation of (9) is achieved by replacing the beamformer's energy over whole direction area with the energy only on sidelobe:

$$\max_{\mathbf{w}} \frac{\int_{\Omega_{PB}} \int_{\Theta_{ML}} \left| \mathbf{w}^H \mathbf{a}(\omega, \theta) \right|^2 d\theta\, d\omega}{\int_{\Omega_{PB}} \int_{\Theta_{SL}} \left| \mathbf{w}^H \mathbf{a}(\omega, \theta) \right|^2 d\theta\, d\omega}, \tag{10}$$

which is equivalent to

$$\max_{\mathbf{w}} \frac{\mathbf{w}^H \mathbf{A}_{ML} \mathbf{w}}{\mathbf{w}^H \mathbf{A}_{SL} \mathbf{w}}, \tag{11}$$

where

$$\begin{aligned} \mathbf{A}_{ML} &= \int_{\Omega_{PB}} \int_{\Theta_{ML}} \mathbf{a}(\omega, \theta)\, \mathbf{a}(\omega, \theta)^H d\theta\, d\omega \\ \mathbf{A}_{SL} &= \int_{\Omega_{PB}} \int_{\Theta_{SL}} \mathbf{a}(\omega, \theta)\, \mathbf{a}(\omega, \theta)^H d\theta\, d\omega. \end{aligned} \tag{12}$$

Maximum energy approach does not guarantee a smooth constant response over different frequencies [2]. Linear constraints $\mathbf{C}^H \mathbf{w} = \mathbf{f}$ can be added to get a smooth response at mainlobe by directly constraining the response of the beamformer over different frequencies [2]. $\mathbf{C}$ is $NJ \times J$ constraint matrix and $\mathbf{f}$ is the $J \times 1$ response vector. Thus, (11) becomes

$$\begin{aligned} &\max_{\mathbf{w}} \quad \frac{\mathbf{w}^H \mathbf{A}_{ML} \mathbf{w}}{\mathbf{w}^H \mathbf{A}_{SL} \mathbf{w}} \\ &\text{subject to} \quad \mathbf{C}^H \mathbf{w} = \mathbf{f}. \end{aligned} \tag{13}$$

Equation (13) can be transformed into another form:

$$\begin{aligned} &\max_{\widehat{\mathbf{w}}} \quad \frac{\widehat{\mathbf{w}}^H \widehat{\mathbf{A}}_{ML} \widehat{\mathbf{w}}}{\widehat{\mathbf{w}}^H \widehat{\mathbf{A}}_{SL} \widehat{\mathbf{w}}} \\ &\text{subject to} \quad \widehat{\mathbf{C}}^H \widehat{\mathbf{w}} = \mathbf{0}, \end{aligned} \tag{14}$$

where

$$\begin{aligned} \widehat{\mathbf{w}} &= \left[\mathbf{w}^H, -1 \right]^H, \\ \widehat{\mathbf{A}}_{ML} &= \begin{bmatrix} \mathbf{A}_{ML} & \mathbf{0} \\ \mathbf{0} & 0 \end{bmatrix}, \\ \widehat{\mathbf{A}}_{SL} &= \begin{bmatrix} \mathbf{A}_{SL} & \mathbf{0} \\ \mathbf{0} & 0 \end{bmatrix}, \\ \widehat{\mathbf{C}} &= \left[\mathbf{C}^H \ \ \mathbf{f} \right]^H. \end{aligned} \tag{15}$$

It is shown in the constraint of (14) that $\widehat{\mathbf{w}}$ locates in the null space of $\widehat{\mathbf{C}}$. Equation (14) can be transformed into a standard maximum energy problem by introducing a matrix $\mathbf{Z}$ whose columns are the bases of the null space of $\widehat{\mathbf{C}}$ and it satisfies $\widehat{\mathbf{w}} = \mathbf{Z}\widetilde{\mathbf{w}}$. Hence,

$$\max_{\widetilde{\mathbf{w}}} \frac{\widetilde{\mathbf{w}}^H \mathbf{Z}^H \widehat{\mathbf{A}}_{ML} \mathbf{Z} \widetilde{\mathbf{w}}}{\widetilde{\mathbf{w}}^H \mathbf{Z}^H \widehat{\mathbf{A}}_{SL} \mathbf{Z} \widetilde{\mathbf{w}}}. \tag{16}$$

Equation (16) is a standard maximum energy problem and can be solved through generalized vector approach. As we can see in the first equation in (15), the first MJ elements of $\widehat{\mathbf{w}}$ can be regarded as $\mathbf{w}$ only when the last element of $\widehat{\mathbf{w}}$ has been scaled to -1.

3. Improved Maximum Energy Approach

3.1. Maximum Energy Approach with Response Variation. RV describes the Euclidean distance between responses at reference frequency and responses at other frequencies within whole angle-frequency region of interest.

In [11–14, 16], a general form of RV is defined:

$$\begin{aligned} \mathrm{RV} &= \int_{\Omega_{PB}} \int_{\Theta} \left| \mathbf{w}^H \mathbf{a}(\omega, \theta) - \mathbf{w}^H \mathbf{a}(\omega_r, \theta) \right|^2 d\theta\, d\omega \\ &= \mathbf{w}^H \mathbf{A}_{RV} \mathbf{w}, \end{aligned} \tag{17}$$

where

$$\begin{aligned} \mathbf{A}_{RV} &= \int_{\Omega_{PB}} \int_{\Theta} \left(\mathbf{a}(\omega, \theta) - \mathbf{a}(\omega_r, \theta) \right) \\ &\quad \cdot \left(\mathbf{a}(\omega, \theta) - \mathbf{a}(\omega_r, \theta) \right)^H d\theta\, d\omega, \end{aligned} \tag{18}$$

where Θ represents the range of angles RV measured.

The introduction of RV factor makes it possible to constrain response directly on the reference frequency. The energy of mainlobe and sidelobe also can be simplified as follows:

$$\mathbf{w}^H \mathbf{a}(\omega_r, \theta_d) = 1 \tag{19}$$

$$\mathbf{A}_{MLr} = \int_{\Theta_{ML}} \mathbf{a}(\omega_r, \theta)\, \mathbf{a}(\omega_r, \theta)^H d\theta \tag{20}$$

$$\mathbf{A}_{SLr} = \int_{\Theta_{SL}} \mathbf{a}(\omega_r, \theta)\, \mathbf{a}(\omega_r, \theta)^H d\theta. \tag{21}$$

Equation (19) constrains the response of beamformer at looking direction θ_d without distortion.

Equation (13) can be transformed into a new form with RV and new constraint:

$$\begin{aligned} &\max_{\mathbf{w}} \quad \frac{\mathbf{w}^H \mathbf{A}_{MLr} \mathbf{w}}{\mathbf{w}^H \left(\mathbf{A}_{SLr} + \alpha \mathbf{A}_{RV} \right) \mathbf{w}} \\ &\text{subject to} \quad \mathbf{w}^H \mathbf{a}(\omega_r, \theta_d) = 1, \end{aligned} \tag{22}$$

where α is a positive trade-off coefficient, usually larger than 1, affecting the frequency invariant (FI) property. The beamformer will achieve better FI property with larger α. However, if the value of α is too large, the whole beam patterns will be degraded.

Equation (22) can be transformed into a similar form to (14):

$$\max_{\mathbf{w}} \frac{\widehat{\mathbf{w}}^H \widehat{\mathbf{A}}_{\mathrm{ML}r} \widehat{\mathbf{w}}}{\widehat{\mathbf{w}}^H \widehat{\mathbf{A}}_{\mathrm{SR}r} \widehat{\mathbf{w}}} \quad \text{subject to} \quad \widehat{\mathbf{C}}^H \widehat{\mathbf{w}} = 0, \tag{23}$$

where

$$\widehat{\mathbf{C}} = \begin{bmatrix} \mathbf{a}(\omega_r, \theta_d) \\ 1 \end{bmatrix}, \quad \widehat{\mathbf{A}}_{\mathrm{ML}r} = \begin{bmatrix} \mathbf{A}_{\mathrm{ML}r} & \mathbf{0} \\ \mathbf{0} & 0 \end{bmatrix}, \quad \widehat{\mathbf{A}}_{\mathrm{SR}r} = \begin{bmatrix} (\mathbf{A}_{\mathrm{SL}r} + \alpha \mathbf{A}_{\mathrm{RV}}) & \mathbf{0} \\ \mathbf{0} & 0 \end{bmatrix}, \quad \widehat{\mathbf{w}} = \begin{bmatrix} \mathbf{w}^H & -1 \end{bmatrix}^H. \tag{24}$$

Similarly, (23) can be transformed into a new standard maximum energy problem:

$$\max_{\widetilde{\mathbf{w}}} \frac{\widetilde{\mathbf{w}}^H \mathbf{Z}^H \widehat{\mathbf{A}}_{\mathrm{ML}r} \mathbf{Z} \widetilde{\mathbf{w}}}{\widetilde{\mathbf{w}}^H \mathbf{Z}^H \widehat{\mathbf{A}}_{\mathrm{SR}r} \mathbf{Z} \widetilde{\mathbf{w}}}. \tag{25}$$

Equation (25) reaches its maximum when $\widetilde{\mathbf{w}}$ is the generalized eigenvector corresponding to the largest generalized eigenvalue of matrix pair $\mathbf{Z}^H \widehat{\mathbf{A}}_{\mathrm{ML}} \mathbf{Z}$ and $\mathbf{Z}^H \widehat{\mathbf{A}}_{\mathrm{SR}} \mathbf{Z}$.

3.2. Design of Null Point

3.2.1. Design of Null with Linear Constraint on Null's Response. Null can be generated by directly imposing a constraint at interference angle θ_i at reference frequency ω_r:

$$\mathbf{w}^H \mathbf{a}(\omega_r, \theta_i) = \varepsilon, \tag{26}$$

where ε is a small positive value, usually smaller than 1, of the response at null point. Null's response in decibels can be computed through $-20 \lg \varepsilon$ dB.

With this constraint, (22) is rewritten as

$$\max_{\mathbf{w}} \frac{\mathbf{w}^H \mathbf{A}_{\mathrm{ML}r} \mathbf{w}}{\mathbf{w}^H (\mathbf{A}_{\mathrm{SL}r} + \alpha \mathbf{A}_{\mathrm{RV}}) \mathbf{w}} \quad \text{subject to} \quad \mathbf{w}^H \mathbf{a}(\omega_r, \theta_d) = 1, \quad \mathbf{w}^H \mathbf{a}(\omega_r, \theta_i) = \varepsilon. \tag{27}$$

Hence, (27) can be transformed into a similar form to (14):

$$\max_{\mathbf{w}} \frac{\widehat{\mathbf{w}}^H \widehat{\mathbf{A}}_{\mathrm{ML}r} \widehat{\mathbf{w}}}{\widehat{\mathbf{w}}^H \widehat{\mathbf{A}}_{\mathrm{SR}r} \widehat{\mathbf{w}}} \quad \text{subject to} \quad \widehat{\mathbf{C}}^H \widehat{\mathbf{w}} = 0, \tag{28}$$

where

$$\widehat{\mathbf{C}} = \begin{bmatrix} \mathbf{a}(\omega_r, \theta_d) & \mathbf{a}(\omega_r, \theta_i) \\ 1 & \varepsilon \end{bmatrix}, \quad \widehat{\mathbf{A}}_{\mathrm{ML}r} = \begin{bmatrix} \mathbf{A}_{\mathrm{ML}r} & \mathbf{0} \\ \mathbf{0} & 0 \end{bmatrix}, \quad \widehat{\mathbf{A}}_{\mathrm{SR}r} = \begin{bmatrix} (\mathbf{A}_{\mathrm{SL}r} + \alpha \mathbf{A}_{\mathrm{RV}}) & \mathbf{0} \\ \mathbf{0} & 0 \end{bmatrix}, \quad \widehat{\mathbf{w}} = \begin{bmatrix} \mathbf{w}^H & -1 \end{bmatrix}^H. \tag{29}$$

And (28) changes into an unconstrained form by introducing a matrix $\mathbf{Z}$ which satisfies $\widehat{\mathbf{w}} = \mathbf{Z}\widetilde{\mathbf{w}}$:

$$\max_{\widetilde{\mathbf{w}}} \frac{\widetilde{\mathbf{w}}^H \mathbf{Z}^H \widehat{\mathbf{A}}_{\mathrm{ML}r} \mathbf{Z} \widetilde{\mathbf{w}}}{\widetilde{\mathbf{w}}^H \mathbf{Z}^H \widehat{\mathbf{A}}_{\mathrm{SR}r} \mathbf{Z} \widetilde{\mathbf{w}}}. \tag{30}$$

Equation (30) is a standard maximum energy problem and reaches its maximum when $\widetilde{\mathbf{w}}$ is the generalized eigenvector corresponding to the largest generalized eigenvalue of matrix pair $\mathbf{Z}^H \widehat{\mathbf{A}}_{\mathrm{ML}r} \mathbf{Z}$ and $\mathbf{Z}^H \widehat{\mathbf{A}}_{\mathrm{SR}r} \mathbf{Z}$.

3.2.2. Design of Null with Expanded Null's Energy. Null's energy at reference frequency is defined as

$$\int_{\Theta_{\mathrm{NL}}} \left| \mathbf{w}^H \mathbf{a}(\omega_r, \theta) \right|^2 d\theta = \mathbf{w}^H \mathbf{A}_{\mathrm{NL}r} \mathbf{w}, \quad \mathbf{A}_{\mathrm{NL}r} = \int_{\Theta_{\mathrm{NL}}} \mathbf{a}(\omega_r, \theta) \mathbf{a}(\omega_r, \theta)^H d\theta. \tag{31}$$

Null is designed for interference, which usually locates in sidelobe: $\Theta_{\mathrm{NL}} \subset \Theta_{\mathrm{SL}}$.

Then, (22) changes to a new form by replacing $\mathbf{A}_{\mathrm{SL}r}$ with $\mathbf{A}_{\mathrm{SL}r} + \beta \mathbf{A}_{\mathrm{NL}r}$:

$$\max_{\mathbf{w}} \frac{\mathbf{w}^H \mathbf{A}_{\mathrm{ML}r} \mathbf{w}}{\mathbf{w}^H (\mathbf{A}_{\mathrm{SL}r} + \beta \mathbf{A}_{\mathrm{NL}r} + \alpha \mathbf{A}_{\mathrm{RV}}) \mathbf{w}} \quad \text{subject to} \quad \mathbf{w}^H \mathbf{a}(\omega_r, \theta_d) = 1, \tag{32}$$

where β is the expansion coefficient which is also a large positive value used to expand null's energy in target function. Commonly, null is lower with larger β. The value of β is related to the actual beamformer, usually bigger than 1000 for a null low enough.

Thus, (32) can be transformed into a new form like (14):

$$\max_{\mathbf{w}} \frac{\widehat{\mathbf{w}}^H \widehat{\mathbf{A}}_{\mathrm{ML}r} \widehat{\mathbf{w}}}{\widehat{\mathbf{w}}^H \widehat{\mathbf{A}}_{\mathrm{SRN}r} \widehat{\mathbf{w}}} \quad \text{subject to} \quad \widehat{\mathbf{C}}^H \widehat{\mathbf{w}} = 0, \tag{33}$$

where

$$\widehat{\mathbf{C}} = \begin{bmatrix} \mathbf{a}(\omega_r, \theta_d) \\ 1 \end{bmatrix},$$

$$\widehat{\mathbf{A}}_{\mathrm{ML}r} = \begin{bmatrix} \mathbf{A}_{\mathrm{ML}r} & \mathbf{0} \\ \mathbf{0} & 0 \end{bmatrix}, \tag{34}$$

$$\widehat{\mathbf{A}}_{\mathrm{SRN}r} = \begin{bmatrix} (\mathbf{A}_{\mathrm{SL}r} + \beta \mathbf{A}_{\mathrm{NL}r} + \alpha \mathbf{A}_{\mathrm{RV}}) & \mathbf{0} \\ \mathbf{0} & 0 \end{bmatrix},$$

$$\widehat{\mathbf{w}} = \begin{bmatrix} \mathbf{w}^H & -1 \end{bmatrix}^H .$$

Similarly, (33) can be transformed into a standard maximum energy problem:

$$\max_{\widetilde{\mathbf{w}}} \frac{\widetilde{\mathbf{w}}^H \mathbf{Z}^H \widehat{\mathbf{A}}_{\mathrm{ML}r} \mathbf{Z} \widetilde{\mathbf{w}}}{\widetilde{\mathbf{w}}^H \mathbf{Z}^H \widehat{\mathbf{A}}_{\mathrm{SRN}r} \mathbf{Z} \widetilde{\mathbf{w}}}. \tag{35}$$

Equation (35) reaches its maximum when $\widetilde{\mathbf{w}}$ is the generalized eigenvector corresponding to the largest generalized eigenvalue of matrix pair $\mathbf{Z}^H \widehat{\mathbf{A}}_{\mathrm{ML}r} \mathbf{Z}$ and $\mathbf{Z}^H \widehat{\mathbf{A}}_{\mathrm{SRN}r} \mathbf{Z}$.

4. Performance Analysis

In this section, we will test and analyze the performance of the methods proposed. The relative coefficients will be discussed as well.

4.1. Experiment 1. The simulations are performed based on the specifications as follows:

(i) A TDL structure with $M = 14$ and $J = 20$ is adopted during the simulation.

(ii) Half of the sampling frequency f_s has been scaled to π and has been sampled to 64 narrow bands; the wide passband is set as $\Omega_{\mathrm{PB}} = [0.5\pi, \pi]$, and the reference frequency is set as $\omega_r = 0.7\pi$.

(iii) Space between elements is half of the wavelength corresponding to the highest frequency of passband.

(iv) Desired signal is coming from 0°, the interference signal comes from 50°, and null is set to be [48°, 52°].

(v) Mainlobe is set as $\Theta_{\mathrm{ML}} = [-15^\circ, 15^\circ]$ and sidelobe is set to be $\Omega_{\mathrm{SL}} = [-90^\circ, -20^\circ] \cup [20^\circ, 90^\circ]$, and the whole direction area is sampled every 1°.

(vi) $\alpha = 10$ and $\varepsilon = 10^{-4}$ in method 1; $\alpha = 200$ and $\beta = 9 \times 10^4$ in method 2.

(vii) To get a real-value $\mathbf{w}$, all the matrices during the computation are the real part of the original matrices.

Beamformer's responses through the 2 proposed methods are shown in Figures 2 and 3, respectively.

Good beam patterns have been achieved with distortionless mainlobes point at 0° and nulls reach −80 dB or lower.

To illustrate the effectiveness of the proposed methods described in (27) and (32), we compare them with the method from [2, 15]. In [15], the design of wideband beamformer was achieved through convex optimization. The weighting function in [15] has been set as $F(\omega, \theta) = 1$, $\delta = 1 \times 10^{-6}$, $\varepsilon = 10^{-4}$.

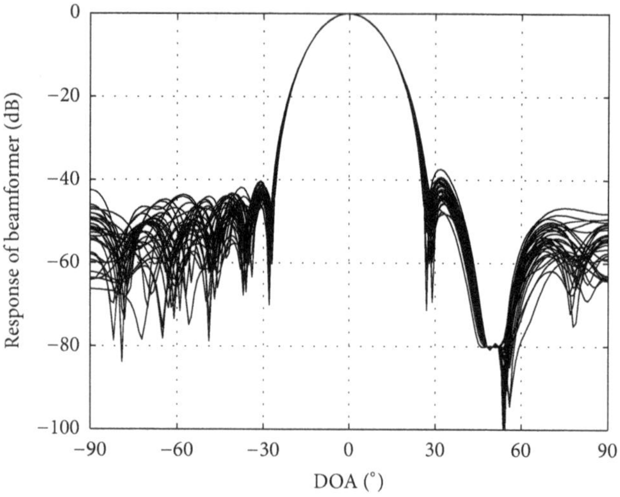

FIGURE 2: Response of beamformer with proposed method 1.

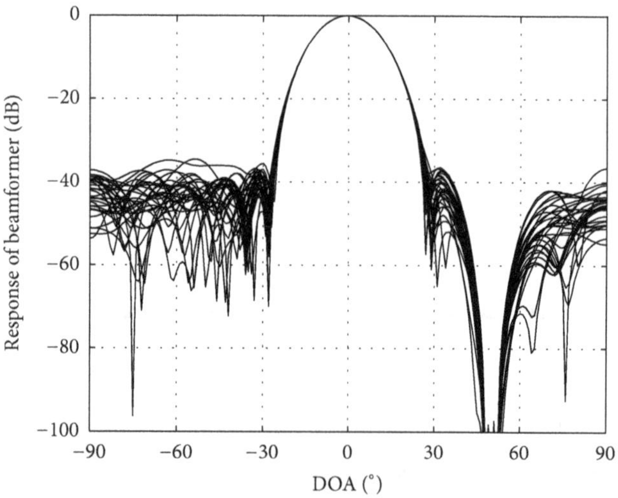

FIGURE 3: Response of beamformer with proposed method 2.

We test these methods versus width of mainlobe over passband, RV value of mainlobe, peak sidelobe level (SLL) of the beam patterns averaged over passband, and suppression values of null point.

The results are shown in Table 1.

The nulls generated by these four methods are constrained to −80 dB, which is very good suppression for interference signal. It can be seen in Table 1 with RV taken into consideration during the design of fixed wideband beamformer that both the proposed methods and the method in [15] achieved constant beam width and smaller RV; however, the values through the proposed methods are smaller. The SLL produced by the proposed methods are more than 10 dB lower than those from the method in [2, 15], indicating better suppression over the sidelobe region.

TABLE 1: Evaluation of four methods.

Measurement	Method in [2]	Method in [15]	Method 1	Method 2
Width of beam	[−12°, 11°]	[−10°, 9°]	[−10°, 9°]	[−10°, 9°]
RV of mainlobe	7.69×10^{-3}	3.27×10^{-8}	1.98×10^{-10}	3.96×10^{-11}
SLL (dB)	−21.06	−27.03	−44.2	−38.21

TABLE 2: Evaluation of the two proposed methods with different α.

α	Method 1		Method 2	
	RV	SLL (dB)	RV	SLL (dB)
1	6.85×10^{-8}	−45.21	1.75×10^{-7}	−41.71
10	1.98×10^{-10}	−44.2	2.21×10^{-9}	−40.99
20	1.76×10^{-10}	−44.04	6.60×10^{-10}	−40.23
50	1.07×10^{-10}	−43.37	7.61×10^{-10}	−38.51
100	6.44×10^{-11}	−42.69	4.04×10^{-11}	−38.39
200	4.81×10^{-11}	−42	3.69×10^{-11}	−38.21
500	9.61×10^{-12}	−40.55	3.17×10^{-11}	−37.41
1000	1.81×10^{-12}	−39.17	4.19×10^{-11}	−37.52

In order to investigate how the RV and SLL change with different α, more simulations have been implemented and the results are shown in Table 2 (all the specifications stay the same except for α).

As we mentioned previously, larger α provide a small RV (better frequency invariant property), but, at the same time, SLL gets higher. One conclusion can be drawn from Table 2: RV will not keep decreasing with the increasing of α (method 2; RV with $\alpha = 1000$ is larger than that with $\alpha = 500$). A good beam pattern should be a balance between the RV and SLL with an appropriate α.

4.2. Experiment 2. Wideband beamformer's performance decreases while the passband gets wider [2]. Wider passband is discussed in this part to test the effectiveness of the proposed methods. The specifications are described as follows:

(i) A TDL structure with $M = 12$ and $J = 12$ is adopted during the simulation.

(ii) Half of the sampling frequency f_s has been scaled to π and has been sampled to 64 narrow bands; the wide passband is set as $\Omega_{PB} = [0.3\pi, \pi]$, and the reference frequency is set as $\omega_r = 0.6\pi$.

(iii) Space between elements is half of the wavelength corresponding to the highest frequency of passband.

(iv) Desired signal is coming from 10°, the interference signal comes from −40°, and null is set to be [−42°, −38°].

(v) Mainlobe is set as $\Theta_{ML} = [-5°, 25°]$ and sidelobe is set to be $\Omega_{SL} = [-90°, -10°] \cup [30°, 90°]$, and the whole direction area is sampled every 1°.

(vi) $\alpha = 10$ and $\varepsilon = 10^{-3}$ in method 1; $\alpha = 20$ and $\beta = 5 \times 10^3$ in method 2.

(vii) To get a real-value $\mathbf{w}$, all the matrices during the computation are the real part of the original matrices.

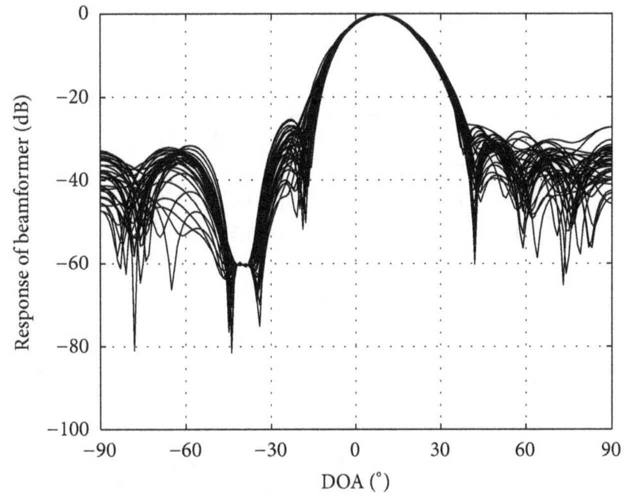

FIGURE 4: Response of beamformer with proposed method 1.

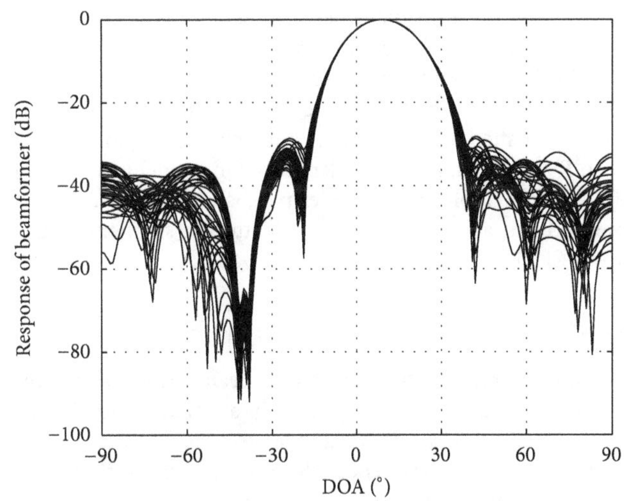

FIGURE 5: Response of beamformer with proposed method 2.

Beamformers' responses through the 2 proposed methods are shown in Figures 4 and 5, respectively.

The beam patterns obtained are shown in Figures 4 and 5. The mainlobes for both beam patterns are distortionless and point at 10° as desired. All realized nulls are −60 dB or less.

Similarly, the proposed methods are compared with the methods from [2, 15]. The weighting function in [15] has been set as $F(\omega, \theta) = 1, \delta = 1\times10^{-5}, \varepsilon = 10^{-3}$. The results are shown in Table 3.

All the nulls reach −60 dB. The performances of four methods decrease when we compare Table 3 with Table 1, due

TABLE 3: Evaluation of four methods.

Measurement	Method in [2]	Method in [15]	Method 1	Method 2
Width of beam	[−3°, 22°]	[0°, 19°]	[0°, 19°]	[0°, 19°]
RV of mainlobe	6.11×10^{-2}	7.74×10^{-7}	2.31×10^{-8}	1.07×10^{-10}
SLL (dB)	−18.21	−22.64	−30.24	−32.81

TABLE 4: Evaluation of the two proposed methods with different α.

α	Method 1		Method 2	
	RV	SLL (dB)	RV	SLL (dB)
1	8.31×10^{-8}	−32.26	2.00×10^{-9}	−34.67
10	2.31×10^{-8}	−30.24	7.07×10^{-11}	−32.48
20	1.64×10^{-8}	−29.69	1.07×10^{-10}	−32.81
50	1.42×10^{-8}	−29	1.53×10^{-10}	−32.1
100	1.86×10^{-8}	−28.37	5.90×10^{-12}	−31.14
200	9.99×10^{-9}	−27.96	5.29×10^{-11}	−31.36
500	2.50×10^{-9}	−25.84	1.32×10^{-12}	−29.85
1000	1.46×10^{-9}	−24.87	1.42×10^{-12}	−29.92

to the wider passband. But the results through the proposed methods are still better than the rest. The mainlobes remain constant over the whole passband, the SLL are more than 8 dB lower than other methods, and RV through our methods is smaller.

More simulations have been implemented to investigate α and the results are shown in Table 4.

Similar to the previous discussion, RV decreases with the increasing of α which also causes an increase in SLL (not strictly satisfied in method 2). A good choice of α should take both RV and SLL into consideration. One conclusion can be drawn from Table 4: for a wider passband, using the second method can achieve better beam patterns (smaller RV and lower SLL).

5. Conclusion

In this paper, a fixed wideband beamformer designing method based on maximum energy approach and RV has been proposed and investigated, as well as two methods to design null for interferences. Simulations show that good frequency invariant property in mainlobes and stable nulls for interference signal can be achieved through the proposed methods and the sidelobe suppression with the proposed methods is also lower.

Competing Interests

The authors declare that they have no competing interests.

References

[1] B. D. Van Veen and K. M. Buckley, "Beamforming: a versatile approach to spatial filtering," *IEEE ASSP Magazine*, vol. 5, no. 2, pp. 4–24, 1988.

[2] W. Liu and S. Weiss, *Wideband Beamforming—Concepts and Techniques*, John Wiley & Sons, 2010.

[3] M. Ström, M. Viberg, and K. Falk, "Robust transceiver design for wideband MIMO radar utilizing a subarray antenna structure," *Signal Processing*, vol. 93, no. 12, pp. 3541–3552, 2013.

[4] S. D. Somasundaram, "Wideband robust capon beamforming for passive sonar," *IEEE Journal of Oceanic Engineering*, vol. 38, no. 2, pp. 308–322, 2013.

[5] Z. G. Feng, K. F. C. Yiu, and S. E. Nordholm, "Placement design of microphone arrays in near-field broadband beamformers," *IEEE Transactions on Signal Processing*, vol. 60, no. 3, pp. 1195–1204, 2012.

[6] S. Tanaka, A. Harada, M. Sawahashi, and F. Adachi, "Experiments on coherent adaptive antenna array diversity for wideband DS-CDMA mobile radio," *IEEE Journal on Selected Areas in Communications*, vol. 18, no. 8, pp. 1495–1504, 2000.

[7] B. S. Huang, Y. L. Huang, P. L. Leu, and S. J. Lee, "Estimation of the rupture velocity and fault length of the 2004 Sumatra–Andaman earthquake using a dense broadband seismic array in Taiwan," *Journal of Asian Earth Sciences*, vol. 40, no. 3, pp. 762–769, 2011.

[8] J. Makhoul, "On the eigenvectors of symmetric Toeplitz matrices," *IEEE Transactions on Acoustics, Speech, and Signal Processing*, vol. 29, no. 4, pp. 868–872, 1981.

[9] D. Korompis, K. Yao, and F. Lorenzelli, "Broadband maximum energy array with user imposed spatial and frequency constraints," in *Proceedings of the IEEE International Conference on Acoustics, Speech and Signal Processing (ICASSP '94)*, vol. 4, pp. 529–532, Adelaide, Australia, April 1994.

[10] S. Doclo and M. Moonen, "Comparison of least-squares and eigenfilter techniques for broadband beamforming," in *Proceedings of the 3rd IEEE Benelux Signal Processing Symposium (SPS '02)*, pp. 73–76, Leuven, Belgium, March 2002.

[11] W. Liu and S. Weiss, "Design of frequency invariant beamformers for broadband arrays," *IEEE Transactions on Signal Processing*, vol. 56, no. 2, pp. 855–860, 2008.

[12] H. Duan, B. P. Ng, C. M. S. See, and J. Fang, "Applications of the SRV constraint in broadband pattern synthesis," *Signal Processing*, vol. 88, no. 4, pp. 1035–1045, 2008.

[13] Y. Zhao, W. Liu, and R. J. Langley, "Subband design of fixed wideband beamformers based on the least squares approach," *Signal Processing*, vol. 91, no. 4, pp. 1060–1065, 2011.

[14] Y. Zhao, W. Liu, and R. J. Langley, "Application of the least squares approach to fixed beamformer design with frequency-invariant constraints," *IET Signal Processing*, vol. 5, no. 3, pp. 281–291, 2011.

[15] Z. Fan and G.-L. Liang, "Broadband beamforming with minimum sidelobe and constant beamwidth based on convex optimization," *Acta Electronica Sinica*, vol. 41, no. 5, pp. 943–948, 2013.

[16] S. Doclo and M. Moonen, "Design of far-field and near-field broadband beamformers using eigenfilters," *Signal Processing*, vol. 83, no. 12, pp. 2641–2673, 2003.

3

A New Generalized Orthogonal Matching Pursuit Method

Liquan Zhao and Yulong Liu

College of Information Engineering, Northeast Electric Power University, Jilin 132012, China

Correspondence should be addressed to Liquan Zhao; zhao_liquan@163.com

Academic Editor: Jar Ferr Yang

To improve the reconstruction performance of the generalized orthogonal matching pursuit, an improved method is proposed. Columns are selected from the sensing matrix by generalized orthogonal matching pursuit, and indices of the columns are added to the estimated support set to reconstruct a sparse signal. Those columns contain error columns that can reduce the reconstruction performance. Therefore, the proposed algorithm adds a backtracking process to remove the low-reliability columns from the selected column set. For any k-sparse signal, the proposed method firstly computes the correlation between the columns of the sensing matrix and the residual vector and then selects s columns that correspond to the s largest correlation in magnitude and adds their indices to the estimated support set in each iteration. Secondly, the proposed algorithm projects the measurements onto the space that consists of those selected columns and calculates the projection coefficient vector. When the size of the support set is larger than k, the proposed method will select k high-reliability indices using a search strategy from the support set. Finally, the proposed method updates the estimated support set using the selected k high-reliability indices. The simulation results demonstrate that the proposed algorithm has a better recovery performance.

1. Introduction

In recent years, a new theory named compressive sensing (CS) [1] has surpassed the limits of the Nyquist sampling rate. Because CS can recover signals at a sampling frequency far lower than the Nyquist sampling rate, CS has aroused tremendous interests over the past few years [2, 3]. CS differs from the traditional Nyquist sampling theory and includes three procedures: sparse representation, nonrelated linear measurement, and signal reconstruction. The reconstruction algorithm aims to recover signals accurately from the measurements, and this step is one of most important parts of CS.

Recently, many reconstruction algorithms have been proposed to obtain the original sparse signal from measurements. Two major classes of reconstruction algorithms are l_1-minimization and greedy pursuit algorithms. Common l_1-minimization approaches include basis pursuit (BP) [4], Gradient projection for sparse reconstruction (GPSR) [5], iterative thresholding (IT) [6], and other algorithms. Those algorithms possess good performance in solving a convex minimization problem, but they have a higher computational complexity.

Greedy algorithms have received increasing attention for their excellent performance and small cost in recovering sparse signals from compressed measurements. A greedy algorithm proposed early on was the matching pursuit algorithm (MP) [7]. Building on the MP algorithm, the orthogonal matching pursuit algorithm (OMP) [8] was proposed to optimize the MP via orthogonalization of the estimate support set. The OMP has become a well-known greedy algorithm with wide application. The regularized orthogonal matching pursuit algorithm (ROMP) [9] was developed to refine the selected columns of the measurement matrix with a regularized rule to improve the speed of OMP. The stage wise orthogonal matching pursuit (StOMP) [10] selects multiple columns in each iteration via a presupposed threshold. The subspace pursuit (SP) [11] and compressive sampling matching pursuit (CoSaMP) [12] proposed similar improvement methods. Both of these algorithms were proposed with the idea of backtracking, and the difference is that SP selects k columns from the sensing matrix for each iteration, while CoSaMP selects $2k$. The generalized orthogonal matching pursuit (GOMP) was proposed by Wang et al. [13, 14]. The algorithm selects S $(S \leq K)$ columns in each iteration.

When $S = 1$, GOMP is identical to OMP. Compared to OMP, which selects only one column in each iteration, GOMP changes the number of columns that are selected in each iteration to improve the computational efficiency and recovery performance. The generalized OMP (GOMP) has received increasing attention in recent years, because the method can enhance the recovery performance of OMP. Several papers have been published on the analysis of the theoretical performance of GOMP [13–17].

2. Compressive Sensing Model

Compressive sensing requires that the target signal is a K-sparse signal. It means that if we regard the signal as a dimensional vector x, there should be at most K no-zero elements in x. However, in practical applications, sparse signals may not exist in all cases. The target signal has to be transformed into a sparse signal based on a set of sparse basis $\Psi = \{\phi_1, \phi_2, \phi_3, \ldots, \phi_N\}$. In this case, x can be defined as

$$x = \sum_{i=1}^{N} \alpha_i \phi_i = \Psi \alpha, \qquad (1)$$

where $\|\alpha\|_0 = K$; $\|\cdot\|_0$ denotes the number of nonzero elements in a vector. Thus, the signal x is equivalently represented by K-sparse vector α under some linear transformation Ψ in some domains. The process of compressive sensing can be regarded as a technique that automatically selects relevant information from signals by a measurement. In the theory, x is translated into M-dimensional measurements y via a matrix multiplication with Φ. We express it as

$$y = \Phi x, \qquad (2)$$

where Φ is defined as the measurement matrix with dimensions $M \times N$. Combining (1) with (2), we can obtain

$$y = \Phi x = \Phi \Psi \alpha = A_{CS} \alpha, \qquad (3)$$

where $A_{CS} = \Phi\Psi$.

In most scenarios, $M \ll N$. Thus, we can interpret that x is compressed into y with a dimension ranging from N to M. Clearly, (3) is an underdetermined equation, and it is difficult to obtain an accurate solution based on the equation. That is to say, it is impossible with traditional methods that obtain the inverse of the matrix Φ to reconstruct the original signal x accurately. In this case, we can obtain x by solving the l_p-minimization problem:

$$\begin{aligned} \widehat{\alpha} = \min \quad & \|\alpha\|_p \\ \text{s.t.} \quad & y = A_{CS}\alpha. \end{aligned} \qquad (4)$$

Several methods exist for solving this problem. When $p = 1$, the problem is a l_1-minimization problem, which can be solved by using a convex optimization algorithm. When $p = 0$, the problem is a l_0-minimization problem, which can be solved using a greedy algorithm. An appropriate condition for exact recovery is that the matrix A_{CS} satisfies the condition of restricted isometry property (RIP) condition [1].

Definition 1. A sensing Matrix A_{CS} is said to satisfy the RIP condition with the smallest number of the K-restricted isometry constant δ_K ($\delta_K \in (0,1)$), if

$$(1 - \delta_K) \|\alpha\|^2 \le \|A_{CS}\alpha\| \le (1 + \delta_K) \|\alpha\|^2 \qquad (5)$$

holds for any K-sparse vector $\alpha \in R^{N\times 1}$ with $\|\alpha\|_0 \le K$.

3. GOMP Algorithm

Greedy algorithms are used widely to recover signals in CS due to their simple algorithms and low computational complexity. Two methods exist for improving the OMP algorithm. The first method is based on the idea of backtracking, as in the subspace pursuit algorithm (SP) [11]. The second method selects more than one atom in each iteration, as in GOMP. The computational complexity of the backtracking method is higher, but it yields higher accuracy in most cases. In this section, we will introduce the GOMP algorithm.

GOMP firstly computes the correlation between the columns of the sensing matrix Φ and the residual vector r^{k-1} by $\Phi' r^{k-1}$ and then selects S columns that correspond to the S largest correlation in magnitude adding their indices to the estimated support set Λ^k in each iteration. Next the projection coefficient vector $\widehat{x}_{\Lambda^k}$ of measurements y onto space of span(Φ_{Λ^k}) is obtained using the least square method (LS). The residual r^k is revised by subtracting $\Phi_{\Lambda^k}\widehat{x}_{\Lambda^k}$ from y. These operations are repeated until either the iteration number reaches the maximum $k = \min\{K, M/S\}$, or the l_2-norm of the residual falls below a threshold ε. For ease of understanding, we describe the GOMP algorithm in Algorithm 1 according to [13, 14].

In Algorithm 1, we observe that the only difference between GOMP and OMP is that the GOMP selects more than one atom in each iteration. The GOMP algorithm selects S ($S \ge 1$) atoms in each iteration. When $S = 1$, the GOMP is identical to OMP.

4. Improved GOMP Algorithm

For the usage of greedy algorithms, it is important to generate an estimate of the correct support set. We assume the correct support is T, and the estimate of T is Λ^k. The goal is to determine the indices for Λ^k that are most similar to T. GOMP selects S columns from the sensing matrix Φ and added the indices of these columns to the estimated support set Λ^k to reconstruct a sparse signal. Those columns contain columns that be selected by error. Although GOMP allows for a number of error indices in Λ^k, these error indices will lead to the size of Λ^k larger than K. This increases the algorithm complexity greatly in the process of estimation of x_{Λ^k} and residual update in the GOMP.

In the process of estimation of x_{Λ^k}, the projection coefficient vector of the measurements onto the space of span(Φ_{Λ^k}) can be expressed as

$$\widehat{x}_{\Lambda^k} = {\Phi_{\Lambda^k}}^{\dagger} y, \qquad (6)$$

Input: measurements $y \in R^{M\times N}$,
sensing matrix $\Phi \in R^{M\times N}$,
Sparsity K,
Initialize: number of indices of columns for each selection S $(S \leq K)$.
iteration count $k = 0$,
residual vector $r^0 = y$,
estimated support set $\Lambda^0 = \varnothing$.
While $\|r^k\| > \varepsilon$ and $k < \min\{K, M/S\}$ do
$k = k + 1$;
(Identification)
Select S largest entries (in magnitude) from $\Phi' r^{k-1}$.
Then record the indices $\{\varphi(i)\}_{i=1,2,3,\ldots,S}$ corresponding to the entries.
(Augmentation)
$\Lambda^k = \Lambda^{k-1} \cup \{\varphi(1), \varphi(2), \ldots, \varphi(S)\}$.
(Estimation of x_{Λ^k})
$\hat{x}_{\Lambda^k} = \arg\min_{\sup p(u)=\Lambda^k} \|y - \Phi u\|_2$.
(Residual Update)
$r^k = y - \Phi_{\Lambda^k}\hat{x}_{\Lambda^k}$.
End
Output The estimated support $\widehat{\Lambda} = \arg\min_{T:\|T\|=K} \|\hat{x}_{\Lambda^k} - \hat{x}_T\|_2$ and signal $\hat{x}_{\widehat{\Lambda}} = \Phi^{\dagger}_{\widehat{\Lambda}} y$.

ALGORITHM 1: GOMP algorithm.

where $\Phi_{\Lambda^k}{}^{\dagger} = (\Phi^T_{\Lambda^k}\Phi_{\Lambda^k})^{-1}\Phi^T_{\Lambda^k}$; y are the measurements of the signal vector x. The cost of (6) is

$$4S^2km + \left(-2S^2 + 5N\right)m + 2S^3k^2 + \left(-4S^3 + 5S^2\right)k + 3S^3 - S^2 - S, \tag{7}$$

where S represents the number of indices of columns for each selection, k is the number of iterations, and m is the dimensionality of measurements y [13].

In the process of estimation of residual update, the residuals in GOMP can be expressed as

$$r^k = y - \Phi_{\Lambda^k}\hat{x}_{\Lambda^k}, \tag{8}$$

where y represents the measurements, Φ_{Λ^k} is the estimated support set, and $\hat{x}_{\Lambda^k}$ is the projection coefficient vector of the measurements onto the space of the estimated support set. The cost of $\Phi_{\Lambda^k}\hat{x}_{\Lambda^k}$ is

$$(2kS - 1)\,m, \tag{9}$$

where k is the number of iterations and S is the number of indices of columns for each selection.

GOMP selects S indices in each iteration and adds them to Λ^k. On the k-th iteration, kS indices will be selected, and the dimensionality of Φ_{Λ^k} is $m \times kS$. If all the indices in the estimated support set Λ^k are correct, Λ^k is identical to T and $kS \approx K$. Although a selection rule exists to ensure that newly added indices belong to the correct support T in GOMP, it is unavoidable that the error indices are selected and once selected, the error indices of the selected columns will remain in the support set throughout the remainder of the reconstruction process. It means $kS > K$. Generally speaking, when a large S is selected, the GOMP should exert higher efficiency than a small S. However, as S increases, the probability of the selected error indices increases as well. When a large number of error indices are selected, it leads to a great increase in kS, further increasing the cost of the algorithm.

To overcome this problem, we propose a method to improve the performance of GOMP. The proposed method will retain the size of the estimated support set as K and update the indices using the backtracking method to reduce the number of error indices, when the number of indices is greater than K. Even if an index is deemed reliable in some iteration, when it is considered unreliable in subsequent iterations, the index will be removed from the estimated support set. This can reduce the number of error indices in the estimated support set.

The main difference between the proposed and the conventional GOMP algorithm is that the new algorithm can add or remove the index from the estimated support set at any stage of the recovery process. The proposed algorithm increases the reliability of the indices of the estimated support set and furthermore reduces the cost of the reconstruction process. The process of proposed algorithm is expressed in Algorithm 2.

In Algorithm 2, we observe that the proposed method adds a backtracking step to the GOMP algorithm. When S is small, it has a high accuracy for the identification process. When $S = 1$, GOMP is identical to OMP. At this step, GOMP has the highest accuracy in the identification step.

Input: measurements $y \in R^{M\times N}$,
sensing matrix $\Phi \in R^{M\times N}$,
Sparsity K,
Initialize: number of indices of columns for each selection S $(S \le K)$.
iteration count $k = 0$,
residual vector $r^0 = y$,
estimated support set $\Lambda^0 = \varnothing$.
While $\|r^k\| > \varepsilon$ and $k < \min\{K, M/S\}$ do
$k = k + 1$;
(Identification)
Select S largest entries (in magnitude) from $\Phi' r^{k-1}$. Then record the indices $\{\varphi(i)\}_{i=1,2,3,\ldots,S}$ corresponding to the entries.
(Augmentation)
$\Lambda^k = \Lambda^{k-1} \cup \{\varphi(1), \varphi(2), \ldots, \varphi(S)\}$.
(Estimation of x_{Λ^k}**)**
$x_{\Lambda^k} = \arg\min_{\sup p(u)=\Lambda^k} \|y - \Phi u\|_2$.
(Backtracking) While $\|x_{\Lambda^k}\|_0 \ge K$, select K largest elements of $|x_{\Lambda^k}|$. Then recording the indices corresponding to the elements, and renew the Λ_k with those indices.
(Residual Update) $r^k = y - \Phi_{\Lambda^k} \widehat{x}_{\Lambda^k}$.
End
Output The estimated support $\widehat{\Lambda} = \arg\min_{T:\|T\|=K} \|\widehat{x}_{\Lambda^k} - \widehat{x}_T\|_2$ and signal $\widehat{x}_{\widehat{\Lambda}} = \Phi^{\dagger}_{\widehat{\Lambda}} y$.

ALGORITHM 2: Improved method of GOMP.

As S increases, the probability of the selected error indices increases as well. The proposed method adds a backtracking process to remove those error indices in subsequent iterations. When $S \le K$, the proposed method requires more than one iteration to create a K-sized support set. When $kS < K$, the support set cannot contain K correct indices, and executing the backtracking step is unnecessary and will cause wasteful computation. Therefore, we designed the algorithm to execute the backtracking step when $kS \ge K$. The dimensionality of x_{Λ^k} is $kS \times 1$ before executing the backtracking step. So the improved algorithm determines whether to execute the backtracking step by the size of x_{Λ^k}. Experimental evidence demonstrated that our changes improved the performance. We will describe our simulation in the next section.

5. Simulation and Discussion

In this part, we will demonstrate the performance of the proposed algorithm with sparse signals. We used the same sparse signal sources as in GOMP with K-sparse to compare the performances of the different algorithms. The components of the sensing matrix were generated randomly with Gaussian distribution, and the size of the sensing matrix was 128×256. We used MATLAB 7.0 with a quad-core 64-bit processor in a Windows 10 environment. We executed each algorithm 1000 times and recorded the probability of the exact reconstructions. We set the threshold $\varepsilon = 10^{-6}$ used in [13] in both GOMP and the proposed method.

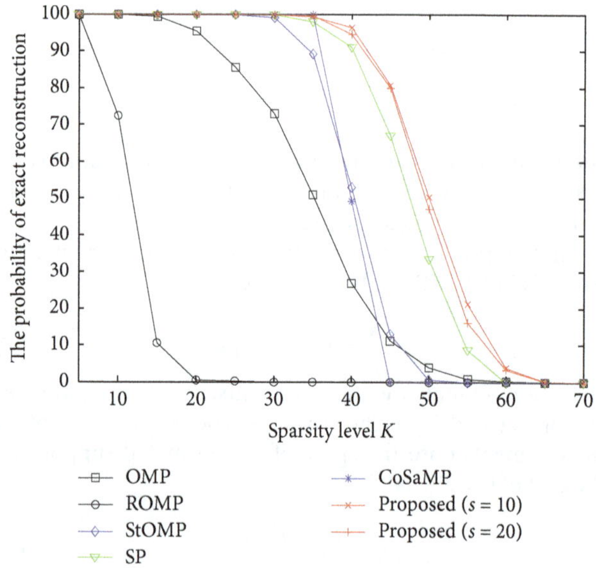

FIGURE 1: Reconstruction performance for K-sparse Gaussian signal vector as a function of sparsity K.

In Figure 1, we compare the probability of successful reconstruction of the proposed method with the ROMP [9], OMP [8], SP [11], CoSaMP [12], and StOMP [10] algorithms. It is evident from Figure 1 that the reconstruction performance of ROMP decreased rapidly with a rise in K.

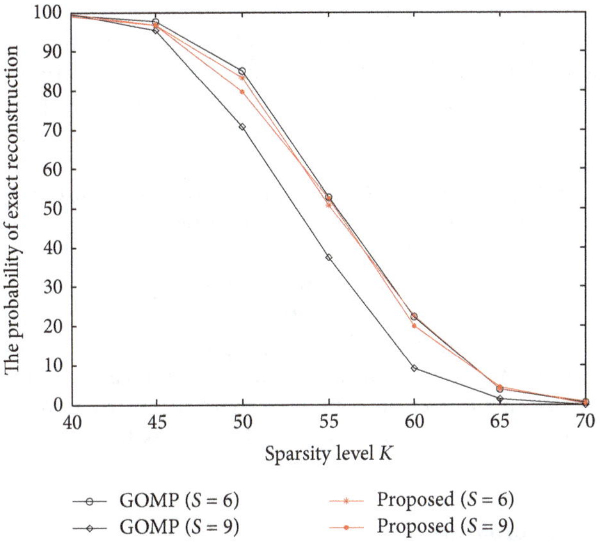

FIGURE 2: Reconstruction performance of GOMP and the proposed method with a small S value.

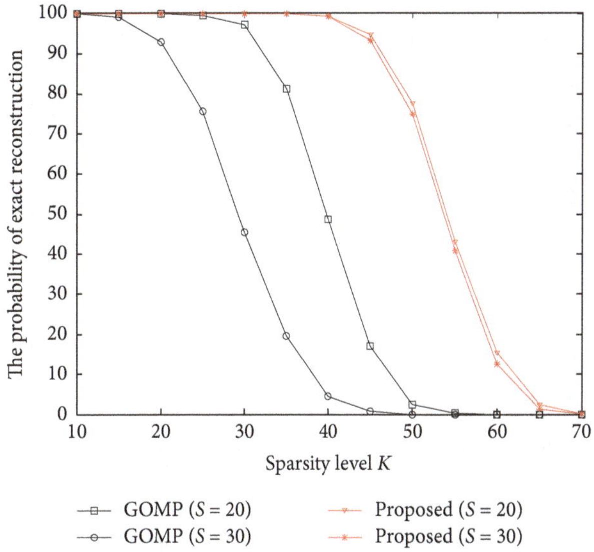

FIGURE 3: Reconstruction performance of GOMP and the proposed method with a large S value.

When $K = 20$, the recovery probability was near zero. The probability of successful recovery of OMP decreased at a smaller rate than that of ROMP, and its critical sparsity was reached at $K = 55$. CoSaMP and StOMP had similar performances. Both retained a high recovery performance when $K \leq 40$. The probability of successful recovery of SP was higher than for ROMP, OMP, CoSaMP, and StOMP with the same K. For our proposed method, we tested the performance with different S values, representing the number of selected columns in each iteration. We set $S = 10$ and $S = 20$. Figure 1 indicates that the smaller the value of S, the higher the probability of successful recovery of the proposed algorithm with the same K. Compared to the other algorithms, under the same conditions, the proposed method with $S = 10$ had the highest probability of successful recovery, followed by the proposed method with $S = 20$, and followed by the SP algorithm. The proposed algorithm with different S values had a higher probability than the other algorithms.

In Figures 2 and 3, we compare the performances of the proposed method and GOMP with different values for the parameter S. In Figure 2, we selected smaller S values to compare the algorithms. All algorithms resulted in high levels of probability of recovery when $K \leq 40$. The probability of exact recovery decreased with an increase in K and reached zero when $K = 70$. Both the proposed algorithm and GOMP had a higher probability of exact recovery with a smaller S value. When $S = 6$, the two algorithms had nearly the same probability values. When $S = 9$, the proposed method had a higher probability of recovery. The probability of exact reconstruction of both algorithms was 100% when $K < 40$ and zero when $K > 70$ for $S = 6$ and $S = 9$, and the difference between two algorithms mainly concentrated on the sparsity level from 40 to 70. Therefore, we compared two algorithms with $K = 40$ to $K = 70$ in Figure 2.

In Figure 3, we selected a larger value of S to compare the proposed algorithm and GOMP. The probability of exact reconstruction of both algorithms was 100% for $K < 10$ and zero for $K > 70$, so we compared the two algorithms with $K = 10$ to $K = 70$ for $S = 20$ and $S = 30$ in Figure 3.

When $K = 40$ and $S = 20$, the probability of exact recovery of GOMP was 48.7%, while the probability value for the proposed method was 99.4%. When $K = 40$ and $S = 30$, the resulting probability values were 4.5% for GOMP and 99.4% for the proposed method. These results indicated that, for different S values, the proposed method had a higher probability of exact recovery than the GOMP method.

Figures 2 and 3 demonstrate that, for different S values, the curves of probability for exact recovery were closer to each other for the proposed method than the GOMP method. This indicated that the proposed method had a more stable performance than the GOMP method for different S values.

Based on the analysis and comparison, we determined that even though the proposed algorithm had an additional backtracking process compared to GOMP, the proposed method demonstrated an excellent performance with regard to running time. In order to compare running times for all the algorithms, we ran each algorithm one thousand times to calculate the average running time. The computing environment was the same as for the determination of the probability of exact reconstruction.

In Figure 2, we can see that the probability of extract reconstruction of both algorithms was below 100% for $K > 40$ with $S = 6$ and $S = 9$, and the effective running time should be computed for successful reconstructing. Therefore, we just compared the running time with $K < 40$ for two methods. The difference was not obvious with $K < 5$ in running time for both methods with $S = 6$ and $S = 9$, so we compared the running time with $K > 5$ and $K < 40$ in Figure 4.

In Figure 3, we can see that probability of extract reconstruction of GOMP algorithm with $S = 20$ was below 100% for $K < 20$, and with $S = 30$ for $K < 10$, and the probability of extract reconstruction of proposed algorithm with $S = 20$ and

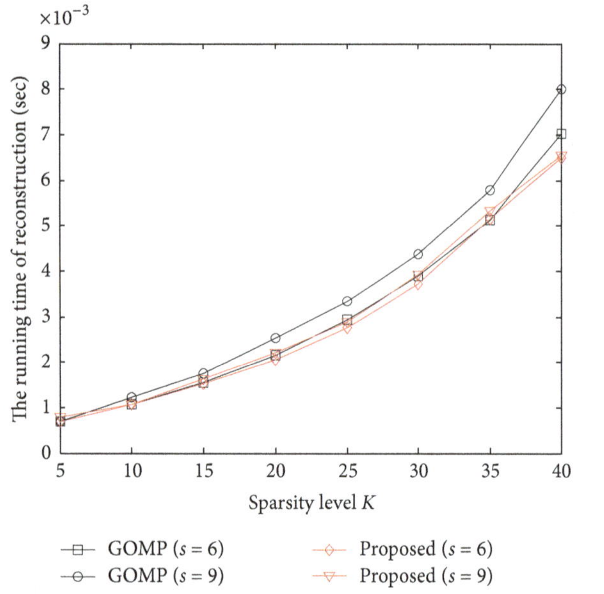

FIGURE 4: The average running time of GOMP and the proposed method with a small S value.

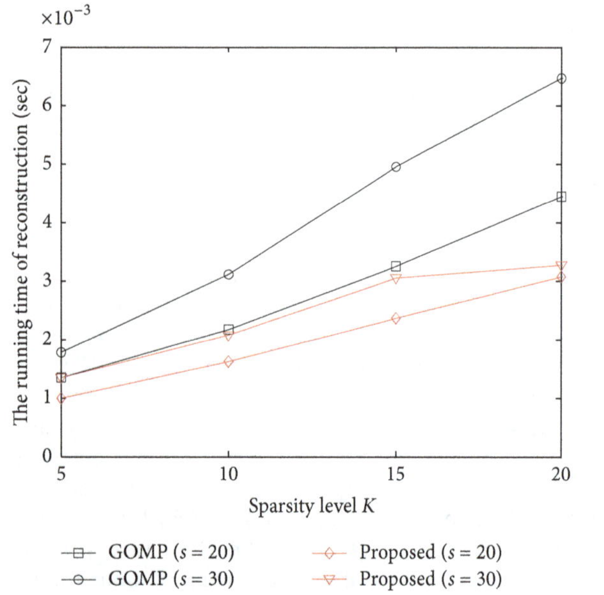

FIGURE 5: The average running time of GOMP and the proposed method with a large S value.

$S = 30$ all were below 100% for $K < 40$. The effective running time should be computed for successful reconstructing, so we just compared the running time of both algorithms for $S = 20$ with $K < 20$, and $S = 30$ with $K < 10$. The difference was not obvious in running time for both methods with smaller K. Based on the above analysis, we just compared their running time with $K > 5$ and $K < 20$ for $S = 20$, and $K > 5$ and $K < 10$ for $S = 30$. In order to unify the sparsity level scope, we set K from 5 to 20 in Figure 5.

In Figure 4, we set $S = 6$ and $S = 9$ to compare the running times for GOMP and the proposed method. The running times for GOMP with $S = 6$ and the proposed method with $S = 6$ and $S = 9$ were nearly identical. The running times under those conditions were less than those for GOMP with $S = 9$. When $K = 40$, the running time of the proposed method with $S = 6$ and $S = 9$ was less than that of GOMP with $S = 6$.

In Figure 5, we set $S = 20$ and $S = 30$. From Figure 5, we can observe that the running time of the proposed algorithms was less than GOMP for $S = 20$ with $K = 5$ to 20, and for $S = 30$ with $K = 5$ to 10. This shows that the proposed algorithm has faster speed for reconstructing.

These results show that the proposed method resulted not only in better performance with regard to running time, but also in a higher probability of exact reconstruction. The advantage is more obvious when larger S is taken in both algorithms.

6. Conclusion

In this paper, a novel method for sparse signal reconstruction has been proposed. The proposed method adds a backtracking step to maintain a K-sized estimated support set, avoiding the extra computation cost for an oversized estimated set. At the same time, the proposed method can increase the reliability of the estimated support set by removing the low-reliability columns from the estimated support set. The proposed reconstruction algorithm performs well at not only reconstructing the sparse signal (i.e., when K is small), but also the less sparse signal (i.e., when K is large). The simulation results showed that the proposed method had a better performance than GOMP with regard to both the probability of exact reconstruction and running time, especially with larger S values. In future research, we want to improve on the proposed algorithm to optimize the greedy algorithm further.

Conflicts of Interest

The authors declare that they have no conflicts of interest.

Acknowledgments

This work was supported by the National Natural Science Foundation of China (61271115) and the Foundation of Jilin Educational Committee (2015235).

References

[1] D. L. Donoho, "Compressed sensing," *Institute of Electrical and Electronics Engineers. Transactions on Information Theory*, vol. 52, no. 4, pp. 1289–1306, 2006.

[2] R. Zhao and H. N. Yu, "Compressed sensing, based over-complete dictionary learning algorithm for images," *Journal of Northeast Dianli University*, vol. 32, no. 4, pp. 44–47, 2012.

[3] H. N. Yu and L. Li, "Compressed sensing," *Journal of Northeast Dianli University*, vol. 32, no. 1, pp. 79–83, 2011.

[4] S. S. Chen, D. L. Donoho, and M. A. Saunders, "Atomic decomposition by basis pursuit," *SIAM Journal on Scientific Computing*, vol. 20, no. 1, pp. 33–61, 1998.

[5] M. A. T. Figueiredo, R. D. Nowak, and S. J. Wright, "Gradient projection for sparse reconstruction: Application to compressed

sensing and other inverse problems," *IEEE Journal on Selected Topics in Signal Processing*, vol. 1, no. 4, pp. 586–597, 2007.

[6] I. Daubechies, M. Defrise, and C. de Mol, "An iterative thresholding algorithm for linear inverse problems with a sparsity constraint," *Communications on Pure and Applied Mathematics*, vol. 57, no. 11, pp. 1413–1457, 2004.

[7] S. G. Mallat and Z. Zhang, "Matching pursuits with time-frequency dictionaries," *IEEE Transactions on Signal Processing*, vol. 41, no. 12, pp. 3397–3415, 1993.

[8] J. A. Tropp and A. C. Gilbert, "Signal recovery from random measurements via orthogonal matching pursuit," *Institute of Electrical and Electronics Engineers. Transactions on Information Theory*, vol. 53, no. 12, pp. 4655–4666, 2007.

[9] D. Needell and R. Vershynin, "Signal recovery from incomplete and inaccurate measurements via regularized orthogonal matching pursuit," *IEEE Journal on Selected Topics in Signal Processing*, vol. 4, no. 2, pp. 310–316, 2010.

[10] D. L. Donoho, Y. Tsaig, I. Drori, and J.-L. Starck, "Sparse solution of underdetermined systems of linear equations by stagewise orthogonal matching pursuit," *Institute of Electrical and Electronics Engineers. Transactions on Information Theory*, vol. 58, no. 2, pp. 1094–1121, 2012.

[11] W. Dai and O. Milenkovic, "Subspace pursuit for compressive sensing signal reconstruction," *Institute of Electrical and Electronics Engineers. Transactions on Information Theory*, vol. 55, no. 5, pp. 2230–2249, 2009.

[12] D. Needell and J. A. Tropp, "CoSaMP: iterative signal recovery from incomplete and inaccurate samples," *Applied and Computational Harmonic Analysis. Time-Frequency and Time-Scale Analysis, Wavelets, Numerical Algorithms, and Applications*, vol. 26, no. 3, pp. 301–321, 2009.

[13] J. Wang, S. Kwon, and B. Shim, "Generalized orthogonal matching pursuit," *IEEE Transactions on Signal Processing*, vol. 60, no. 12, pp. 6202–6216, 2012.

[14] J. Wang, S. Kwon, P. Li, and B. Shim, "Recovery of sparse signals via generalized orthogonal matching pursuit: a new analysis," *IEEE Transactions on Signal Processing*, vol. 64, no. 4, pp. 1076–1089, 2016.

[15] S. Satpathi, R. L. Das, and M. Chakraborty, "Improving the bound on the RIP constant in generalized orthogonal matching pursuit," *IEEE Signal Processing Letters*, vol. 20, no. 11, pp. 1074–1077, 2013.

[16] Y. Shen, B. Li, W. Pan, and J. Li, "Analysis of generalised orthogonal matching pursuit using restricted isometry constant," *Electronics Letters*, vol. 50, no. 14, pp. 1020–1022, 2014.

[17] B. Li, Y. Shen, Z. Wu, and J. Li, "Sufficient conditions for generalized orthogonal matching pursuit in noisy case," *Signal Processing*, vol. 108, pp. 111–123, 2015.

4

Rao-Blackwellized Gaussian Sum Particle Filtering for Multipath Assisted Positioning

Markus Ulmschneider, Christian Gentner, Thomas Jost, and Armin Dammann

German Aerospace Center (DLR), Institute of Communications and Navigation, Muenchner Str. 20, 82334 Wessling, Germany

Correspondence should be addressed to Markus Ulmschneider; Markus.Ulmschneider@dlr.de

Academic Editor: Huimin Chen

In multipath assisted positioning, multipath components arriving at a receiver are regarded as being transmitted by a virtual transmitter in a line-of-sight condition. As the locations and clock offsets of the virtual and physical transmitters are in general unknown, simultaneous localization and mapping (SLAM) schemes can be applied to simultaneously localize a user and estimate the states of physical and virtual transmitters as landmarks. Hence, multipath assisted positioning enables localizing a user with only one physical transmitter depending on the scenario. In this paper, we present and derive a novel filtering approach for our multipath assisted positioning algorithm called Channel-SLAM. Making use of Rao-Blackwellization, the location of a user is tracked by a particle filter, and each landmark is represented by a sum of Gaussian probability density functions, whose parameters are estimated by unscented Kalman filters. Since data association, that is, finding correspondences among landmarks, is essential for robust long-term SLAM, we also derive a data association scheme. We evaluate our filtering approach for multipath assisted positioning by simulations in an urban scenario and by outdoor measurements.

1. Introduction

The amount of available and potential services requiring precise localization of a user has steadily increased over the recent years. Global navigation satellite systems (GNSSs) can often satisfy the demands for localization in scenarios where the receiver has a clear view of the sky. However, if the view of the sky is obstructed, such as indoors, in urban canyons, or in tunnels, the positioning performance of GNSSs may be drastically decreased, or no positioning solution may be obtained at all [1]. Reasons for this include a low received signal power due to signal blocking or shadowing and multipath propagation.

In contrast to GNSS signals, many kinds of terrestrial signals are likely to have a good coverage in GNSS denied places. In particular, cellular radio frequency (RF) signals are designed to be reliably available at least in populated areas, and they may be used as signals of opportunity (SoOs) for positioning. However, also terrestrial signals experience multipath propagation. Multipath propagation biases range estimates if standard correlator based methods are used. Various approaches to handle the multipath problem have been addressed in the literature, for example, in [2]. Advanced methods such as maximum likelihood (ML) mitigation algorithms try to estimate the channel impulse response (CIR) and to mitigate the influence of multipath components (MPCs) on the line-of-sight (LoS) path [3].

The idea of multipath assisted positioning is contrary, though. Instead of regarding multipath propagation as ill, the spatial information of MPCs on the receiver position is exploited. In [4], the information of MPCs is used in a fingerprinting scheme. Going one step further, each MPC can be regarded as being transmitted by a virtual transmitter in a pure LoS condition, and the virtual transmitters can be used to locate the user. Such an approach is called multipath assisted positioning.

The authors of [5, 6] derived some theoretical bounds for multipath assisted positioning. Multipath assisted positioning schemes have, for example, been applied in radar applications [7], using ultrawideband (UWB) [8, 9] or 5G [10] systems and in cooperative systems [11].

If the locations of physical transmitters and reflecting and scattering objects are known, the locations of virtual transmitters can be calculated based on geometrical considerations. The authors of [12] assume the room layout to be known and focus on the association among virtual transmitters and reflecting walls. In a general setting, however, the scenario is unknown to the user.

The authors of [13, 14] have presented a multipath assisted positioning scheme named Channel-SLAM that does not rely on prior information on the scenario. Instead, the locations of the physical and virtual transmitters are estimated simultaneously with the user position in a simultaneous localization and mapping (SLAM) [15, 16] approach. In general, SLAM describes the simultaneous estimation of a user position and the locations of landmarks. In Channel-SLAM, the landmarks are the physical and virtual transmitters. Previous extensions to Channel-SLAM include mapping of the user positions [17], the consideration of vehicular applications [18], and data association methods [19, 20], for example.

Nonlinearities in the prediction and update equations of the Bayesian recursive estimation framework prohibit the use of optimal algorithms such as the Kalman filter, since the integrals involved in the estimation process cannot be solved in closed form or become intractable. A popular alternative is the extended Kalman filter (EKF) [21], which linearizes the nonlinear terms using a first-order Taylor series expansion. However, such a linearization can introduce large errors in the estimation process [22]. The unscented Kalman filter (UKF) [23, 24] uses a nonlinear transformation to deal with nonlinearities and outperforms the EKF in a wide range of applications [22, 25].

UKF methods have found their way into localization problems, for example, in [27, 28]. The authors of [29] propose Gaussian sum cubature filters. In [30, 31], the authors consider a Rao-Blackwellization scheme for SLAM with a particle filter for the user state and UKFs for the landmark states, where the measurement model is based on linearization, though.

The current Channel-SLAM algorithm uses a Rao-Blackwellized particle filter to estimate the user state and the location of transmitters simultaneously. Hence, both the user state probability density function (PDF) and the transmitter state PDFs are represented by a large set of particles, tending to result in a high memory occupation. This paper is an extension of [32], where we proposed a novel estimation approach for Channel-SLAM scheme based on Rao-Blackwellization and performed first simulations. We refer to this new estimation method as Rao-Blackwellized Gaussian sum particle filter (RBGSPF). In the RBGSPF, the user position is tracked by a sequential importance resampling (SIR) particle filter, while the physical and virtual transmitter state PDFs are represented by Gaussian mixture models estimated by UKFs. This parametrized representation of the transmitter states is a key enabler for exchanging maps of transmitters among users, since the amount of data that has to be communicated among users can be decreased drastically compared to the nonparametric representation with particles. Such an exchange of maps may be performed directly among users or via a central entity, for example, in form of local dynamic maps (LDMs) in an intelligent transportation system (ITS) context. In this paper, we provide a full and detailed derivation of our novel algorithm. In particular, we derive the calculation of the particle weights in the user particle filter given the representation of the transmitters in the UKF framework. Since data association is an essential feature for the accuracy in long-term SLAM, we also derive a data association method based on [33]. We evaluate our algorithm by both simulations in an urban scenario and outdoor measurements.

The remainder of this paper is structured as follows. Section 2 describes the fundamental idea behind multipath assisted positioning and Channel-SLAM. In Section 3, we briefly summarize some concepts of nonlinear Kalman filtering. The derivation of the RBGSPF is presented in Section 4, and a solution to data association is presented in Section 5. After the experimental results in Section 6, Section 7 concludes the paper.

Throughout the paper, we use the following notation:

(i) As indices, i stands for a user particle, j denotes a transmitter or a signal component, ℓ is a component in a Gaussian mixture model, and m stands for a sigma point.

(ii) $(\cdot)^T$ denotes the transpose of a matrix or vector.

(iii) $\mathbb{1}_n$ denotes the identity matrix of dimension $n \times n$.

(iv) $\mathbb{0}_n$ and $\mathbb{0}_{m\times n}$ denote the zero matrices of dimensions $n \times n$ and $m \times n$, respectively.

(v) $\mathcal{N}(\mathbf{x}; \boldsymbol{\mu}, \mathbf{C})$ denotes the PDF of a normal distribution in $\mathbf{x}$ with mean $\boldsymbol{\mu}$ and covariance $\mathbf{C}$.

(vi) c_0 denotes the speed of light.

(vii) $\|\cdot\|$ denotes the Euclidean norm of a vector.

2. Multipath Assisted Positioning

2.1. Virtual Transmitters. The idea of virtual transmitters is illustrated in Figure 1. The physical transmitter Tx transmits an RF signal. A mobile user equipped with an RF receiver receives the transmitted signal via three different propagation paths.

In the first case, the signal is reflected at the reflecting surface. The user treats the corresponding impinging MPC as being transmitted by the virtual transmitter vTx1 in a pure LoS condition. The location of vTx1 is the location of the physical transmitter Tx mirrored at the reflecting surface. When the user moves along the trajectory, the reflection point at the wall moves as well. However, the location of vTx1 is static. The two transmitters Tx and vTx1 are inherently perfectly synchronized.

In the second case, the signal from the physical transmitter is scattered by a point scatterer and then received by the user. We define the effect of scattering such that the energy of an electromagnetic wave impinging against the scatterer is distributed uniformly in all directions [34]. The user regards the scattered signal as a LoS signal from the virtual transmitter vTx2, which is located at the scatterer location. If the signal is scattered, the physical and the virtual

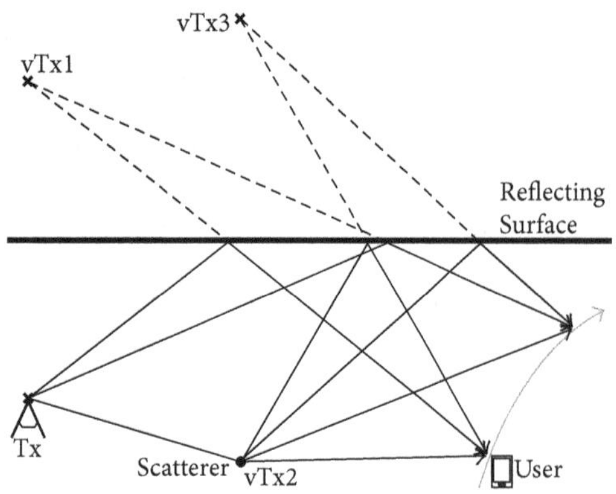

FIGURE 1: Signals from the physical transmitter Tx are received at the two user positions via different propagation paths. Each MPC arriving at the receiver is regarded as being transmitted by a virtual transmitter in a pure LoS condition. The propagation paths correspond to a reflection at the wall (vTx1), a scattering at a point scatterer (vTx2), and a scattering followed by a reflection at the wall (vTx3).

transmitter are not time synchronized: the virtual transmitter has an additional delay offset to the physical transmitter corresponding to the propagation time of the signal traveling from the physical to the virtual transmitter.

In the third case, the signal is first scattered at the scatterer and then reflected at the surface. The user treats this signal as being sent from the virtual transmitter vTx3. The location of vTx3 is the location of vTx2, that is, the scatterer location, mirrored at the reflecting surface. Accordingly, the concept of single reflections and scatterings can be generalized in a straightforward manner to the case of multiple reflections and scatterings by applying the first two cases iteratively. In case the signal undergoes only reflections, the physical and the virtual transmitters are inherently time synchronized. If the signal is scattered at least once, the delay offset corresponds to the actual propagation time of the signal from the physical transmitter to the last scatterer the signal interacts with. Therefore, in Figure 1, the virtual transmitters vTx2 and vTx3 have the same delay offset towards the physical transmitter. Note that a delay offset can be interpreted as a clock offset.

Throughout the paper, we consider the physical transmitter and the environment to be static. Hence, the virtual transmitters are static as well.

2.2. Recursive Bayesian Estimation. Recursive Bayesian estimation [35] is a method to recursively estimate the evolution of a state vector $\mathbf{x}$, where the state evolution is modeled as

$$\mathbf{x}_k = \mathbf{f}_k\left(\mathbf{x}_{k-1}, \mathbf{v}_{k-1}\right). \quad (1)$$

The index k denotes the time instant, the function $\mathbf{f}_k(\cdot)$ is assumed to be known, and $\mathbf{v}_{k-1}$ denotes a sample of the process noise with covariance matrix $\mathbf{Q}$. The state is related to the measurement $\mathbf{z}_k$ by

$$\mathbf{z}_k = \mathbf{h}_k\left(\mathbf{x}_k, \mathbf{n}_k\right), \quad (2)$$

where $\mathbf{h}_k(\cdot)$ is again a known function and $\mathbf{n}_k$ is a sample of the measurement noise with covariance matrix $\mathbf{R}$. Recursive Bayesian estimation works in two steps, the prediction and the update step. The corresponding PDFs can be calculated recursively by

$$\mathrm{p}\left(\mathbf{x}_k \mid \mathbf{z}_{1:k-1}\right) = \int \mathrm{p}\left(\mathbf{x}_k \mid \mathbf{x}_{k-1}\right)\mathrm{p}\left(\mathbf{x}_{k-1} \mid \mathbf{z}_{1:k-1}\right) d\mathbf{x}_{k-1} \quad (3)$$

for the prediction step and by

$$\mathrm{p}\left(\mathbf{x}_k \mid \mathbf{z}_{1:k}\right) = \frac{1}{c_k}\mathrm{p}\left(\mathbf{z}_k \mid \mathbf{x}_k\right)\mathrm{p}\left(\mathbf{x}_k \mid \mathbf{z}_{1:k-1}\right), \quad (4)$$

for the update step, where c_k is a constant and $\mathbf{z}_{1:k}$ denotes the measurements from time instant 1 to k. The state transition prior $\mathrm{p}(\mathbf{x}_k \mid \mathbf{x}_{k-1})$ and the measurement likelihood $\mathrm{p}(\mathbf{z}_k \mid \mathbf{x}_k)$ are obtained from the movement model in (1) and the measurement model in (2), respectively.

2.3. Channel-SLAM. In the following, we will revise the Channel-SLAM algorithm from [13, 17]. Figure 2 gives an overview of the two stages of Channel-SLAM. In the first stage, the parameters of the signal components received by the user via different propagation paths are estimated. The resulting estimates are used as measurement input in the second stage, where the states of the physical and virtual transmitters and the user position are estimated simultaneously in a SLAM scheme. Further sensors, such as an inertial measurement unit (IMU), may be included in the second stage. The locations of both the physical and virtual transmitters are assumed to be unknown. Thus, Channel-SLAM does not differentiate between physical and virtual transmitters, and the term transmitter comprises both physical and virtual transmitters in the following. Each signal component arriving at the receiver corresponds to one transmitter.

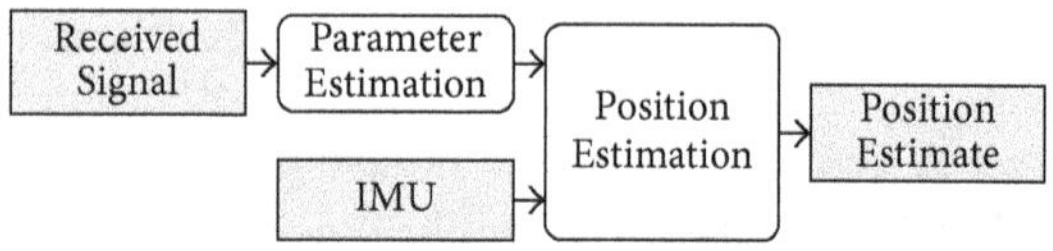

FIGURE 2: Based on the received signal, the parameters of the propagation paths are estimated in the first step by the KEST algorithm. In the second step, the estimates serve as measurements for estimating the positions of the user and the physical and virtual transmitters. In addition, user heading rates of change measurements from an IMU are incorporated in the second step.

The RF propagation channel between the physical transmitter and the user equipped with a receiver is assumed to be a linear and time-variant multipath channel.

The received signal is modeled as a superposition of signal components of the transmit signal $s(t)$, where the jth signal component is defined by a complex amplitude $a_j(t_k)$ and a delay $d_j(t_k)$ at time t_k. The signal received by the user at time instant t_k is

$$y\left(\tau, t_k\right) = \sum_j a_j\left(t_k\right) s\left(\tau - d_j\left(t_k\right)\right) + n(\tau), \quad (5)$$

where $n(\tau)$ is a sample from a colored noise sequence incorporating both dense multipath components (DMCs) and additive Gaussian noise. The channel is assumed to be constant during the short time interval when the received signal is sampled at time instant k.

The physical transmitter continuously broadcasts the signal $s(t)$ that is known to the user. At the user side, the parameters of the signal components arriving at the receiver are estimated. Such parameters can in general be the complex amplitude, time of arrival (ToA), angle of arrival (AoA), or Doppler shift, depending on the available hardware and the scenario. For the signal parameter estimation, we use the KEST algorithm [36]. The KEST estimator works in two stages. In an inner stage, a ML parameter estimator such as Space-Alternating Generalized Expectation-Maximization (SAGE) [37] estimates the parameters of the signal components jointly on a snapshot basis. An outer stage tracks these estimated parameters over time with a Kalman filter and keeps track of the number of signal components. The KEST estimator is in general able to handle the DMCs in the noise term in (5). However, DMC handling is not implemented in our evaluations, leading to a model mismatch in KEST and hence to a higher variance in the parameter estimation. However, we do not expect many DMCs in our evaluation scenarios. In an indoor scenario, for example, DMCs need to be considered [38].

In the literature, there are alternatives to the KEST estimator. For example, the authors of [39] track signal component parameters based on an EKF, though the authors of [40] showed that the KEST estimator is more robust in resolving signal components that are close to each other in the state space. In [41], an EKF is used as well for parameter estimation, while the position estimation is based on the time difference of arrival (TDoA) of virtual transmitters. The authors of [42] consider the linearization of the observation model in the EKF a major drawback that might lead to a tracking loss.

In the second stage of Channel-SLAM, we use only the delays, that is, ToAs, and AoAs, estimated by KEST as measurement inputs. Hence, after sampling the received signal, the KEST estimates at time instant k are comprised in the vector

$$\mathbf{z}_k = \left[\mathbf{d}_k^T \ \boldsymbol{\theta}_k^T\right]^T, \tag{6}$$

where

$$\mathbf{d}_k = \left[d_{1,k} \ \cdots \ d_{N_{\mathrm{TX}},k}\right]^T \tag{7}$$

are the ToA estimates for the N_{TX} signal components, or transmitters, and

$$\boldsymbol{\theta}_k = \left[\theta_{1,k} \ \cdots \ \theta_{N_{\mathrm{TX}},k}\right]^T \tag{8}$$

are the corresponding AoA estimates. Note that the number of signal components and thus transmitters may change over time. Nevertheless, for notational convenience, we omit the time instant index k in N_{TX}.

In the second stage of Channel-SLAM, the user state $\mathbf{x}_{\mathrm{u},k}$ is estimated simultaneously with the state of the transmitters $\mathbf{x}_{\mathrm{TX},k}$. The entire state vector is hence

$$\begin{aligned}\mathbf{x}_k &= \left[\mathbf{x}_{\mathrm{u},k}{}^T \ \mathbf{x}_{\mathrm{TX},k}{}^T\right]^T \\ &= \left[\mathbf{x}_{\mathrm{u},k}{}^T \ \mathbf{x}_{\mathrm{TX},k}^{\langle 1\rangle}{}^T \ \cdots \ \mathbf{x}_{\mathrm{TX},k}^{\langle N_{\mathrm{TX}}\rangle}{}^T\right]^T,\end{aligned} \tag{9}$$

where $\mathbf{x}_{\mathrm{TX},k}^{\langle j\rangle}$ is the state of the jth transmitter. As we consider a two-dimensional scenario, the user state at time instant k is defined by

$$\mathbf{x}_{\mathrm{u},k} = \left[x_k \ y_k \ v_{x,k} \ v_{y,k}\right]^T = \left[\mathbf{p}_{\mathrm{u},k}^T \ \mathbf{v}_{\mathrm{u},k}^T\right]^T, \tag{10}$$

where the user position is defined by $\mathbf{p}_{\mathrm{u},k} = \left[x_k \ y_k\right]^T$ and the user velocity by $\mathbf{v}_{\mathrm{u},k} = \left[v_{x,k} \ v_{y,k}\right]^T$. Each transmitter is defined by its location $\mathbf{p}_{\mathrm{TX},k} = \left[x_{\mathrm{TX},k} \ y_{\mathrm{TX},k}\right]^T$ and a clock offset $\tau_{0,k}$ at time instant k. The state vector of the jth transmitter is hence defined by

$$\mathbf{x}_{\mathrm{TX},k}^{\langle j\rangle} = \left[x_{\mathrm{TX},k}^{\langle j\rangle} \ y_{\mathrm{TX},k}^{\langle j\rangle} \ \tau_{0,k}^{\langle j\rangle}\right]^T = \left[\mathbf{p}_{\mathrm{TX},k}^{\langle j\rangle}{}^T \ \tau_{0,k}^{\langle j\rangle}\right]^T. \tag{11}$$

Our goal is to find the minimum mean square error (MMSE) estimator for $\mathbf{x}_k$, which is defined as

$$\hat{\mathbf{x}}_k = \int \mathbf{x}_k \mathrm{p}\left(\mathbf{x}_k \mid \mathbf{z}_{1:k}\right) \mathrm{d}\mathbf{x}_k, \tag{12}$$

where $\mathbf{z}_{1:k}$ denotes all measurements up to time instant k.

We use a recursive Bayesian estimation scheme as in Section 2.2 to estimate the posterior PDF $\mathrm{p}(\mathbf{x}_k \mid \mathbf{z}_{1:k})$. This posterior can be factorized as

$$\begin{aligned}\mathrm{p}\left(\mathbf{x}_k \mid \mathbf{z}_{1:k}\right) &= \mathrm{p}\left(\mathbf{x}_{\mathrm{TX},k}, \mathbf{x}_{\mathrm{u},k} \mid \mathbf{z}_{1:k}\right) \\ &= \mathrm{p}\left(\mathbf{x}_{\mathrm{TX},k} \mid \mathbf{x}_{\mathrm{u},k}, \mathbf{z}_{1:k}\right) \mathrm{p}\left(\mathbf{x}_{\mathrm{u},k} \mid \mathbf{z}_{1:k}\right).\end{aligned} \tag{13}$$

The signal components arriving at the receiver are assumed to be independent of each other; that is, we assume they interact with distinct objects. Assuming independence among the measurements for distinct transmitters (on the one hand, the parameters of the signal components are estimated jointly by the KEST algorithm, and hence these estimates might be correlated between signal components and between the parameters; on the other hand, the correlation is likely to have effect only on a short term basis as KEST estimates are unbiased when observed over a longer time), the first factor in (13) can be factorized further as

$$\mathrm{p}\left(\mathbf{x}_{\mathrm{TX},0:k} \mid \mathbf{x}_{\mathrm{u},0:k}, \mathbf{z}_{1:k}\right) = \prod_{j=1}^{N_{\mathrm{TX}}} \mathrm{p}\left(\mathbf{x}_{\mathrm{TX},0:k}^{\langle j\rangle} \mid \mathbf{x}_{\mathrm{u},0:k}, \mathbf{z}_{1:k}^{\langle j\rangle}\right). \tag{14}$$

With the above factorization, the transmitter states are estimated independently from each other.

As we consider a static scenario, the virtual transmitters are static as well, and the transition prior for the jth transmitter is calculated as

$$\mathrm{p}\left(\mathbf{x}_{\mathrm{TX},k}^{\langle j\rangle} \mid \mathbf{x}_{\mathrm{TX},k-1}^{\langle j\rangle}\right) = \delta\left(\mathbf{x}_{\mathrm{TX},k}^{\langle j\rangle} - \mathbf{x}_{\mathrm{TX},k-1}^{\langle j\rangle}\right), \tag{15}$$

where $\delta(\cdot)$ denotes the Dirac distribution.

For the prediction of the user, additional sensors such as an IMU carried by the user may be integrated into the movement model. Within this paper, we assume only heading change rate measurements from a gyroscope to be available, though, and no knowledge on the user speed.

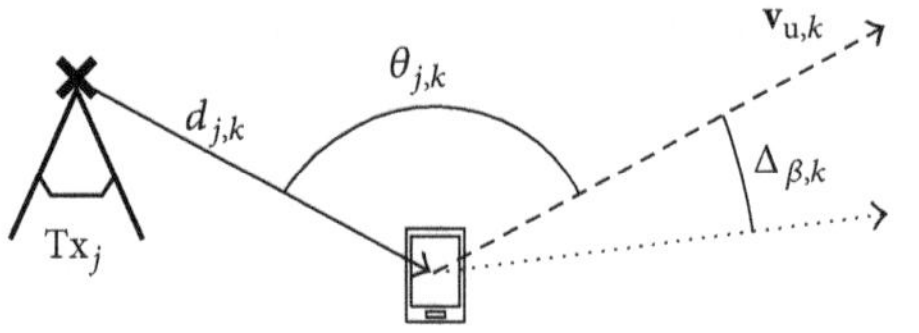

FIGURE 3: The user moves in the direction $\mathbf{v}_{u,k}$ at time instant k. The heading change rate from the IMU is $\Delta_{\beta,k}$. The ToA and AoA measurements for the signal from the jth transmitter are $d_{j,k}$ and $\theta_{j,k}$, respectively, where $\theta_{j,k}$ describes the angle between the heading direction $\mathbf{v}_{u,k}$ of the user and the arriving signal.

With the gyroscope heading change rate $\Delta_{\beta,k}$, we predict the movement of the user by

$$\mathbf{x}_{u,k} = \begin{bmatrix} \mathbb{1}_2 & T_k \mathbb{1}_2 \\ \mathbb{0}_2 & \mathbf{R}\left(\Delta_{\beta,k}\right) \end{bmatrix} \mathbf{x}_{u,k-1} = \mathbf{F}_{u,k}\mathbf{x}_{u,k-1}, \quad (16)$$

where T_k denotes the time between instants $k-1$ and k. The two-dimensional rotation matrix $\mathbf{R}(\Delta_{\beta,k})$ is defined as

$$\mathbf{R}\left(\Delta_{\beta,k}\right) = \begin{bmatrix} \cos\left(\Delta_{\beta,k} + w_k\right) & -\sin\left(\Delta_{\beta,k} + w_k\right) \\ \sin\left(\Delta_{\beta,k} + w_k\right) & \cos\left(\Delta_{\beta,k} + w_k\right) \end{bmatrix}, \quad (17)$$

where w_k is the heading noise which is distributed following a von Mises distribution. Hence, the function $\mathbf{f}_k$ in (1) can be expressed in our case as

$$\mathbf{f}_k\left(\mathbf{x}_{k-1}, \mathbf{v}_{k-1}\right) = \begin{bmatrix} \mathbf{F}_{u,k} & \mathbb{0}_{4\times 3N_{TX}} \\ \mathbb{0}_{3N_{TX}\times 4} & \mathbb{1}_{3N_{TX}} \end{bmatrix} \mathbf{x}_{k-1} + \mathbf{v}_{k-1}, \quad (18)$$

where the process noise covariance matrix $\mathbf{Q}$ is diagonal.

As depicted in Figure 3, an AoA measurement for a transmitter j describes the angle $\theta_{j,k}$ between the user heading direction $\mathbf{v}_{u,k}$ and the incoming signal from the transmitter. The measurement noise for the ToA and AoA measurements is assumed to be zero-mean Gaussian distributed with variances $\sigma^2_{d,j}$ and $\sigma^2_{\theta,j}$, respectively, for the jth transmitter. Also, we assume no cross-correlation between the single ToA and AoA measurements. The likelihood for the measurement vector $\mathbf{z}_k$ conditioned on the state vector $\mathbf{x}_k$ is therefore the product

$$\mathrm{p}\left(\mathbf{z}_k \mid \mathbf{x}_k\right) = \prod_{j=1}^{N_{TX}} \mathcal{N}\left(d_{j,k}; \widehat{d}_{j,k}, \sigma^2_{d,j}\right) \mathcal{N}\left(\theta_{j,k}; \widehat{\theta}_{j,k}, \sigma^2_{\theta,j}\right), \quad (19)$$

where the predicted ToA between the user and the jth transmitter is

$$\widehat{d}_{j,k} = \frac{1}{c_0}\left\|\mathbf{p}_{u,k} - \mathbf{p}^{\langle j\rangle}_{TX,k}\right\| + \tau^{\langle j\rangle}_{0,k}, \quad (20)$$

and the predicted AoA is calculated as

$$\widehat{\theta}_{j,k} = \operatorname{atan2}\left(y_k - y^{\langle j\rangle}_{TX,k}, x_k - x^{\langle j\rangle}_{TX,k}\right) - \operatorname{atan2}\left(v_{y,k}, v_{x,k}\right). \quad (21)$$

The function $\operatorname{atan2}(y, x)$ calculates the four-quadrant inverse tangent function. It returns the counterclockwise angle between the positive x-axis and the point given by the coordinates (x, y).

3. Nonlinear Kalman Filtering

3.1. Unscented Transform. If a random variable $\mathbf{x}$ is transformed by a function $\mathbf{g}(\cdot)$ such that $\mathbf{y} = \mathbf{g}(\mathbf{x})$, the statistics of $\mathbf{y}$ cannot always be calculated in closed form. Monte Carlo (MC) methods try to estimate the statistics of $\mathbf{y}$ from a set of randomly chosen sample points of $\mathbf{x}$ that undergo the transformation $\mathbf{g}(\cdot)$. For the unscented transform, a set of the so-called sigma points is propagated through the function $\mathbf{g}(\cdot)$ to obtain transformed sigma points yielding the statistics of $\mathbf{y}$. However, the sigma points are not chosen randomly, but in a deterministic manner, which is the fundamental difference to MC methods.

Based on the unscented transform, numerical approximations of integrals can be derived. In particular, for the case when the integrand is a product of an arbitrary function $\mathbf{g}(\mathbf{x})$ of the integration variable $\mathbf{x}$ and a Gaussian PDF $\mathcal{N}(\mathbf{x}; \boldsymbol{\mu}_\mathbf{x}, \mathbf{C}_\mathbf{x})$ an integration rule of the form

$$\int \mathbf{g}(\mathbf{x})\, \mathcal{N}\left(\mathbf{x}; \boldsymbol{\mu}_\mathbf{x}, \mathbf{C}_\mathbf{x}\right) \mathrm{d}\mathbf{x} \approx \sum_{m=1}^{N_{sig}} \omega_m \mathbf{g}\left(\mathbf{X}_m\right) \quad (22)$$

can be applied, where $\mathbf{X}_m$ is the mth of the N_{sig} sigma points with its associated weight ω_m. The idea of the UKF is to approximate the posterior PDF in recursive Bayesian estimation by a Gaussian PDF. Hence, the integral in the prediction step is approximated by the integration rule in (22). The authors of [43] provide further insight into sigma point methods and their relation to Gaussian process quadrature.

3.2. Choice of Sigma Points. In the literature, different sets of sigma points have been proposed for the unscented transform [44]. Let $\mathbf{X}_m$ be the mth sigma point and ω_m its associated weight. The dimension, mean, and covariance of the random variable $\mathbf{x}$ are denoted by N, $\boldsymbol{\mu}_\mathbf{x}$, and $\mathbf{C}_\mathbf{x}$, respectively. In [23], the sigma points and their weights are defined for some $\kappa \in \mathbb{R}$ as

$$\begin{aligned} \mathbf{X}_0 &= \boldsymbol{\mu}_\mathbf{x}, \quad \omega_0 = \frac{\kappa}{\kappa + N}, \\ \mathbf{X}_m &= \boldsymbol{\mu}_\mathbf{x} + \left(\sqrt{(N+\kappa)\,\mathbf{C}_\mathbf{x}}\right)_m, \quad \omega_m = \frac{1}{2(\kappa + N)}, \\ \mathbf{X}_{m+N} &= \boldsymbol{\mu}_\mathbf{x} - \left(\sqrt{(N+\kappa)\,\mathbf{C}_\mathbf{x}}\right)_m, \\ & \omega_{m+N} = \frac{1}{2(\kappa + N)}, \end{aligned} \quad (23)$$

where $m = 1, \ldots, N$, $(\mathbf{A})_m$ denotes the mth row or column of the matrix $\mathbf{A}$, and $(N+\kappa)\mathbf{C}_\mathbf{x}$ is factorized into

$$(N+\kappa)\,\mathbf{C}_\mathbf{x} = \sqrt{(N+\kappa)\,\mathbf{C}_\mathbf{x}}\sqrt{(N+\kappa)\,\mathbf{C}_\mathbf{x}}^T. \quad (24)$$

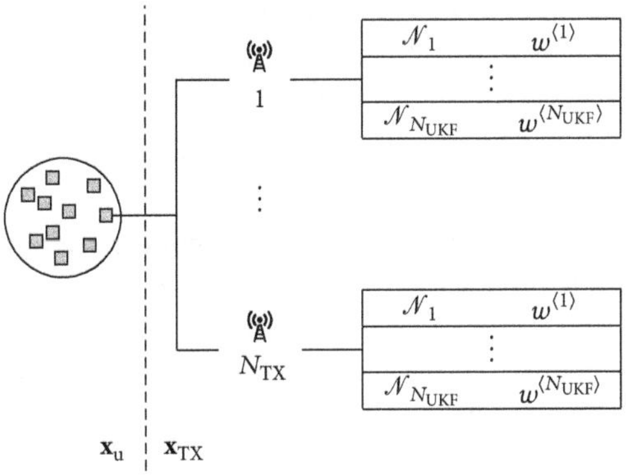

FIGURE 4: Structure of the RBGSPF representation: the user state $\mathbf{x}_{\text{u}}$ is represented by a number of particles. Each particle estimates the transmitters' states on its own. Each of the N_{TX} transmitter states is represented by a sum of N_{UKF} Gaussian distributions $\mathcal{N}_\ell$ with associated weights $w^{\langle \ell \rangle}$.

This definition leads to $N_{\text{sig}} = 2N + 1$ sigma points. The authors of [45] presented the cubature Kalman filter (CKF) with an intuitive derivation of the choice of sigma points and their weights. The CKF differs from the UKF only in the choice of the sigma points. Its derivation is based on the fact that the approximation of an integral using the unscented transform as in (22) is exact for $\mathbf{g}(\mathbf{x})$ being a monomial of an order not greater than some integer d. The resulting sigma points are the points in (23) for $\kappa = 0$. Since the weight of the first sigma point is zero, there are only $2N$ effective points. Although the derivation in [45] gives useful insight into the UKF, the same choice of sigma points had been proposed in [44] before.

The Appendix summarizes the equations for the prediction and update steps of the UKF. If the state transition model in (1) or the measurement model in (2) are linear or if Gaussian noise is assumed in the state transition or measurement model, methods from [46] can be applied to decrease the computational complexity of the UKF.

4. Derivation of the Gaussian Sum Particle Filter

4.1. The Rao-Blackwellized Gaussian Sum Particle Filter. The factorization in (13) allows for estimating the user state independently from the transmitter states. For the estimation of the user state in the RBGSPF, we use a SIR particle filter [26, 47]. The single transmitter states $\mathbf{x}_{\text{TX},k}^{\langle j \rangle}$ are estimated independently from each other following (14). Each transmitter state is represented by a Gaussian mixture model or Gaussian sum model [48]. The posterior PDF of each of the N_{UKF} Gaussian components in a Gaussian mixture is estimated by a UKF. The structure of the resulting RBGSPF representation is shown in Figure 4.

A particle filter is a MC based method, where the posterior PDF is represented by a number of random samples, called particles, with associated weights. The user posterior PDF is approximated as

$$\mathrm{p}\left(\mathbf{x}_{\text{u},k} \mid \mathbf{z}_{1:k}\right) = \sum_{i=1}^{N_p} w_k^{\langle i \rangle} \delta\left(\mathbf{x}_{\text{u},k} - \mathbf{x}_{\text{u},k}^{\langle i \rangle}\right), \tag{25}$$

where $\mathbf{x}_{\text{u},k}^{\langle i \rangle}$ is the ith user particle, $w_k^{\langle i \rangle}$ its associated weight, and N_p the number of particles in the particle filter. From the structure of (14), we see that the transmitter states are estimated for each user particle independently.

The posterior distribution of the state of each transmitter is approximated by a Gaussian mixture model. In a Gaussian mixture model, a PDF is described as a sum of weighted Gaussian PDFs, each described by a mean and a covariance. Hence, the posterior PDF of the state of the jth transmitter of the ith user particle is represented as [48]

$$\begin{aligned}\mathrm{p}\left(\mathbf{x}_{\text{TX},k}^{\langle i,j \rangle} \mid \mathbf{x}_{\text{u},k}^{\langle i \rangle}, \mathbf{z}_{1:k}^{\langle j \rangle}\right) = &\sum_{\ell=1}^{N_{\text{UKF}}} w_k^{\langle i,j,\ell \rangle} \\ &\times \mathcal{N}\left(\mathbf{x}_{\text{TX},k}^{\langle i,j \rangle}; \mathbf{x}_{\text{TX},k|k}^{\langle i,j,\ell \rangle}, \mathbf{P}_{k|k}^{\langle i,j,\ell \rangle}\right),\end{aligned} \tag{26}$$

where $\mathbf{z}_{1:k}^{\langle j \rangle}$ are the measurements for the jth transmitter and $\mathbf{x}_{\text{TX},k|k}^{\langle i,j,\ell \rangle}$, $\mathbf{P}_{k|k}^{\langle i,j,\ell \rangle}$, and $w_k^{\langle i,j,\ell \rangle}$ are the mean, the covariance matrix, and the weight, respectively, of the ℓth Gaussian component of the Gaussian mixture for the jth transmitter. Both $\mathbf{x}_{\text{TX},k|k}^{\langle i,j,\ell \rangle}$ and $\mathbf{P}_{k|k}^{\langle i,j,\ell \rangle}$ are obtained from the update step of the corresponding UKF. Similarly, the likelihood for the measurement of the jth transmitter of the ith user particle is

$$\begin{aligned}\mathrm{p}\left(\mathbf{z}_k^{\langle j \rangle} \mid \mathbf{x}_{\text{u},k}^{\langle i \rangle}, \mathbf{x}_{\text{TX},k}^{\langle i,j \rangle}\right) = &\sum_{\ell=1}^{N_{\text{UKF}}} w_k^{\langle i,j,\ell \rangle} \\ &\times \mathcal{N}\left(\mathbf{z}_k^{\langle j \rangle}; \widehat{\mathbf{z}}_k^{\langle i,j,\ell \rangle}, \mathbf{R}_k^{\langle j \rangle}\right),\end{aligned} \tag{27}$$

where $\mathbf{R}_k^{\langle j \rangle}$ is the measurement noise covariance matrix for the jth transmitter. The number N_{UKF} of Gaussian components might differ between transmitters, user particles, and time instants. However, for notational convenience, we drop the particle, transmitter, and time instant indices of N_{UKF}. The predicted measurement of the ℓth Gaussian component for the jth transmitter of the ith user particle,

$$\widehat{\mathbf{z}}_k^{\langle i,j,\ell \rangle} = \left[\widehat{d}_k^{\langle i,j,\ell \rangle} \; \widehat{\theta}_k^{\langle i,j,\ell \rangle}\right]^T, \tag{28}$$

consists of the predicted ToA measurement

$$\widehat{d}_k^{\langle i,j,\ell \rangle} = \frac{1}{c_0}\left\|\mathbf{p}_{\text{u},k}^{\langle i \rangle} - \mathbf{p}_{\text{TX},k}^{\langle i,j,\ell \rangle}\right\| + \tau_{0,k}^{\langle i,j,\ell \rangle} \tag{29}$$

and the predicted AoA measurement

$$\begin{aligned}\widehat{\theta}_k^{\langle i,j,\ell \rangle} &= \operatorname{atan2}\left(y_k^{\langle i \rangle} - y_{\text{TX},k}^{\langle i,j,\ell \rangle}, x_k^{\langle i \rangle} - x_{\text{TX},k}^{\langle i,j,\ell \rangle}\right) \\ &\quad - \operatorname{atan2}\left(v_{y,k}^{\langle i \rangle}, v_{x,k}^{\langle i \rangle}\right),\end{aligned} \tag{30}$$

where the ith user particle is

$$\mathbf{x}_{\text{u},k}^{\langle i \rangle} = \left[x_k^{\langle i \rangle} \; y_k^{\langle i \rangle} \; v_{x,k}^{\langle i \rangle} \; v_{y,k}^{\langle i \rangle}\right]^T = \left[\mathbf{p}_{\text{u},k}^{\langle i \rangle \, T} \; \mathbf{v}_{\text{u},k}^{\langle i \rangle \, T}\right]^T, \tag{31}$$

and the mean of the corresponding ℓth Gaussian component of the jth transmitter is

$$\mathbf{x}_{\mathrm{TX},k|k}^{\langle i,j,\ell\rangle} = \left[x_{\mathrm{TX},k}^{\langle i,j,\ell\rangle} \; y_{\mathrm{TX},k}^{\langle i,j,\ell\rangle} \; \tau_{0,k}^{\langle i,j,\ell\rangle}\right]^T = \left[\mathbf{p}_{\mathrm{TX},k}^{\langle i,j,\ell\rangle T} \; \tau_{0,k}^{\langle i,j,\ell\rangle}\right]^T. \quad (32)$$

In the prediction step of the user particle filter, new particles are sampled based on the transition prior $\mathrm{p}(\mathbf{x}_{\mathrm{u},k} \mid \mathbf{x}_{\mathrm{u},k-1})$. Hence, the particle $\mathbf{x}_{\mathrm{u},k}^{\langle i\rangle}$ is drawn as

$$\mathbf{x}_{\mathrm{u},k}^{\langle i\rangle} = \mathbf{f}_{\mathrm{u},k}\left(\mathbf{x}_{\mathrm{u},k-1}^{\langle i\rangle}, \mathbf{v}_{\mathrm{u},k-1}\right), \quad (33)$$

where the function $\mathbf{f}_{\mathrm{u},k}(\cdot)$ describes the user movement model, and $\mathbf{v}_{\mathrm{u},k-1}$ is a noise sample drawn from the user process noise PDF. For the prediction and the update step of a Gaussian component of a transmitter's state, the equations of the UKF are summarized in the Appendix.

4.2. Derivation of the Particle Weight Calculation. In the following, we will derive the calculation of the particle weights in the user particle filter and of the weights for the Gaussian components in the Gaussian mixture models used to describe the PDFs of the transmitter states. As the importance density of the SIR particle filter is the state transition prior and resampling of the particles is performed at every time instant, the weight for the ith particle at time instant k is given by [26]

$$w_k^{\langle i\rangle} \propto \mathrm{p}\left(\mathbf{z}_k \mid \mathbf{x}_{\mathrm{u},0:k}^{\langle i\rangle}, \mathbf{z}_{1:k-1}\right). \quad (34)$$

This expression can be written as

$$\begin{aligned} w_k^{\langle i\rangle} &\propto \int \mathrm{p}\left(\mathbf{z}_k \mid \mathbf{x}_{\mathrm{u},0:k}^{\langle i\rangle}, \mathbf{x}_{\mathrm{TX},k}^{\langle i\rangle}, \mathbf{z}_{1:k-1}\right) \\ &\quad \times \mathrm{p}\left(\mathbf{x}_{\mathrm{TX},k}^{\langle i\rangle} \mid \mathbf{x}_{\mathrm{u},0:k}^{\langle i\rangle}, \mathbf{z}_{1:k-1}\right) d\mathbf{x}_{\mathrm{TX},k}^{\langle i\rangle} \\ &\propto \int \mathrm{p}\left(\mathbf{z}_k \mid \mathbf{x}_{\mathrm{u},k}^{\langle i\rangle}, \mathbf{x}_{\mathrm{TX},k}^{\langle i\rangle}\right) \\ &\quad \times \mathrm{p}\left(\mathbf{x}_{\mathrm{TX},k}^{\langle i\rangle} \mid \mathbf{x}_{\mathrm{u},k}^{\langle i\rangle}, \mathbf{z}_{k-1}\right) d\mathbf{x}_{\mathrm{TX},k}^{\langle i\rangle}, \end{aligned} \quad (35)$$

where we use the assumption of a first-order hidden Markov model. With the assumption that the measurements are independent for different transmitters, (35) can be expressed as

$$\begin{aligned} w_k^{\langle i\rangle} &\propto \prod_{j=1}^{N_{\mathrm{TX}}} \int \mathrm{p}\left(\mathbf{z}_k^{\langle j\rangle} \mid \mathbf{x}_{\mathrm{u},k}^{\langle i\rangle}, \mathbf{x}_{\mathrm{TX},k}^{\langle i,j\rangle}\right) \\ &\quad \times \mathrm{p}\left(\mathbf{x}_{\mathrm{TX},k}^{\langle i,j\rangle} \mid \mathbf{x}_{\mathrm{u},k}^{\langle i\rangle}, \mathbf{z}_{k-1}^{\langle j\rangle}\right) d\mathbf{x}_{\mathrm{TX},k}^{\langle i,j\rangle}. \end{aligned} \quad (36)$$

Furthermore, using the Gaussian mixture model representation from (26) and (27) and assuming Gaussian measurement noise, the integrand can be expressed as a sum of weighted Gaussian PDFs; namely,

$$\mathrm{p}\left(\mathbf{z}_k^{\langle j\rangle} \mid \mathbf{x}_{\mathrm{u},k}^{\langle i\rangle}, \mathbf{x}_{\mathrm{TX},k}^{\langle i,j\rangle}\right) \mathrm{p}\left(\mathbf{x}_{\mathrm{TX},k}^{\langle i,j\rangle} \mid \mathbf{x}_{\mathrm{u},k}^{\langle i\rangle}, \mathbf{z}_{k-1}^{\langle j\rangle}\right) = \sum_{\ell=1}^{N_{\mathrm{UKF}}} w_{k-1}^{\langle i,j,\ell\rangle} \mathrm{p}_{\mathrm{z},k}^{\langle i,j,\ell\rangle} \mathrm{p}_{\mathrm{TX},k}^{\langle i,j,\ell\rangle}. \quad (37)$$

In (37), we defined for notational brevity

$$\mathrm{p}_{\mathrm{z},k}^{\langle i,j,\ell\rangle} = \mathcal{N}\left(\mathbf{z}_k^{\langle j\rangle}; \widehat{\mathbf{z}}_k^{\langle i,j,\ell\rangle}, \mathbf{R}_k^{\langle j\rangle}\right), \quad (38)$$

$$\mathrm{p}_{\mathrm{TX},k}^{\langle i,j,\ell\rangle} = \mathcal{N}\left(\mathbf{x}_{\mathrm{TX},k}^{\langle i,j,\ell\rangle}; \mathbf{x}_{\mathrm{TX},k|k}^{\langle i,j,\ell\rangle}, \mathbf{P}_{k|k}^{\langle i,j,\ell\rangle}\right), \quad (39)$$

where $\mathbf{R}_k^{\langle j\rangle}$ is the measurement noise covariance matrix for the jth transmitter and $\mathbf{x}_{\mathrm{TX},k}^{\langle i,j,\ell\rangle}$ denotes the state of the ℓth Gaussian component of the jth transmitter of the ith user particle. As we assume no correlation among the ToA and AoA measurements, we have

$$\mathbf{R}_k^{\langle j\rangle} = \begin{bmatrix} \sigma_{d,j}^2 & 0 \\ 0 & \sigma_{\theta,j}^2 \end{bmatrix}. \quad (40)$$

Inserting (37) into (36) leads to

$$w_k^{\langle i\rangle} \propto \prod_{j=1}^{N_{\mathrm{TX}}} \sum_{\ell=1}^{N_{\mathrm{UKF}}} w_{k-1}^{\langle i,j,\ell\rangle} \int \mathrm{p}_{\mathrm{z},k}^{\langle i,j,\ell\rangle} \mathrm{p}_{\mathrm{TX},k}^{\langle i,j,\ell\rangle} d\mathbf{x}_{\mathrm{TX},k}^{\langle i,j,\ell\rangle}. \quad (41)$$

The predicted measurement $\widehat{\mathbf{z}}_k^{\langle i,j,\ell\rangle}$ in (38), defined in (28), is a nonlinear function of $\mathbf{x}_{\mathrm{TX},k}^{\langle i,j,\ell\rangle}$. We express the measurement likelihood in (38) explicitly as a function $\mathbf{g}(\cdot)$ of $\mathbf{x}_{\mathrm{TX},k}^{\langle i,j,\ell\rangle}$, resulting in

$$\begin{aligned} \mathbf{g}\left(\mathbf{x}_{\mathrm{TX},k}^{\langle i,j,\ell\rangle}\right) &= \mathrm{p}_{\mathrm{z},k}^{\langle i,j,\ell\rangle} \\ &= \mathcal{N}\left(d_{j,k}; \widehat{d}_k^{\langle i,j,\ell\rangle}, \sigma_{d,j}^2\right) \\ &\quad \times \mathcal{N}\left(\theta_{j,k}; \widehat{\theta}_k^{\langle i,j,\ell\rangle}, \sigma_{\theta,j}^2\right). \end{aligned} \quad (42)$$

Due to the nonlinearity in (42), stemming from (29) and (30), the integral in (41) cannot be solved analytically. Instead, we use the approximation from (22) to calculate the integral as

$$\int \mathbf{g}\left(\mathbf{x}_{\mathrm{TX},k}^{\langle i,j,\ell\rangle}\right) \mathcal{N}\left(\mathbf{x}_{\mathrm{TX},k}^{\langle i,j,\ell\rangle}; \mathbf{x}_{\mathrm{TX},k|k}^{\langle i,j,\ell\rangle}, \mathbf{P}_{k|k}^{\langle i,j,\ell\rangle}\right) d\mathbf{x}_{\mathrm{TX},k}^{\langle i,j,\ell\rangle} \approx \sum_{m=1}^{N_{\mathrm{sig}}} \omega_m \mathbf{g}\left(\mathbf{X}_m\right), \quad (43)$$

where the sigma points $\mathbf{X}_m$ and their weights ω_m can be calculated by (23), where

$$\boldsymbol{\mu}_{\mathbf{x}} = \mathbf{x}_{\mathrm{TX},k|k}^{\langle i,j,\ell\rangle}, \qquad \mathbf{C}_{\mathbf{x}} = \mathbf{P}_{k|k}^{\langle i,j,\ell\rangle}, \quad (44)$$

and N is the dimension of a transmitter's state; that is, $N = 3$. Finally, the weight of the ith particle is calculated as

$$w_k^{\langle i\rangle} \propto \prod_{j=1}^{N_{\mathrm{TX}}} \sum_{\ell=1}^{N_{\mathrm{UKF}}} w_{k-1}^{\langle i,j,\ell\rangle} \sum_{m=1}^{N_{\mathrm{sig}}} \omega_m \mathbf{g}\left(\mathbf{X}_m\right). \quad (45)$$

It follows directly from (45) that the weights of the Gaussian components are updated by

$$w_k^{\langle i,j,\ell\rangle} \propto w_{k-1}^{\langle i,j,\ell\rangle} \sum_{m=1}^{N_{\mathrm{sig}}} \omega_m \mathbf{g}\left(\mathbf{X}_m\right). \quad (46)$$

Input: $\mathbf{x}_{\mathrm{u},k-1}^{\langle i\rangle}$, $\mathbf{x}_{\mathrm{TX},k-1}^{\langle i,j,\ell\rangle}$ and $w_{k-1}^{\langle i,j,\ell\rangle}$ for $i = 1,\ldots,N_p$, $j = 1,\ldots,N_{\mathrm{TX}}$, $\ell = 1,\ldots,N_{\mathrm{UKF}}$, $\mathbf{z}_k$
Output: $\mathbf{x}_{\mathrm{u},k}^{\langle i\rangle}$, $\mathbf{x}_{\mathrm{TX},k}^{\langle i,j,\ell\rangle}$ and $w_{k}^{\langle i,j,\ell\rangle}$ for $i = 1,\ldots,N_p$, $j = 1,\ldots,N_{\mathrm{TX}}$, $\ell = 1,\ldots,N_{\mathrm{UKF}}$
for $i = 1,\ldots,N_p$ **do**
  draw new user particle $\mathbf{x}_{\mathrm{u},k}^{\langle i\rangle}$ using (33);
  if *any new signal components detected* **then**
    initialize the new transmitter(s) based on $\mathbf{z}_k$;
  if *track of any signal components lost* **then**
    delete the corresponding transmitter(s);
  for $j = 1,\ldots,N_{\mathrm{TX}}$ **do**
    for $\ell = 1,\ldots,N_{\mathrm{UKF}}$ **do**
      perform UKF prediction and update to calculate $\mathbf{x}_{\mathrm{TX},k}^{\langle i,j,\ell\rangle}$ using the UKF equations in the Appendix;
      calculate the weight $w_k^{\langle i,j,\ell\rangle}$ with (46);
      if $w_k^{\langle i,j,\ell\rangle} < \rho$ **then**
        prune this Gaussian component: set $w_k^{\langle i,j,\ell\rangle} = 0$;
  calculate the weight $w_k^{\langle i\rangle}$ with (45);
for $i = 1,\ldots,N_p$ **do**
  for $j = 1,\ldots,N_{\mathrm{TX}}$ **do**
    normalize the weights $w_k^{\langle i,j,\ell\rangle}$ for $\ell = 1,\ldots,N_{\mathrm{UKF}}$;
normalize the weights $w_k^{\langle i\rangle}$ for $i = 1,\ldots,N_p$;
resample the user particles $\mathbf{x}_{\mathrm{u},k}^{\langle i\rangle}$ [26];

Algorithm 1: RBGSPF for time instant $k > 0$.

Note that the weights $w_k^{\langle i\rangle}$ and $w_k^{\langle i,j,\ell\rangle}$ in (45) and (46) of the user particles and the Gaussian components, respectively, are not yet normalized. Since resampling is performed at every time instant in the SIR particle filter, the user particle weights $w_k^{\langle i\rangle}$ do not depend on the weights $w_{k-1}^{\langle i\rangle}$ from the previous time instant [26].

4.3. Merging and Pruning of Gaussian Components. When the KEST estimator detects a new signal component, a new transmitter is initialized for each user particle based on the ToA and AoA measurement for that new transmitter at the current time instant. The posterior PDF of the new transmitter is represented by a number of Gaussian PDFs, whose means are initially placed on a grid dependent on the measurement. The number of Gaussian components depends on the measurement as well.

As the user travels through a scenario, the means, covariances, and weights of the Gaussian components of a transmitter's state posterior PDF change over time depending on the available measurements. The mean and covariance of a Gaussian component may be regarded as a hypothesis and a corresponding uncertainty, respectively, for the state of a transmitter. If the weight of a Gaussian component becomes smaller, the hypothesis for that state of the transmitter becomes less likely. Hence, if the weight of a Gaussian component falls below a threshold ρ, the Gaussian component is pruned; that is, its weight $w_{k-1}^{\langle i,j,\ell\rangle}$ is set to zero. If the means of two Gaussian components get very close to each other, they may be merged in order to reduce the computational complexity.

The final algorithm for one time instant $k > 0$ of the RBGSPF is summarized in Algorithm 1. For a particle filter resampling algorithm, we refer to [26]. Note again that we have dropped the particle and transmitter indices in N_{UKF}.

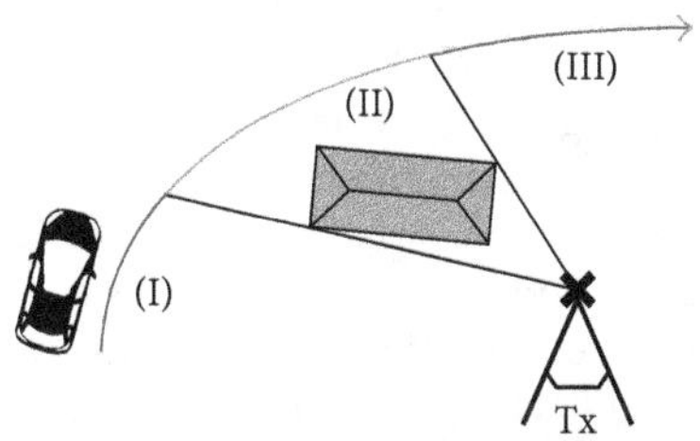

Figure 5: A user moves along the trajectory. The LoS signal to the transmitter in Region (I) is lost in Region (II) temporarily due to blocking by an obstacle and received again in Region (III).

5. Data Association

Data association is of crucial importance for robust long-term SLAM. It describes the correspondences among landmarks, which are transmitters in multipath assisted positioning. In Figure 5, a user travels along its trajectory. In Region (I), the LoS signal from the transmitter Tx can be tracked. This signal is lost in Region (II) and regained in Region (III). However, KEST is not able to retrack a former path. Hence, when the user enters Region (III), KEST detects a new signal component, and consequently a new transmitter is initialized. However, the transmitter is the same as that which had been observed in Region (I).

We define the set of transmitters that had been observed previously but are not detected any more, as *old* transmitters. When a new signal component is detected by KEST, a new transmitter has to be initialized. Consequently, two cases may arise:

(1) the new signal component corresponds indeed to a new transmitter, or

(2) the new signal component corresponds to an old transmitter that had been observed before already.

Data association is the decision on the above two cases when a new signal component is detected. In the first case, a new transmitter is initialized for the newly detected signal component. In the second case, the newly detected signal component is associated with a previously observable, that is, old, transmitter.

In [33], a multiple hypothesis tracking (MHT) association method was introduced and derived for FastSLAM, where the user state is represented by a particle filter and each landmark state by an EKF. In [19], the same method has been derived for a Rao-Blackwellized particle filter. In the following, we will derive the method for the RBGSPF derived in Section 4.

Each user particle decides for associations individually and thus carries a hypothesis for associations. Hence, association decisions are hard decisions for each particle. Regarding the ensemble of user particles, though, there are many different hypotheses on associations in the user state estimate, and the association method can be regarded as a soft decision method. Consequently, the state vector of the user is increased by data association.

In the following, we describe how to make an association decision for a single user particle, where we omit the particle index i in the association variables for notational brevity. The value of the association variable n_k denotes an association of the new transmitter with the old transmitter n_k. We denote the marginalized likelihood of the measurement of the new transmitter that is to be initialized at time instant k by p_{n_k} assuming that the new transmitter is associated with the old transmitter n_k. The set of association decisions up to time instant $k-1$ is denoted by N_{k-1}. From [33], we have

$$\begin{aligned} p_{n_k} &= \mathrm{p}\left(\mathbf{z}_k \mid n_k, N_{k-1}, \mathbf{x}_{\mathrm{u},k}^{\langle i\rangle}, \mathbf{z}_{1:k-1}\right) \\ &= \int \mathrm{p}\left(\mathbf{z}_k \mid \mathbf{x}_{\mathrm{TX},k}^{\langle i,n_k\rangle}, n_k, N_{k-1}, \mathbf{x}_{\mathrm{u},k}^{\langle i\rangle}, \mathbf{z}_{1:k-1}\right) \\ &\quad \times \mathrm{p}\left(\mathbf{x}_{\mathrm{TX},k}^{\langle i,n_k\rangle} \mid n_k, N_{k-1}, \mathbf{x}_{\mathrm{u},k}^{\langle i\rangle}, \mathbf{z}_{1:k-1}\right) \mathrm{d}\mathbf{x}_{\mathrm{TX},k}^{\langle i,n_k\rangle}, \end{aligned} \tag{47}$$

where $\mathbf{x}_{\mathrm{TX},k}^{\langle i,n_k\rangle}$ denotes the state vector of the n_kth transmitter for the ith user particle.

Assuming a first-order hidden Markov model, the first integrand in (47) can be simplified to

$$\begin{aligned} &\mathrm{p}\left(\mathbf{z}_k \mid \mathbf{x}_{\mathrm{TX},k}^{\langle i,n_k\rangle}, n_k, N_{k-1}, \mathbf{x}_{\mathrm{u},k}^{\langle i\rangle}, \mathbf{z}_{1:k-1}\right) \\ &\quad = \mathrm{p}\left(\mathbf{z}_k \mid \mathbf{x}_{\mathrm{TX},k}^{\langle i,n_k\rangle}, n_k, N_{k-1}, \mathbf{x}_{\mathrm{u},k}^{\langle i\rangle}\right). \end{aligned} \tag{48}$$

Since we use a Gaussian mixture model to represent the single transmitter states, the second integrand in (47) can be rewritten as

$$\begin{aligned} &\mathrm{p}\left(\mathbf{x}_{\mathrm{TX},k}^{\langle i,n_k\rangle} \mid n_k, N_{k-1}, \mathbf{x}_{\mathrm{u},k}^{\langle i\rangle}, \mathbf{z}_{1:k-1}\right) \\ &\quad = \sum_{\ell=1}^{N_{\mathrm{UKF}}} w_k^{\langle i,n_k,\ell\rangle} \mathcal{N}\left(\mathbf{x}_{\mathrm{TX},k}^{\langle i,n_k\rangle}; \mathbf{x}_{\mathrm{TX},k|k-1}^{\langle i,n_k,\ell\rangle}, \mathbf{P}_{k|k-1}^{\langle i,n_k,\ell\rangle}\right). \end{aligned} \tag{49}$$

Inserting (48) and (49) into (47) yields

$$p_{n_k} = \sum_{\ell=1}^{N_{\mathrm{UKF}}} w_k^{\langle i,n_k,\ell\rangle} \int \mathrm{p}_{\mathrm{z},k}^{\langle i,n_k,\ell\rangle} \mathrm{p}_{\mathrm{TX},k}^{\langle i,n_k,\ell\rangle} \mathrm{d}\mathbf{x}_{\mathrm{TX},k}^{\langle i,n_k,\ell\rangle}, \tag{50}$$

where $\mathrm{p}_{\mathrm{z},k}^{\langle i,n_k,\ell\rangle}$ is defined as in (38), and

$$\mathrm{p}_{\mathrm{TX},k|k-1}^{\langle i,n_k,\ell\rangle} = \mathcal{N}\left(\mathbf{x}_{\mathrm{TX},k}^{\langle i,n_k,\ell\rangle}; \mathbf{x}_{\mathrm{TX},k|k-1}^{\langle i,n_k,\ell\rangle}, \mathbf{P}_{k|k-1}^{\langle i,n_k,\ell\rangle}\right). \tag{51}$$

Similar to (42), we define

$$\begin{aligned} \mathbf{g}\left(\mathbf{x}_{\mathrm{TX},k}^{\langle i,n_k,\ell\rangle}\right) &= \mathrm{p}_{\mathrm{z},k}^{\langle i,n_k,\ell\rangle} \\ &= \mathcal{N}\left(d_{n_k,k}; \hat{d}_k^{\langle i,n_k,\ell\rangle}, \sigma_{d,n_k}^2\right) \\ &\quad \times \mathcal{N}\left(\theta_{n_k,k}; \hat{\theta}_k^{\langle i,n_k,\ell\rangle}, \sigma_{\theta,n_k}^2\right), \end{aligned} \tag{52}$$

which is nonlinear in $\mathbf{x}_{\mathrm{TX},k}^{\langle i,n_k,\ell\rangle}$, and rewrite the integral in (50) as

$$\begin{aligned} \int \mathrm{p}_{\mathrm{z},k}^{\langle i,n_k,\ell\rangle} \mathrm{p}_{\mathrm{TX},k|k-1}^{\langle i,n_k,\ell\rangle} \mathrm{d}\mathbf{x}_{\mathrm{TX},k}^{\langle i,n_k,\ell\rangle} &= \int \mathbf{g}\left(\mathbf{x}_{\mathrm{TX},k}^{\langle i,n_k,\ell\rangle}\right) \\ &\quad \times \mathcal{N}\left(\mathbf{x}_{\mathrm{TX},k}^{\langle i,n_k,\ell\rangle}; \mathbf{x}_{\mathrm{TX},k|k-1}^{\langle i,n_k,\ell\rangle}, \mathbf{P}_{k|k-1}^{\langle i,n_k,\ell\rangle}\right) \mathrm{d}\mathbf{x}_{\mathrm{TX},k}^{\langle i,n_k,\ell\rangle}. \end{aligned} \tag{53}$$

Approximating the integral using (22) and inserting it into (50) finally yield

$$p_{n_k} = \sum_{\ell=1}^{N_{\mathrm{UKF}}} w_k^{\langle i,n_k,\ell\rangle} \sum_{m=1}^{N_{\mathrm{sig}}} \omega_m \mathbf{g}\left(\mathbf{X}_m\right). \tag{54}$$

The sigma points $\mathbf{X}_m$ and their weights ω_m can again be calculated by (23) with

$$\begin{aligned} \boldsymbol{\mu}_{\mathbf{x}} &= \mathbf{x}_{\mathrm{TX},k|k-1}^{\langle i,n_k,\ell\rangle}, \\ \mathbf{C}_{\mathbf{x}} &= \mathbf{P}_{k|k-1}^{\langle i,n_k,\ell\rangle}. \end{aligned} \tag{55}$$

The authors of [33] propose two ways to come to an association decision, a ML method and data association sampling (DAS). The probability for making no association is defined and denoted by p_0. The set of indices of old transmitters that have not yet been and hence might be associated is denoted by Γ_k.

For the ML association method, the association of the new transmitter with the old transmitter n_k is chosen to be

$$\hat{n}_{\mathrm{ML},k} = \arg\max_{n_k \in \Gamma_k \cup \{0\}} p_{n_k}. \tag{56}$$

In DAS, an association is sampled based on the likelihoods $c p_{n_k}$ for $n_k \in \Gamma_k \cup \{0\}$, where c is a normalization constant.

If a detected transmitter is associated with an old transmitter, the new transmitter can be initialized with the

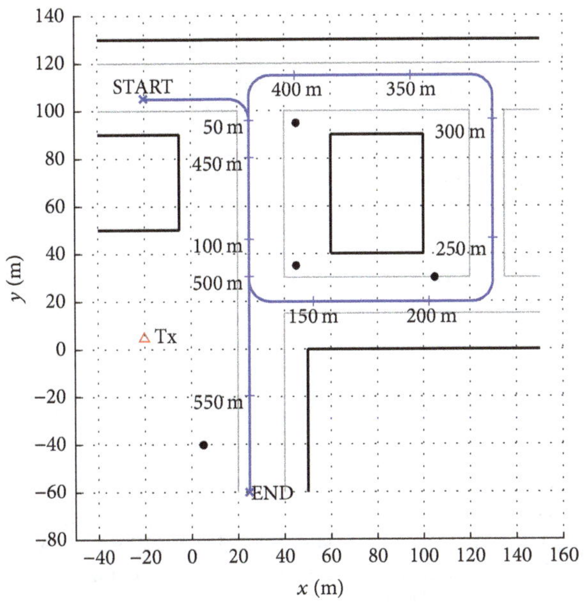

FIGURE 6: The simulation scenario with thick black lines representing reflecting walls and black circles representing objects that scatter the RF signals. The physical transmitter is marked by the red upward triangle labeled Tx. The user travels along the blue line from START to END with one loop around the central building.

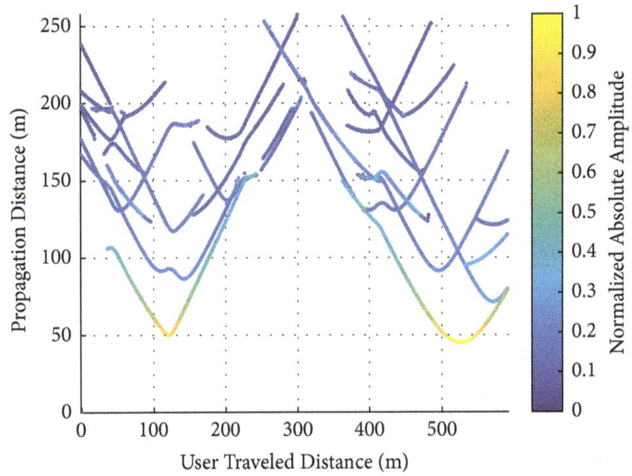

FIGURE 7: The results of the KEST estimator for the simulations showing the propagation distances of signal components versus the user traveled distance. The propagation distances are the ToA multiplied by the speed of light. Only signal components that are observable for a traveled distance of at least 35 m are shown. The color indicates the normalized amplitude in linear domain.

posterior PDF of the associated old transmitter. Thus, data association has to be incorporated in Line (4) of Algorithm 1.

The above method describes how to take association decisions if no more than one new transmitter is initialized at one time instant, that is, if no more than one new signal component is detected by KEST at a time instant. In case of multiple transmitters being initialized at the same time instant, a greedy algorithm [20] may be applied.

6. Evaluations

In the following, we evaluate the RBGSPF derived in Section 4 by means of simulations and actual outdoor measurements. For the evaluations, we implemented a square-root version of the cubature Kalman filter as in [45] for numerical stability. The sigma points are the ones in (23) for $\kappa = 0$. Since the movement model of the transmitters is linear and we assume Gaussian noise, the prediction step can be calculated analytically. For the description of a prediction step of a square-root version of the conventional Kalman filter, we refer to [48].

6.1. Simulations in an Urban Scenario. A top view of the urban simulation scenario is depicted in Figure 6. The thick black lines represent walls, for example, from buildings, that reflect RF signals, and the black circles are objects such as traffic light poles acting as scatterers. There is one physical transmitter in the scenario marked by the red upward triangle labeled Tx. The user travels with a constant speed of 10 m/s along the blue line with a loop around the central building. The initial and final user positions are labeled START and END, respectively. The traveled distances of the user are marked for every 50 m.

The transmitter continuously broadcasts a signal that is known to the user and has a rectangular shape in frequency domain with a center frequency of 1.5 GHz and a bandwidth of 100 MHz. As we know the environment, a CIR and the received signal can be modeled for every user position with a simple ray-tracing approach. We incorporate first- and second-order reflections and scattering, that is, single and double reflections and/or scattering. The power loss for the signal being reflected is 3 dB and 6 dB when the signal is scattered at a point scatterer. The average signal-to-noise ratio (SNR) at the user is 7 dB.

The user is equipped with an RF receiver and a two-dimensional, rectangular antenna array consisting of nine elements. Hence, both the ToA and the AoA estimates from KEST are incorporated in the estimation of the user and the transmitters' states. Based on the received signal, KEST estimates the ToAs and AoAs every 50 ms.

The results of the KEST estimator are plotted in Figure 7. It shows the propagation distance, which is the ToA multiplied by the speed of light, of the signal components versus the traveled distance of the user. Each continuous line represents one signal component and its evolution as the user travels through the scenario. The color of each line indicates the normalized absolute value of the amplitude of the corresponding signal component in linear domain. Since signal components that are observable for a long time can contribute much better to Channel-SLAM than components which are observable only for a short time, only signal components that are observable for a user traveled distance of at least 35 m are plotted and used. Using all detected signal components would dramatically increase the computational complexity and hardly increase the positioning performance.

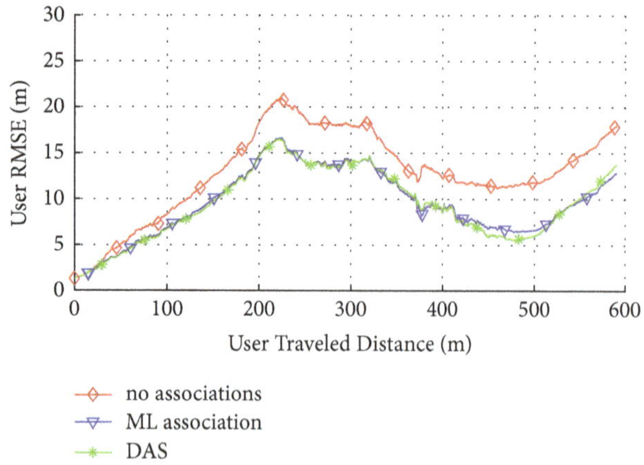

FIGURE 8: The RMSE of the user position versus the user traveled distance for the simulations. The red curve shows the RMSE if no associations among transmitters are made, the blue curve if the ML method for associations is applied, and the green curve for using DAS.

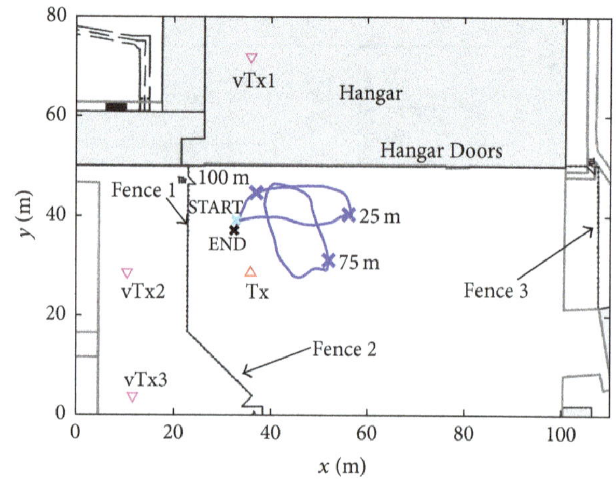

FIGURE 9: Top view of the measurement scenario in front of a hangar. The physical transmitter location is marked by the red triangle labeled Tx. The user travels along the blue track from START to END. The expected virtual transmitter locations are marked by the magenta downward triangles.

In Channel-SLAM, the user position is estimated relative to the physical and virtual transmitters in the scenario. Thus, to create a local coordinate system, the initial state of the user is assumed to be known. However, no prior knowledge on any transmitter is assumed. In the Rao-Blackwellized particle filter, the number of user particles in the particle filter is 4000, while the number of Gaussian components for each transmitter depends on the first ToA measurement for that transmitter.

The root mean square error (RMSE) of the user position versus its traveled distance is plotted in Figure 8. The red curve shows the RMSE if no associations among transmitters are made; that is, every signal component that is detected by the KEST algorithm is assumed to be a new transmitter. The RMSEs with the ML association method and DAS from Section 5 being applied are plotted in blue and green, respectively. Since the particle filter is a MC based method, all RMSE curves are averaged over 100 simulations.

As we assume the starting position of the user to be known, all three curves start with a low RMSE that increases linearly during the first 200 m as expected. The increase of the RMSE is less due to a bias in the position estimate but more due to an increasing uncertainty, that is, variance, about the user position. After approximately 200 m, the RMSE tends to decrease for all three curves. As more and more transmitters are observed, the weight for some user particles becomes small, and these particles are unlikely to be resampled. Towards the end of the track, the geometrical delusion of precision (GDOP) causes an increase in the RMSE, since most of the transmitters are observed from the same direction. After a traveled distance of around 370 m, several transmitters that had been observed in the beginning are observed again, and correspondences among them can be found. If data association methods are used, the RMSE decreases particularly in that region. The ML method and DAS show a similar performance. Note that there are several reasons for which associations among transmitters can be found. Examples are signal blocking or the geometry of the environment causing virtual transmitters to be observable only from certain regions. In addition, when KEST loses and regains track of a signal component if its received power fluctuates or if another signal component arrives at the receiver with a very small difference in delay, transmitters may be discarded and initialized again at a later point, and associations among them may be found. This explains the increasing positioning performance gain after approx. 50 m using data association.

6.2. Outdoor Measurements. In addition to the simulations as described above, we performed outdoor measurements on an airfield. A top view of the measurement scenario is depicted in Figure 9. The grey area is an airplane hangar with solid metallic doors. The user track with a total length of 112.5 m is plotted in blue. The user walked along the track starting from the light blue cross labeled START to the black cross labeled END. The traveled distance of the user is marked after 25 m, 75 m, and 100 m. There is one physical transmitter marked by the red upward triangle labeled Tx. The user is in LoS to the physical transmitter throughout the entire track.

In the scenario, we have three fences labeled Fence 1, Fence 2, and Fence 3. We expect these fences and the hangar door to reflect the RF signal emitted by the physical transmitter. Hence, we expect a virtual transmitter for each of the fences and for the hangar door following Section 2.1. The virtual transmitter corresponding to the reflection of the signal at Fence 1 is the magenta downward triangle labeled vTx2. It is located at the physical transmitter position mirrored at Fence 1. Likewise, the location of the virtual transmitter corresponding to Fence 2 is labeled as vTx3. For the reflection of the signal at the hangar doors, we expect the virtual transmitter located at the magenta triangle labeled as vTx1. The expected virtual transmitter corresponding to Fence 3, vTx4, is outside of the boundaries of Figure 9.

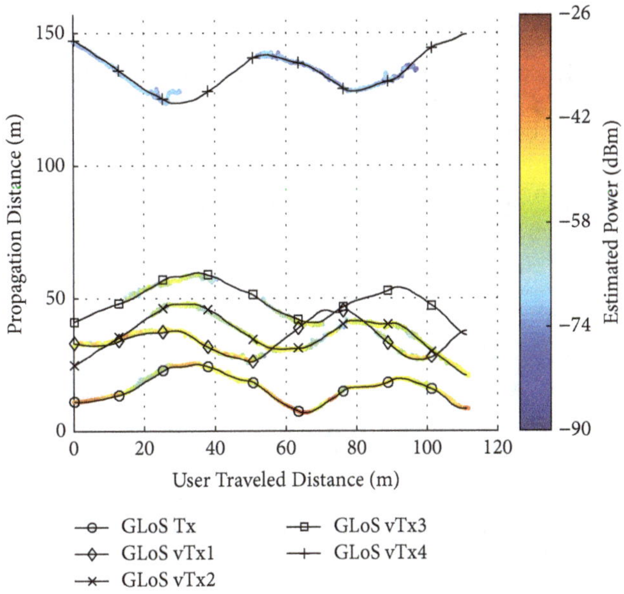

Figure 10: The results of the KEST estimator for the outdoor measurements showing the propagation distances, that is, the ToAs multiplied by the speed of light, of signal components, versus the user traveled distance. Only signal components that are observable for long traveled distance are shown. The color indicates the estimated received power for the signal components in dB.

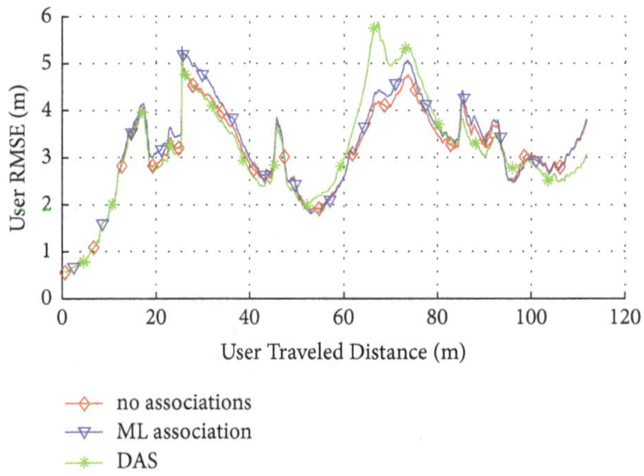

Figure 11: The RMSE of the user position versus the user traveled distance for the outdoor measurements. The red curve shows the RMSE if no associations among transmitters are made, the blue curve if the ML method for associations is applied, and the green curve for using DAS.

The Medav RUSK broadband channel sounder [49] was used to perform the measurements. The transmit signal is a multitone signal with a center frequency of 1.51 GHz and a bandwidth of 100 MHz. The signal has 1281 subcarriers with equal gains and a total transmit power of 10 mW.

The user was equipped with an RF receiver, recording a snapshot of the received signal every 1.024 ms. For later evaluation, the user carried a prism mounted next to the receiver antenna that was tracked by a tachymeter (Leica Geosystems TCRP1200) to obtain the ground truth of the user location in centimeter accuracy. In addition, the user carried an Xsens MTI-G-700 IMU. Only heading change rate measurements were used from the IMU.

On both transmitter and receiver side, single antennas were used. Hence, no AoA information about the imping-ing signal components can be used for Channel-SLAM. Instead, only ToA estimates from KEST are incorporated. The likelihood function in (19) is adapted accordingly for the evaluation.

The results of the KEST estimator for the outdoor measurements are plotted in Figure 10. The colors indicate the power estimated by KEST in dBm. As for the simulations, only signal components that are observable for a long user traveled distance are plotted and used. In addition, the ground truth geometrical line-of-sight (GLoS) propagation distances from the physical and the expected virtual transmitters as in Figure 9 to the user are plotted by black lines. They match the KEST estimates very well, justifying the signal model in (5) without considering DMCs in KEST for the measurement scenario.

The RMSE of the user position versus its traveled distance for the outdoor measurements is plotted in Figure 11, where the RMSE is averaged over 50 particle filter simulations. As for Figure 8, the red curve denotes the RMSE with no association method applied; the blue and green curves show the RMSE if the ML method and DAS, respectively, are incorporated for data association.

The user is always in a LoS condition to the physical transmitter vTx, and the corresponding LoS signal component is tracked by the user throughout the track, as becomes evident in Figure 10. Likewise, the signal component corresponding to the virtual transmitter vTx2 can be tracked after a traveled distance of approximately 22 m until the end.

The almost continuous presence of the signals from these two transmitters is reflected in the user RMSE in Figure 11. The RMSE without data association methods applied increases in the beginning but then stays constant in the order of 3-4 m with some fluctuations. This is due to the fact that once the variance on the states of transmitters vTx and vTx2 has decreased far enough, they serve as reliable anchors throughout the track. Hence, they prevent the uncertainty about the user state from increasing further, although we measure only the ToA for each signal component in the outdoor measurement scenario.

For the same reason, the data association methods cannot really improve the user positioning performance in the outdoor measurements. From another point of view, a correct data association is inherently made for vTx and vTx2 throughout the track, since once these transmitters have been initialized they stay observable throughout the track. However, if a user was to go through the same scenario a second time with prior information of the transmitter states as estimated during the first run, data association would improve the positioning performance as correspondences among transmitters estimated during the first and second run could be found and exploited.

In contrast, the user RMSE in Figure 8 without data association in the simulations keeps increasing, since there are constantly new transmitters showing up and current transmitters disappear. As mentioned above, the uncertainty about transmitters is high upon initialization, since the measurements obtained from KEST are of fewer dimensions than the transmitter states. In addition, the current user uncertainty adds up to the transmitter uncertainty. No transmitter can be tracked throughout the scenario, and the overall uncertainty keeps increasing. In the simulations, data association relates new transmitters with previously observable transmitters and decreases the uncertainty about the states of transmitters drastically. Consequently, also the uncertainty about and hence the RMSE of the user state decrease.

7. Conclusion

Within this paper, we derived a novel filtering approach for Channel-SLAM. Using Rao-Blackwellization, the user state is represented by a number of particles and estimated by a particle filter. The states of the landmarks, which are the physical and virtual transmitters in Channel-SLAM, are represented by a sum of Gaussian PDFs, where each Gaussian component PDF is filtered by a UKF. The approach can be applied to SLAM problems in general.

We evaluated our approach in simulations in an urban scenario as well as with outdoor measurement data, where we could track a user's position with only one physical transmitter whose location was unknown. For the simulations in the urban scenario, the user RMSE was always below 21 m. With the presented data association methods applied, it was always below 16.5 m. For the measurements on an airfield, the user RMSE was in the order of 3-4 m.

Appendix

A. UKF Prediction and Update Equations

A.1. Prediction Step

(1) Given the Gaussian state PDF $p(\mathbf{x}_{k-1} \mid \mathbf{z}_{k-1}) = \mathcal{N}(\hat{\mathbf{x}}_{k-1|k-1}, \mathbf{P}_{k-1|k-1})$, calculate the set of N_{sig} sigma points $\mathbf{X}_{m,k-1|k-1}$ and their weights ω_m for $m = 1, \ldots, N_{\text{sig}}$, for example, using (23).

(2) Propagate the sigma points through the movement model:

$$\mathbf{X}^*_{m,k|k-1} = \mathbf{f}_k\left(\mathbf{X}_{m,k-1|k-1}\right). \tag{A.1}$$

(3) Calculate the predicted state:

$$\hat{\mathbf{x}}_{k|k-1} = \sum_{m=1}^{N_{\text{sig}}} \omega_m \mathbf{X}^*_{m,k|k-1}. \tag{A.2}$$

(4) Calculate the predicted error covariance:

$$\mathbf{P}_{k|k-1} = \sum_{m=1}^{N_{\text{sig}}} \omega_m \mathbf{X}^*_{m,k|k-1} \mathbf{X}^{*\,T}_{m,k|k-1} - \hat{\mathbf{x}}_{k|k-1}\hat{\mathbf{x}}^T_{k|k-1} + \mathbf{Q}_{k-1}. \tag{A.3}$$

A.2. Update Step

(1) Given the Gaussian state PDF $p(\mathbf{x}_k \mid \mathbf{z}_{k-1}) = \mathcal{N}(\hat{\mathbf{x}}_{k|k-1}, \mathbf{P}_{k|k-1})$, calculate the set of N_{sig} predicted sigma points $\mathbf{X}_{m,k|k-1}$ and their weights ω_m as for the prediction.

(2) Propagate the predicted sigma points through the measurement function:

$$\mathbf{Z}_{m,k|k-1} = \mathbf{h}_k\left(\mathbf{X}_{m,k|k-1}\right). \tag{A.4}$$

(3) Calculate the predicted measurement:

$$\hat{\mathbf{z}}_{k|k-1} = \sum_{m=1}^{N_{\text{sig}}} \omega_m \mathbf{Z}_{m,k|k-1}. \tag{A.5}$$

(4) Calculate the estimated innovation covariance matrix:

$$\mathbf{P}^{\langle zz \rangle}_{k|k-1} = \sum_{m=1}^{N_{\text{sig}}} \omega_m \mathbf{Z}_{m,k|k-1} \mathbf{Z}_{m,k|k-1}^{\;T} - \hat{\mathbf{z}}_{k|k-1}\hat{\mathbf{z}}^T_{k|k-1} + \mathbf{R}_{k-1}. \tag{A.6}$$

(5) Calculate the cross-covariance matrix:

$$\mathbf{P}^{\langle xz \rangle}_{k|k-1} = \sum_{m=1}^{N_{\text{sig}}} \omega_m \mathbf{X}_{m,k|k-1} \mathbf{Z}_{m,k|k-1}^{\;T} - \hat{\mathbf{x}}_{k|k-1}\hat{\mathbf{z}}^T_{k|k-1}. \tag{A.7}$$

(6) Calculate the Kalman gain:

$$\mathbf{W}_k = \mathbf{P}^{\langle xz \rangle}_{k|k-1} \mathbf{P}^{\langle zz \rangle\,-1}_{k|k-1}. \tag{A.8}$$

(7) Calculate the updated state estimate:

$$\hat{\mathbf{x}}_{k|k} = \hat{\mathbf{x}}_{k|k-1} + \mathbf{W}_k\left(\mathbf{z}_k - \hat{\mathbf{z}}_{k|k-1}\right). \tag{A.9}$$

(8) Calculate the updated error covariance:

$$\mathbf{P}_{k|k} = \mathbf{P}_{k|k-1} - \mathbf{W}_k \mathbf{P}^{\langle zz \rangle}_{k|k-1} \mathbf{W}^T_k. \tag{A.10}$$

Conflicts of Interest

The authors declare that there are no conflicts of interest regarding the publication of this article.

Acknowledgments

This work was partially supported by the European Union's Project HIGHTS (High Precision Positioning for Cooperative-ITS Applications) MG-3.5a-2014-636537 and the DLR project Navigation 4.0.

References

[1] J. Li, Y. Zhao, J. Zhang, R. Jiang, C. Tao, and Z. Tan, "Radio channel measurements and analysis at 2.4/5GHz in subway tunnels," *China Communications*, vol. 12, no. 1, pp. 36–45, 2015.

[2] M. Lentmaier and B. Krach, "Maximum likelihood multipath estimation in comparison with conventional delay lock loops," in *Proceedings of the Institute of Navigation - 19th International Technical Meeting of the Satellite Division, ION GNSS 2006*, pp. 1741–1751, usa, September 2006.

[3] F. Antreich, J. A. Nossek, and W. Utschick, "Maximum likelihood delay estimation in a navigation receiver for aeronautical applications," *Aerospace Science and Technology*, vol. 12, no. 3, pp. 256–267, 2008.

[4] M. Triki, D. T. M. Slock, V. Rigal, and P. François, "Mobile terminal positioning via power delay profile fingerprinting: Reproducible validation simulations," in *Proceedings of the 2006 IEEE 64th Vehicular Technology Conference, VTC-2006 Fall*, pp. 2840–2844, September 2006.

[5] Y. Shen and M. Z. Win, "On the use of multipath geometry for wideband cooperative localization," in *Proceedings of the 2009 IEEE Global Telecommunications Conference, GLOBECOM 2009*, December 2009.

[6] P. Setlur and N. Devroye, "Multipath exploited Bayesian and Cramér-Rao bounds for single-sensor target localization," *EURASIP Journal on Advances in Signal Processing*, vol. 2013, no. 1, article no. 53, 2013.

[7] P. Setlur, G. E. Smith, F. Ahmad, and M. G. Amin, "Target localization with a single sensor via multipath exploitation," *IEEE Transactions on Aerospace and Electronic Systems*, vol. 48, no. 3, pp. 1996–2014, 2012.

[8] P. Meissner, E. Leitinger, M. Fröhle, and K. Witrisal, "Accurate and robust indoor localization systems using ultra-wideband signals," in *Proceedings of the in European Conference on Navigation (ENC)*, 2013.

[9] E. Leitinger, P. Meissner, M. Lafer, and K. Witrisal, "Simultaneous localization and mapping using multipath channel information," in *Proceedings of the IEEE International Conference on Communication Workshop, ICCW 2015*, pp. 754–760, June 2015.

[10] K. Witrisal, P. Meissner, E. Leitinger et al., "High-accuracy localization for assisted living: 5G systems will turn multipath channels from foe to friend," *IEEE Signal Processing Magazine*, vol. 33, no. 2, pp. 59–70, 2016.

[11] M. Froehle, E. Leitinger, P. Meissner, and K. Witrisal, "Cooperative multipath-assisted indoor navigation and tracking (Co-MINT) using UWB signals," in *Proceedings of the 2013 IEEE International Conference on Communications Workshops, ICC 2013*, pp. 16–21, Hungary, June 2013.

[12] A. O'Connor, P. Setlur, and N. Devroye, "Single-sensor RF emitter localization based on multipath exploitation," *IEEE Transactions on Aerospace and Electronic Systems*, vol. 51, no. 3, pp. 1635–1651, 2015.

[13] C. Gentner and T. Jost, "Indoor positioning using time difference of arrival between multipath components," in *Proceedings of the 2013 International Conference on Indoor Positioning and Indoor Navigation, IPIN 2013*, Montbeliard, France, October 2013.

[14] C. Gentner, R. Pöhlmann, M. Ulmschneider, T. Jost, and A. Dammann, "Multipath Assisted Positioning for Pedestrians," in *Proceedings of the in Proceedings of the ION GNSS+*, Tampa, FL, USA, 2015.

[15] H. Durrant-Whyte and T. Bailey, "Simultaneous localization and mapping: Part I," *IEEE Robotics and Automation Magazine*, vol. 13, no. 2, pp. 99–108, 2006.

[16] T. Bailey and H. Durrant-Whyte, "Simultaneous localization and mapping (SLAM): Part II," *IEEE Robotics and Automation Magazine*, vol. 13, no. 3, pp. 108–117, 2006.

[17] C. Gentner, B. Ma, M. Ulmschneider, T. Jost, and A. Dammann, "Simultaneous localization and mapping in multipath environments," in *Proceedings of the IEEE/ION Position, Location and Navigation Symposium, PLANS 2016*, pp. 807–815, Georgia, April 2016.

[18] M. Ulmschneider, R. Raulefs, C. Gentner, and M. Walter, "Multipath assisted positioning in vehicular applications," in *Proceedings of the 13th Workshop on Positioning, Navigation and Communication, WPNC 2016*, Germany, October 2016.

[19] M. Ulmschneider, C. Gentner, T. Jost, and A. Dammann, "Association of Transmitters in Multipath-Assisted Positioning," in *Proceedings of the GLOBECOM 2017 - 2017 IEEE Global Communications Conference*, pp. 1–7, Singapore, December 2017.

[20] M. Ulmschneider, C. Gentner, T. Jost, and A. Dammann, "Multiple hypothesis data association for multipath-assisted positioning," in *Proceedings of the 2017 14th Workshop on Positioning, Navigation and Communications (WPNC)*, pp. 1–6, Bremen, October 2017.

[21] Y. Bar-Shalom, X. Li, and T. Kirubarajan, *Estimation with Applications to Tracking and Navigation*, John Wiley & Sons, New York, NY, USA, 2001.

[22] E. A. Wan and R. van der Merwe, "The unscented Kalman filter for nonlinear estimation," in *Proceedings of the IEEE Adaptive Systems for Signal Processing, Communications, and Control Symposium (AS-SPCC '00)*, pp. 153–158, Lake Louise, Canada, October 2000.

[23] S. J. Julier, J. K. Uhlmann, and H. F. Durrant-Whyte, "A new approach for filtering nonlinear systems," in *Proceedings of the American Control Conference*, pp. 1628–1632, June 1995.

[24] S. J. Julier and J. K. Uhlmann, "A new extension of the Kalman filter to nonlinear systems," in *Proceedings of the Signal Processing, Sensor Fusion, and Target Recognition VI*, vol. 3068 of *Proceedings of SPIE*, pp. 182–193, Orlando, Fla, USA, 1997.

[25] S. J. Julier and J. K. Uhlmann, "Unscented filtering and nonlinear estimation," *Proceedings of the IEEE*, vol. 92, no. 3, pp. 401–422, 2004.

[26] M. S. Arulampalam, S. Maskell, N. Gordon, and T. Clapp, "A tutorial on particle filters for online nonlinear/non-Gaussian Bayesian tracking," *IEEE Transactions on Signal Processing*, vol. 50, no. 2, pp. 174–188, 2002.

[27] R. Martinez-Cantin and J. A. Castellanos, "Unscented SLAM for large-scale outdoor environments," in *Proceedings of the IEEE IRS/RSJ International Conference on Intelligent Robots and Systems (IROS '05)*, pp. 3427–3432, Edmonton, Canada, August 2005.

[28] S. Thrun, W. Burgard, and D. Fox, *Probabilistic Robotics, ser. Intelligent robotics and autonomous agents*, MIT Press, 2005.

[29] P. H. Leong, S. Arulampalam, T. A. Lamahewa, and T. D. Abhayapala, "A Gaussian-sum based cubature Kalman filter for bearings-only tracking," *IEEE Transactions on Aerospace and Electronic Systems*, vol. 49, no. 2, pp. 1161–1176, 2013.

[30] C. Kim, R. Sakthivel, and W. K. Chung, "Unscented FastSLAM: a robust algorithm for the simultaneous localization and mapping problem," in *Proceedings of the IEEE International Conference on Robotics and Automation (ICRA '07)*, pp. 2439–2445, Roma, Italy, April 2007.

[31] C. Kim, R. Sakthivel, and W. K. Chung, "Unscented FastSLAM: a robust and efficient solution to the SLAM problem," *IEEE Transactions on Robotics*, vol. 24, no. 4, pp. 808–820, 2008.

[32] M. Ulmschneider, C. Gentner, R. Faragher, and T. Jost, "Gaussian mixture filter for multipath assisted positioning," in *Proceedings of the 29th International Technical Meeting of the*

Satellite Division of the Institute of Navigation, ION GNSS 2016, pp. 1263–1269, usa, September 2016.

[33] S. Thrun, M. Montemerlo, D. Koller, B. Wegbreit, J. Nieto, and E. Nebot, "FastSLAM: An Efficient Solution to the Simultaneous Localization And Mapping Problem with Unknown Data Association," *Journal of Machine Learning Research*, vol. 4, no. 3, pp. 380–407, 2004.

[34] T. S. Rappaport, *Wireless Communications: Principles and Practice, ser. Prentice Hall communications engineering and emerging technologies*, Prentice Hall PTR, 2002.

[35] S. Särkkä, *Bayesian Filtering and Smoothing*, Cambridge University Press, Cambridge, UK, 2013.

[36] T. Jost, W. Wang, U.-C. Fiebig, and F. Perez-Fontan, "Detection and tracking of mobile propagation channel paths," *IEEE Transactions on Antennas and Propagation*, vol. 60, no. 10, pp. 4875–4883, 2012.

[37] B. H. Fleury, M. Tschudin, R. Heddergott, D. Dahlhaus, and K. I. Pedersen, "Channel parameter estimation in mobile radio environments using the SAGE algorithm," *IEEE Journal on Selected Areas in Communications*, vol. 17, no. 3, pp. 434–450, 1999.

[38] J. Poutanen, *enGeometry-based radio channel modeling: Propagation analysis and concept development [Ph.D. thesis]*, Aalto University, 2011.

[39] J. Salmi, A. Richter, and V. Koivunen, "Detection and tracking of MIMO propagation path parameters using state-space approach," *IEEE Transactions on Signal Processing*, vol. 57, no. 4, pp. 1538–1550, 2009.

[40] W. Wang, T. Jost, and A. Dammann, "Estimation and modelling of NLoS time-variant multipath for localization channel model in mobile radios," in *Proceedings of the 53rd IEEE Global Communications Conference, GLOBECOM 2010*, USA, December 2010.

[41] M. Zhu, J. Vieira, Y. Kuang, K. Åström, A. F. Molisch, and F. Tufvesson, "Tracking and positioning using phase information from estimated multi-path components," in *Proceedings of the IEEE International Conference on Communication Workshop, ICCW 2015*, pp. 712–717, June 2015.

[42] G. E. Kirkelund, G. Steinböck, X. Yin, and B. Fleury, "Tracking of the temporal behaviour of path components in the radio channel - A comparison between methods," in *Proceedings of the 2008 Annual IEEE Student Paper Conference, AISPC*, Denmark, February 2008.

[43] S. Särkkä, J. Hartikainen, L. Svensson, and F. Sandblom, "On the relation between Gaussian process quadratures and sigma-point methods," *Journal of Advances in Information Fusion*, vol. 11, no. 1, pp. 31–46, 2016.

[44] D. Simon, *Optimal State Estimation: Kalman, H∞, and Nonlinear Approaches*, Wiley-Interscience, 2006.

[45] I. Arasaratnam and S. Haykin, "Cubature Kalman filters," *Institute of Electrical and Electronics Engineers Transactions on Automatic Control*, vol. 54, no. 6, pp. 1254–1269, 2009.

[46] M. Briers, S. R. Maskell, and R. Wright, "A rao-blackwellised unscented Kalman filter," in *Proceedings of the 6th International Conference on Information Fusion, FUSION 2003*, pp. 55–61, Australia, July 2003.

[47] N. J. Gordon, D. J. Salmond, and A. F. M. Smith, "Novel approach to nonlinear/non-gaussian Bayesian state estimation," *IEE proceedings F*, vol. 140, no. 2, pp. 107–113, 1993.

[48] B. Anderson and J. Moore, *Optimal Filtering, ser. Dover Books on Electrical Engineering*, Dover Publications, 2012.

[49] R. S. Thoma, M. Landmann, G. Sommerkorn, and A. Richter, "Multidimensional high-resolution channel sounding in mobile radio," in *Proceedings of the 21st IEEE Instrumentation and Measurement Technology Conference, IMTC/04*, pp. 257–262, ita, May 2004.

5

SSCM: An Unambiguous Acquisition Algorithm for CBOC Modulated Signal

Yongxin Feng, Shunchao Fei, Fang Liu, and Bo Qian

Communication and Network Institute, Shenyang Ligong University, Shenyang 110159, China

Correspondence should be addressed to Yongxin Feng; fengyongxin@263.net

Academic Editor: Tho Le-Ngoc

Composite binary offset carrier (CBOC) signal has been widely researched in GNSS. The main ingredient of CBOC signal is BOC(1, 1) signal. Usually, the acquisition method for BOC(1, 1) signal is used to capture CBOC signal, while the research of special acquisition method for CBOC signal is rare. In this letter, according to the principle and characteristics of CBOC signal, a special side-peak cancellation method (SSCM) is proposed and simulated. In this method, two special auxiliary signals are introduced. And the local reference signals are obtained by multiplying the data channel signal and pilot channel signal by the auxiliary signals. The cross-correlation results from the received pilot signal and the two local pilot signals with different auxiliary signals will subtract from one another. Then, side peaks of correlation function and in-band noise of pilot channel are suppressed, while the data channel has the same operation results. At last the outputs of pilot channel and data channel will be added up to make full use of the received signal power. By this way, the acquisition efficiency, accuracy, and adaptability to low signal-to-noise ratio (SNR) conditions for CBOC signal have been improved, alongside possible solution when the GNSS receiver works in a critical environment.

1. Introduction

With the development of wireless devices, radio environment is becoming more and more complex. Global Navigation Satellite Systems (GNSS) have to share the crowded frequency source and work in terrible environment with multipath or interference. Then, in 2004, the BOC modulation was proposed by the European Union (EU) and the US, which can be used for modernized civil Global Position System (GPS) signal on L1 band and Galileo Open Service (OS) on E1 band [1]. The new BOC modulation reduces the interference level caused by the existing GPS L1 C/A signal, since it splits the power spectral away from the center frequency. In 2007, in order to improve the interoperability and compatibility between the PRN code tracking accuracy and navigation systems, MBOC modulation has been recommended by the GPS-Galileo working group. Multiplexed Binary Offset Carrier (MBOC) [2] signal is denoted as the optimization modulation method instead of the initial BOC(1, 1) modulation, which can restrain multipath.

The MBOC modulated signal can be produced by CBOC or time-multiplexed BOC (TMBOC) signals. This new modulation allocates a wide band signal BOC(6, 1) in E1/L1 band without interfering with other existing signals and realizes the compatibility and interoperability between GPS and Galileo system. CBOC modulated signal can get more high-frequency components on the power spectral density (PSD) which improves the performance of tracking accuracy and antimultipath capability. However, some drawbacks have been noted, especially associated with the multiple side peaks of autocorrelation function (ACF) causing the ambiguity problem. In order to remove the side peaks, several acquisition algorithms have been proposed and introduced in the past few years, including BPSK-like technique [3], ASPeCT technique [4], pseudo correlation function (PCF) technique [5–7], subcarrier phase cancellation (SCPC) technique [8], and unambiguous acquisition algorithms without auxiliary signals [9]. Nevertheless, most of the methods are designed for BOC modulated signal. Therefore, a novel acquisition algorithm for the CBOC signal is proposed.

The main contributions of this paper are enumerated below:

(i) An optimized acquisition technique named SSCM is proposed to remove side peaks and then the ambiguity problem can be restrained.

(ii) CBOC modulated signal is an important part of modern navigation systems. Some acquisition methods only use part of signal power to capture signal, while SSCM make full use of CBOC signal power. Then the acquisition performance is improved.

(iii) In order to overcome the complicated wireless environment, this method is suitable when the GNSS receiver works in a serious condition, which is based on the antijamming performance of modern GNSS signal.

The remainder of this paper is organized as follows. In Section 2, before expound SSCM, some related work has been undertook. In Section 3, properties of the CBOC modulated signal are shown. In Section 4, SSCM with noise reduction for CBOC modulated signal is presented in detail. Section 5 simulates and compares the proposed algorithm with preexisting acquisition algorithms. Lastly, Section 6 concludes the paper.

2. Related Work

Several techniques have been proposed in the last ten years to overcome ambiguity problem. The BPSK-like method regards BOC or CBOC modulated signal as the sum of binary phase shift keying (BPSK) signals, using filtering, shifting, and integration to change the ACF shape into BPSK-like correlation function shape. Then, an unambiguous correlation function with a decreasing precision is obtained. However, the structure of BPSK-like method is complex and expensive to implement. The in-band noise of intermediate frequency (IF) signal will have an effect on acquisition. In [4], ASPeCT method has a good performance within solving the ambiguity tracking problem, but also the single peak feature and the inhibiting of side peaks are unsatisfactory. Additionally, the BOC(6, 1) component will be regarded as noise in the ASPeCT method. Then, it is possible that some side peak will be larger than main peak, and false lock will happen. According to literature [5], PCF method uses similar locally BOC modulated signals to obtain unambiguous function. This method shows a good adaptability. However, some disadvantages such as lower main peak and energy loss are unavoidable. The subcarrier phase cancellation (SCPC) technique generates an in-phase local subcarrier signal and a quadra-phase local subcarrier signal, which are used to correlate with the received filtered signal. The outputs of two correlation channels are combined to remove side peaks, and an unambiguous correlation function is obtained. However, the suppression of the in-band noise and the utilization of two channels of CBOC modulated signal have not been fully considered. When the in-band noise is serious and no special care is taken, false acquisition or biased tracking will occur. In order to remove the ambiguities of ACF and unwanted replicas of the signal spectrum, a quick unambiguous acquisition algorithm for BOC modulated signals is proposed [10], which exploits a reduced-complexity filter composed of only seven nonzero samples. However, the high-frequency power is sacrificed when the scheme is applied to the CBOC modulated signal. Thus, the advantages of CBOC modulated signal will be wasted. New unambiguous acquisition algorithms, using auxiliary signals, have been proposed in [10–12]. However, all of them are proposed for sine-BOC or cosine-BOC modulated signal in particular. When considering the unique characteristic of CBOC modulated signals, it is more urgent to study the effective acquisition algorithms for CBOC modulated signal. Then, a novel acquisition algorithm for CBOC modulated signal is proposed, named SSCM.

In the new algorithm SSCM, a special auxiliary signal for CBOC modulated signal is introduced to remove side peaks. Moreover, the data channel and pilot channel signals are utilized completely, to ensure the detection probability. Additionally, the subtraction of the correlation channels is used to suppress in-band noise. This way, the side peaks are removed before signal detection. And the acquisition precision will be improved, which accelerates the tracking process.

3. Signal Model

CBOC signal is recommended and produced by BOC(1, 1) and BOC(6, 1) signal. In Galileo E1 OS, CBOC signal has the same power for data and pilot channels. Two different implementations of CBOC are proposed for a 50%/50% power split between data and pilot components. CBOC(6, 1, 1/11) is an important implementation method, where 1/11 denotes the percentage of power of BOC(6, 1) with respect to the total signal CBOC power. The PSD of CBOC(6, 1, 1/11) [13–16] is given by

$$\begin{aligned} G_{\text{CBOC}(6,1,1/11)}(f) = \frac{10}{11} G_{\text{BOC}(1,1)}(f) \\ + \frac{1}{11} G_{\text{BOC}(6,1)}(f), \end{aligned} \tag{1}$$

where $G_{\text{BOC}(1,1)}(f)$ and $G_{\text{BOC}(6,1)}(f)$ can be represented with $G_{\text{BOC}(f_s,f_c)}(f)$, which is the normalized baseband PSD of a BOC modulation, shown as below. From expression (2) of [17], $G_{\text{BOC}(f_s,f_c)}(f)$ can be expressed as below, when $2f_s/f_c$ of BOC(1, 1) and BOC(6, 1) are both even.

$$G_{\text{BOC}(f_s,f_c)}(f) = f_c \left(\frac{\sin(\pi f/2f_s)\sin(\pi f/f_c)}{\pi f \cos(\pi f/2f_s)} \right)^2. \tag{2}$$

The power spectrum and autocorrelation function of CBOC modulated signal are shown in Figures 1 and 2.

The subcarrier formula of BOC(1, 1) and BOC(6, 1) are given as follows:

$$\begin{aligned} \text{SC}_{\text{BOC}(1,1)}(t) &= \begin{cases} \text{sign}\left[\sin\left(\dfrac{2\pi t}{T_c}\right)\right], & 0 \le t \le T_c \\ 0, & \text{elsewhere}, \end{cases} \\ \text{SC}_{\text{BOC}(6,1)}(t) &= \begin{cases} \text{sign}\left[\sin\left(\dfrac{12\pi t}{T_c}\right)\right], & 0 \le t \le T_c \\ 0, & \text{elsewhere}, \end{cases} \end{aligned} \tag{3}$$

where T_c is the code chip duration. Figure 3 shows the schematic diagram of CBOC(6, 1, 1/11) modulated signal.

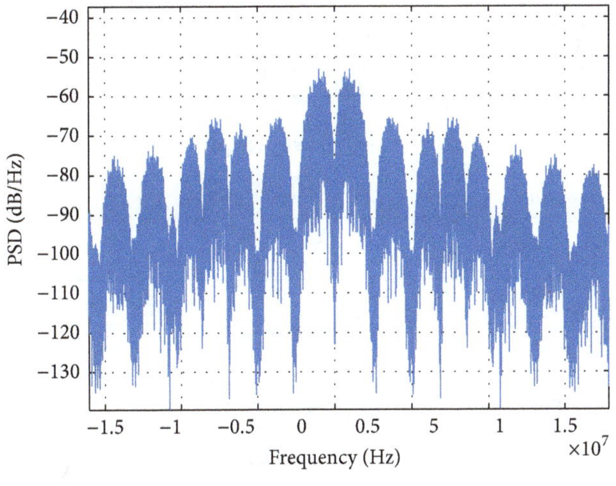

FIGURE 1: PSD of CBOC signal.

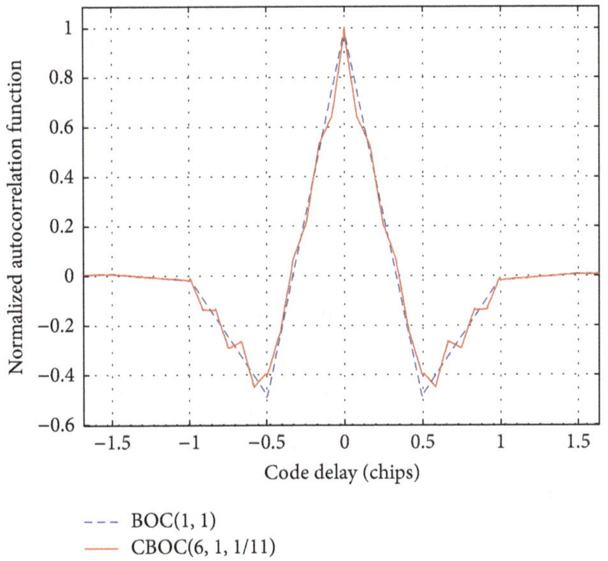

FIGURE 2: Normalized ACFs of BOC(1, 1) and CBOC(6, 1, 1/11) signal.

4. Proposed Acquisition Algorithm

CBOC(6, 1, 1/11) is one of modulations for Galileo E1 signal [18], including E1-B and E1-C channels. CBOC(6, 1, 1/11) of baseband can be expressed mathematically as follows:

$$s(t) = s_{\text{E1-B}}(t) - s_{\text{E1-C}}(t), \tag{4}$$

$$s_{\text{E1-B}}(t) = \left(\frac{1}{\sqrt{2}}\right) \cdot e_{\text{E1-B}}(t) \cdot \left(P \cdot \text{sc}_{\text{E1-B},a}(t) + Q \cdot \text{sc}_{\text{E1-B},b}(t)\right), \tag{5}$$

$$s_{\text{E1-C}}(t) = \left(\frac{1}{\sqrt{2}}\right) \cdot e_{\text{E1-C}}(t) \cdot \left(P \cdot \text{sc}_{\text{E1-C},a}(t) - Q \cdot \text{sc}_{\text{E1-C},b}(t)\right), \tag{6}$$

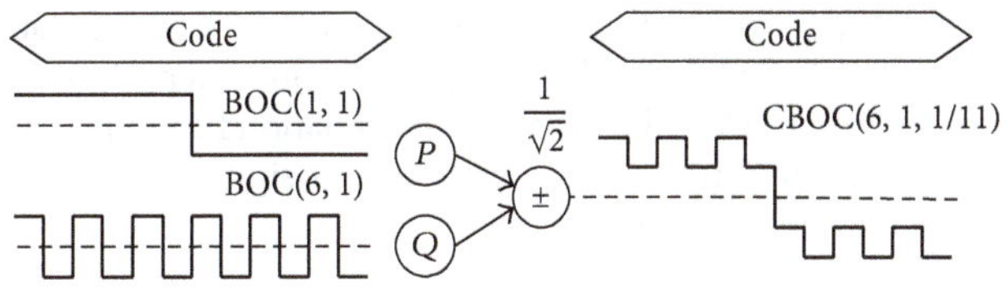

FIGURE 3: CBOC(6, 1, 1/11) signal generation graph.

where $e_{\text{E1-B}}(t)$ and $e_{\text{E1-C}}(t)$ are the binary signal component of data and pilot at the code frequency $f_c(= 1/T_c)$. P equals $\sqrt{10/11}$. Q equals $\sqrt{1/11}$. $\text{sc}_{\text{E1-B},a}(t)$, $\text{sc}_{\text{E1-B},b}(t)$, $\text{sc}_{\text{E1-C},a}(t)$, and $\text{sc}_{\text{E1-C},b}(t)$, respectively, denote the subcarrier, which can be expressed as follows. And f_x represents the subcarrier frequency

$$\text{sc}_x(t) = \text{sign}\left(\sin\left(2\pi f_x t\right)\right). \tag{7}$$

For the case of a CBOC waveform on both data and pilot components, CBOC(6, 1, 1/11) and the rest of received signals of E1 signal can be expressed mathematically as follows:

$$\begin{aligned} s_r(t) = {} & A\left[c_D(t)\, d_D(t)\left(P\text{sc}_1(t) + Q\text{sc}_6(t)\right)\right. \\ & \left. - c_P(t)\left(P\text{sc}_1(t) - Q\text{sc}_6(t)\right)\right] \\ & \cdot \cos\left(2\pi\left(f_{\text{IF}} + f_D\right)t + \theta\right) + n(t), \end{aligned} \tag{8}$$

where c_D and c_P are the data and pilot channels spreading code sequences, d is the navigation message, BOC(1, 1) spreading symbols denoted $\text{sc}_1(t)$ and BOC(6, 1) spreading symbols denoted $\text{sc}_6(t)$, A is the amplitude, f_{IF} and f_D are IF frequency and Doppler frequency, respectively, θ is the unknown carrier phase, and $n(t)$ is the baseband equivalent noise of the received noise that is assumed to be Gaussian.

In order to introduce the method, the baseband signal of CBOC(6, 1, 1/11) is given by

$$\begin{aligned} s(t) = {} & A\left[c_D(t)\, d_D(t)\left(P\text{sc}_1(t) + Q\text{sc}_6(t)\right)\right. \\ & \left. - c_P(t)\left(P\text{sc}_1(t) - Q\text{sc}_6(t)\right)\right]. \end{aligned} \tag{9}$$

Based on the waveform of CBOC(6, 1, 1/11) signal, an auxiliary signal $a_i(t)$ $(i = 1, 2)$ is introduced to signal acquisition process. The local reference signals of pilot and data channels with auxiliary signal are represented in the following equations.

$$S_{D_i}(t) = C_{D-L}(t)\left(P \cdot \text{SC}_1(t) + Q \cdot \text{SC}_6(t)\right) a_i(t), \tag{10}$$

$$S_{P_i}(t) = C_{P-L}(t)\left(P \cdot \text{SC}_1(t) - Q \cdot \text{SC}_6(t)\right) a_i(t), \tag{11}$$

$$a_i(t) = \begin{cases} 1.5, & (n-1)T_c < t < \left(n - \frac{2}{3}\right)T_c \\ 0, & \left(n - \frac{2}{3}\right)T_c < t < \left(n - \frac{1}{3}\right)T_c \\ 1.5 * (-1)^i, & \left(n - \frac{1}{3}\right)T_c < t < nT_c \end{cases} \quad i = 1, 2, \tag{12}$$

where $C_{D-L}(t)$ and $C_{P-L}(t)$ are local PRN code of data channel and pilot channel, T_c is the PRN code interval, and n is the number of T_c. Then the cross-correlation functions can be obtained, multiplying the local reference signals by $S_{\text{base}}(t)$. And τ represents code delay.

$$R_{S/S_{D1}}(\tau) = \sum_{-\infty}^{+\infty} S(t)\, S_{D_1}(t-\tau), \tag{13}$$

$$R_{S/S_{D2}}(\tau) = \sum_{-\infty}^{+\infty} S(t)\, S_{D_2}(t-\tau), \tag{14}$$

$$R_{S/S_{P1}}(\tau) = \sum_{-\infty}^{+\infty} S(t)\, S_{P_1}(t-\tau), \tag{15}$$

$$R_{S/S_{P2}}(\tau) = \sum_{-\infty}^{+\infty} S(t)\, S_{P_2}(t-\tau). \tag{16}$$

Then, the correlation results (13), (14), (15), and (16) can be computed based on equations (10), (11), and (12). $\text{tri}(x/y)$ is a triangular function of width $2y$, centred in $x = 0$, where it has a unity value. It can be used to express the correlation peaks of correlation function.

$$\left|R_{S/S_{D1}}(\tau)\right| = \sum_{i=-11, i\neq 0}^{11} \alpha_i \cdot \text{tri}\left(\frac{\tau + i/12}{1/12}\right), \tag{17}$$

where 1/12 is the bottom width, decided by the subcarrier frequency of BOC(6, 1). And the number of triangular functions is decided by the waveform of S and S_{D1}.

Based on the fixed component and ratio of CBOC(6, 1, 1/11), α_i have fixed values through calculating. And the results of α_i are obtained as follows.

$$\begin{aligned}\alpha_i = \Bigg[& \frac{1+2\sqrt{10/11}}{8}, \frac{9}{44}, \frac{3+2\sqrt{10/11}}{8}, \frac{9}{22}, \frac{1}{2}, \frac{9}{22}, \\ & \frac{12-\sqrt{10}}{44}, \frac{1}{22}, \frac{7+2\sqrt{10}}{88}, \frac{1}{4}, \frac{9}{88}, \frac{9}{88}, \frac{1}{4}, \frac{7+2\sqrt{10}}{88}, \\ & \frac{1}{22}, \frac{12-\sqrt{10}}{44}, \frac{9}{22}, \frac{1}{2}, \frac{9}{22}, \frac{3+2\sqrt{10/11}}{8}, \frac{9}{44}, \\ & \frac{1+2\sqrt{10/11}}{8} \Bigg], \quad i = -11 \sim 11,\ i \neq 0. \end{aligned} \tag{18}$$

Similarly, the correlation results $|R_{S/S_{D2}}(\tau)|$ can be obtained.

$$\left|R_{S/S_{D2}}(\tau)\right| = \sum_{i=-11}^{11} \beta_i * \text{tri}\left(\frac{\tau + i/12}{1/12}\right),$$

$$\begin{aligned}\beta_i = \Bigg[& \frac{1+2\sqrt{10/11}}{8}, \frac{9}{44}, \frac{3+2\sqrt{10/11}}{8}, \frac{9}{22}, \frac{1}{2}, \frac{9}{22}, \\ & \frac{12-\sqrt{10}}{44}, \frac{1}{22}, \frac{25+2\sqrt{10}}{88}, \frac{3}{4}, \frac{63}{88}, 1, \frac{63}{88}, \frac{3}{4}, \\ & \frac{25+2\sqrt{10}}{88}, \frac{1}{22}, \frac{12-\sqrt{10}}{44}, \frac{9}{22}, \frac{1}{2}, \frac{9}{22}, \frac{3+2\sqrt{10/11}}{8}, \\ & \frac{9}{44}, \frac{1+2\sqrt{10/11}}{8} \Bigg], \quad i = -11 \sim 11. \end{aligned} \tag{19}$$

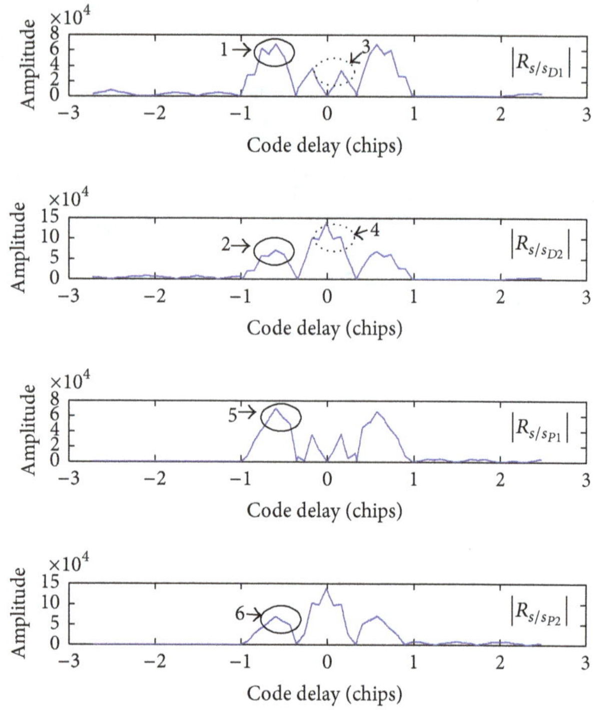

Figure 4: Correlation results of $|R_{S/S_{D1}}(\tau)|$, $|R_{S/S_{D2}}(\tau)|$, $|R_{S/S_{P1}}(\tau)|$, and $|R_{S/S_{P2}}(\tau)|$.

By calculating and comparing, $|R_{S/S_{D1}}(\tau)|$ and $|R_{S/S_{D2}}(\tau)|$ own similar side peaks, while there are seven differences, including $i = [-3, -2, -1, 0, 1, 2, 3]$. The same situation occurs on $|R_{S/S_{P1}}(\tau)|$ and $|R_{S/S_{P2}}(\tau)|$. All the correlation outputs have been simulated and shown in Figure 4.

Looking at Figure 4, we can find that label 1 and label 2 have the same shape as well as label 5 and label 6. The result of the experiment is accordant with the theory. Then, if $|R_{S/S_{D2}}(\tau)|$ is subtracted from $|R_{S/S_{D1}}(\tau)|$, the side peaks will be removed nearly. Furthermore, when label 4 is subtracted from label 3, the result is a main peak, restraining side peaks. The same situation occurs on $|R_{S/S_{P1}}(\tau)|$ and $|R_{S/S_{P2}}(\tau)|$. According to correlation results, outputs of data and pilot channels are obtained.

$$\left|R_{S/S_{D2}}(\tau)\right| - \left|R_{S/S_{D1}}(\tau)\right| = \sum_{i=-3}^{3} \varepsilon_i \cdot \text{tri}\left(\frac{\tau + i/12}{1/12}\right), \tag{20}$$

$$\varepsilon_i = \frac{9}{44}, \frac{1}{2}, \frac{27}{44}, 1, \frac{27}{44}, \frac{1}{2}, \frac{9}{44},$$

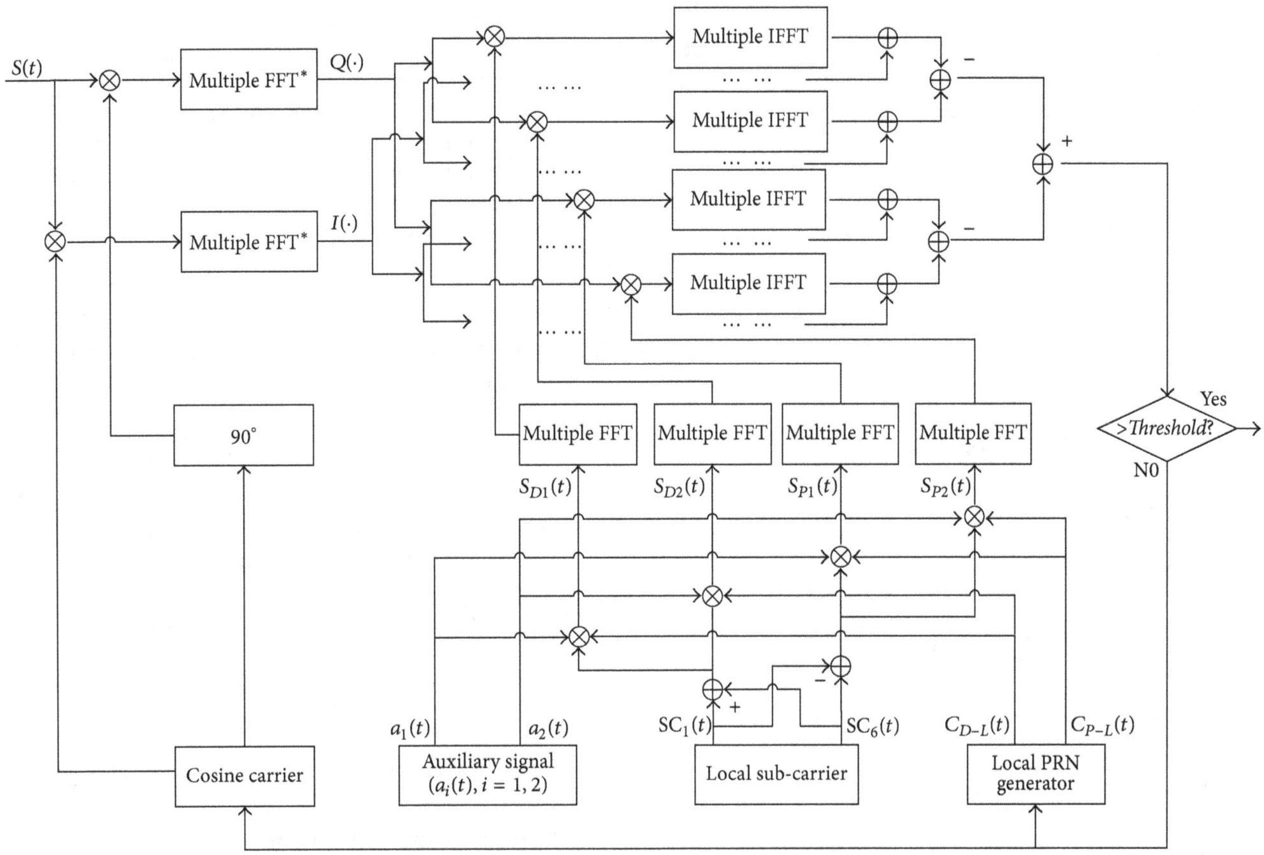

FIGURE 5: Architecture of SSCM algorithm.

$$\left|R_{S/S_{P2}}(\tau)\right| - \left|R_{S/S_{P1}}(\tau)\right| = \sum_{i=-3}^{3} \varepsilon_i \cdot \mathrm{tri}\left(\frac{\tau + i/12}{1/12}\right),$$
$$\varepsilon_i = \frac{9}{44}, \frac{1}{2}, \frac{27}{44}, 1, \frac{27}{44}, \frac{1}{2}, \frac{9}{44}. \tag{21}$$

From (20) and (21), one can see that the correlation result is a linear superposition of several triangular functions, and the two functions are the same. Therefore, a normalized correlation function named special side-peak cancel method (SSCM) can be denoted as follows:

$$\begin{aligned} R_{\text{proposed}}(\tau) &= \left|R_{S/S_{D2}}(\tau)\right| - \left|R_{S/S_{D1}}(\tau)\right| + \left|R_{S/S_{P2}}(\tau)\right| \\ &\quad - \left|R_{S/S_{P1}}(\tau)\right| \\ &= 2 \cdot \sum_{i=-3}^{3} \varepsilon_i \cdot \mathrm{tri}\left(\frac{\tau + i/12}{1/12}\right) \\ &\approx 2 \cdot \mathrm{tri}\left(\frac{\tau}{1/3}\right). \end{aligned} \tag{22}$$

From formula (22), the correlation function of SSCM contains only one main peak, solving the phase ambiguity problem effectively.

Based on subtract operation, the noise of each channel is restrained, while the main peak value of correlation output is uninfluenced. Furthermore, by using $|R_{S/S_{D2}}(\tau)| - |R_{S/S_{D1}}(\tau)|$ and $|R_{S/S_{P2}}(\tau)| - |R_{S/S_{P1}}(\tau)|$ in combination, a bigger main peak can be obtained, and the pilot channel and data channel are both fully used. Then the schematic diagram of SSCM algorithm is shown in Figure 5.

Eight correlators are employed in this architecture. And the in-phase branches are omitted to simplify structure.

In order to analyze detection probability (P_d) and false alarm probability (P_f), a detailed description of the spread spectrum signal acquisition theory [19] is used. In traditional acquisition scheme, the test criterion is given by

$$T = \sum_{i=0}^{M-1} I_i^2 + Q_i^2, \tag{23}$$

where I_i and Q_i are, respectively, in-phase and quadra-phase correlator outputs. M is the number of noncoherent summations. Once the maximum correlation result is larger than a threshold, detection is declared. The main idea SSCM based is to construct local auxiliary signals, the cross-correlation of which with the received signal can be used to remove the undesired side peaks. We choose the test criterion as follows.

$$\begin{aligned} T = \sum_{i=0}^{M-1} \Big[&\left(I_{S/S_{D2},i}^2 + Q_{S/S_{D2},i}^2\right) - \left(I_{S/S_{D1},i}^2 + Q_{S/S_{D1},i}^2\right) \\ &+ \left(I_{S/S_{P2},i}^2 + Q_{S/S_{P2},i}^2\right) - \left(I_{S/S_{P1},i}^2 + Q_{S/S_{P1},i}^2\right) \Big], \end{aligned} \tag{24}$$

where $I_{S/S_{D2},i}$, $Q_{S/S_{D2},i}$, $I_{S/S_{D1},i}$, $Q_{S/S_{D1},i}$, $I_{S/S_{P2},i}$, $Q_{S/S_{P2},i}$, $I_{S/S_{P1},i}$, and $Q_{S/S_{P1},i}$ are in-phase and quadra-phase correlate outputs when the local signals employ the auxiliary signal. All the outputs are shown in the following.

$$\begin{aligned}
I_{S/S_{D1},i} &= \sqrt{\frac{T_{\text{sum}}C}{N_0}} \operatorname{sinc}(\Delta f T_{\text{sum}}) R_{S/S_{D1}}(\Delta\tau)\cos(\Delta\varphi) + n_{I,i}, \\
Q_{S/S_{D1},i} &= \sqrt{\frac{T_{\text{sum}}C}{N_0}} \operatorname{sinc}(\Delta f T_{\text{sum}}) R_{S/S_{D1}}(\Delta\tau)\sin(\Delta\varphi) + n_{Q,i}, \\
I_{S/S_{D2},i} &= \sqrt{\frac{T_{\text{sum}}C}{N_0}} \operatorname{sinc}(\Delta f T_{\text{sum}}) R_{S/S_{D2}}(\Delta\tau)\cos(\Delta\varphi) + n_{I,i}, \\
Q_{S/S_{D2},i} &= \sqrt{\frac{T_{\text{sum}}C}{N_0}} \operatorname{sinc}(\Delta f T_{\text{sum}}) R_{S/S_{D2}}(\Delta\tau)\sin(\Delta\varphi) + n_{Q,i}, \\
I_{S/S_{P1},i} &= \sqrt{\frac{T_{\text{sum}}C}{N_0}} \operatorname{sinc}(\Delta f T_{\text{sum}}) R_{S/S_{P1}}(\Delta\tau)\cos(\Delta\varphi) + n_{I,i}, \\
Q_{S/S_{P1},i} &= \sqrt{\frac{T_{\text{sum}}C}{N_0}} \operatorname{sinc}(\Delta f T_{\text{sum}}) R_{S/S_{P1}}(\Delta\tau)\sin(\Delta\varphi) + n_{Q,i}, \\
I_{S/S_{P2},i} &= \sqrt{\frac{T_{\text{sum}}C}{N_0}} \operatorname{sinc}(\Delta f T_{\text{sum}}) R_{S/S_{P2}}(\Delta\tau)\cos(\Delta\varphi) + n_{I,i}, \\
Q_{S/S_{P2},i} &= \sqrt{\frac{T_{\text{sum}}C}{N_0}} \operatorname{sinc}(\Delta f T_{\text{sum}}) R_{S/S_{P2}}(\Delta\tau)\sin(\Delta\varphi) + n_{Q,i}.
\end{aligned} \tag{25}$$

Here sinc(x) is equal to $\sin(x)/x$. $\Delta\tau$ means the code phase error, T_{sum}, is the coherent integration time, Δf represents the Doppler frequency, C/N_0 means the ratio of carrier and noise, and $\Delta\varphi$ refer to the frequency wipe-off error. $n_{I,i}$ and $n_{Q,i}$ are the correlation calculation results of an additive white Gaussian noise with zero and single-sided noise PSD N_0.

Then the four terms $\sum_{i=0}^{M-1}(I^2_{S/S_{D2},i} + Q^2_{S/S_{D2},i})$, $\sum_{i=0}^{M-1}(I^2_{S/S_{D1},i} + Q^2_{S/S_{D1},i})$, $\sum_{i=0}^{M-1}(I^2_{S/S_{P2},i} + Q^2_{S/S_{P2},i})$, and $\sum_{i=0}^{M-1}(I^2_{S/S_{P1},i} + Q^2_{S/S_{P1},i})$ follow χ^2 distribution with $2M$ degrees of freedom (DOF), in which the noncentrality parameters of the four terms are

$$\begin{aligned}
a_1^2 &= M \cdot T_{\text{sum}} \cdot \frac{C}{N_0} \cdot \operatorname{sinc}^2(\Delta f T_{\text{sum}}) \cdot R^2_{S/S_{D2}}(\Delta\tau), \\
a_2^2 &= M \cdot T_{\text{sum}} \cdot \frac{C}{N_0} \cdot \operatorname{sinc}^2(\Delta f T_{\text{sum}}) \cdot R^2_{S/S_{D1}}(\Delta\tau), \\
a_3^2 &= M \cdot T_{\text{sum}} \cdot \frac{C}{N_0} \cdot \operatorname{sinc}^2(\Delta f T_{\text{sum}}) \cdot R^2_{S/S_{P2}}(\Delta\tau), \\
a_4^2 &= M \cdot T_{\text{sum}} \cdot \frac{C}{N_0} \cdot \operatorname{sinc}^2(\Delta f T_{\text{sum}}) \cdot R^2_{S/S_{P1}}(\Delta\tau).
\end{aligned} \tag{26}$$

And the test criterion without noise can be expressed as

$$\begin{aligned}
T &\approx M \cdot T_{\text{sum}} \cdot \frac{C}{N_0} \cdot \operatorname{sinc}^2(\pi\Delta f T_{\text{sum}}) \cdot \left(R^2_{S/S_{D2}}(\Delta\tau)\right. \\
&\quad \left. - R^2_{S/S_{D1}}(\Delta\tau) + R^2_{S/S_{P2}}(\Delta\tau) - R^2_{S/S_{P1}}(\Delta\tau)\right) = a_1^2 \\
&\quad - a_2^2 + a_3^2 - a_4^2.
\end{aligned} \tag{27}$$

According to the statistical theory [19–21], in order to simplify the computation of theoretical P_{fa} and P_d, the detection probability and false alarm probability of the proposed method have similar form as the GRASS technique [10]. And the Gaussian Q-function is adopted to approximate

$$P_d(V) \approx Q\left(\frac{V - 4M - M(T-4)}{4\sqrt{M(4+T)}}\right), \tag{28}$$

where V denotes threshold of false alarm, which can be calculated as

$$P_{fa}(V) = \exp\left(-\frac{V}{2}\right)\sum_{n=0}^{M-1}\frac{1}{n!}\left(\frac{V}{2}\right)^n. \tag{29}$$

5. Simulation and Analysis

In order to verify the effectiveness and validity of SSCM, it is necessary to analyze the efficiency in terms of correlation function, peak-to-average ratio (PAR) with different noise, detection probability (P_d), false alarm probability (P_f), and processing complexity. In the following comparisons, we assume the following parameters: the center frequency f_c is 48 MHz, sampling frequency f_s is 192 MHz, accumulation time T is 1 ms, and signal-to-noise rate (SNR) range is −35~0 dB.

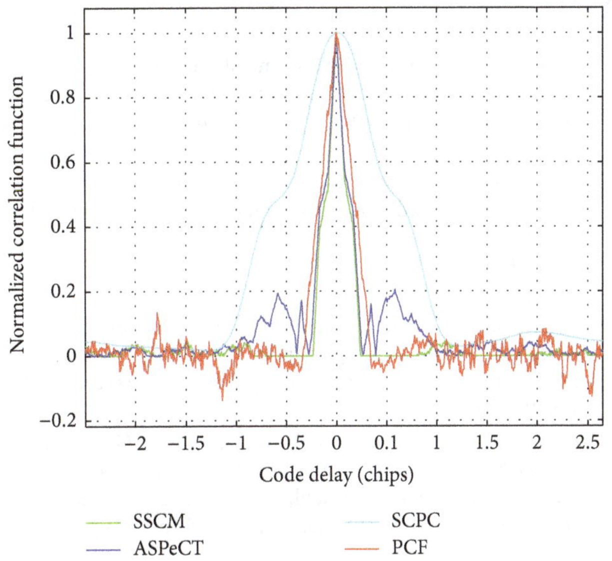

FIGURE 6: Normalized correlation functions of SSCM, ASPeCT, SCPC, and PCF.

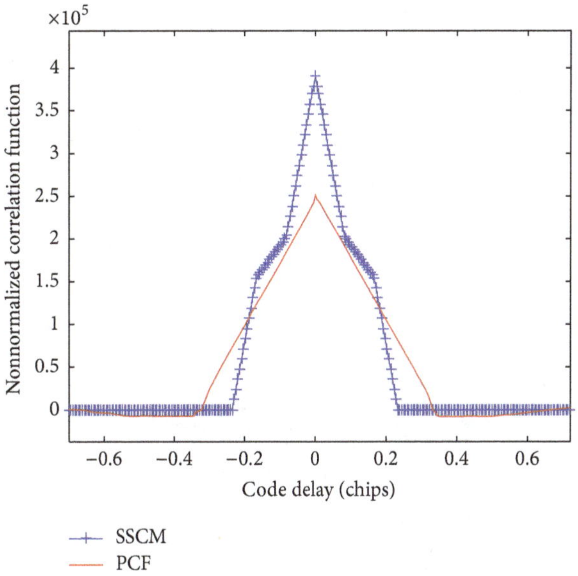

FIGURE 7: Nonnormalized correlation functions of SSCM and PCF.

Figure 6 shows the normalized correlation functions of these four methods. And the SNR is set at −15 dB. SCPC method is similar to that of BPSK, and this method destroys the narrow peak of autocorrelation function (ACF). Therefore, SCPC method is not suitable to CBOC signal. For ASPeCT method, the influence of remaining side peaks still exist. With PRN code delay and Doppler shift, each value is tested in order to get a correlative result. Once the maximum correlation result exceeds the setting threshold, detection will occur. It is probably that one of side peak's energy will be larger than the main peak under the influence of remaining side peaks; thus false acquisition will happen, and this side peak will be locked. In these four methods, PCF method has similar ACF with SSCM method, owning only one main peak. Nevertheless, the two methods differed in power efficiency. From formula (4) one can find that CBOC signal owns data component and pilot component. In PCF method, only one component is utilized, while the two components are fully used in SSCM method. And the main peak of SSCM will be higher than PCF's. Through comparison and analysis, it is found that PCF method owns the best correlation function among PCF, ASPeCT, and SCPC, which is similar to SSCM. Then, take PCF method as example, the performances of PCF and SSCM are compared. In Figure 7, the nonnormalized correlation functions of PCF and SSCM method are shown.

In CBOC(6, 1, 1/11) modulation, data component and pilot component are modulated in the same carrier. And both components have the same Doppler shift and code synchronous message. Therefore, depending on using a combination of the two components, PAR of SSCM method is increased to guarantee P_d under severe environment.

Along with SNR increasing from −35 dB to 0 dB, simulation shows that SSCM method own the best PAR than other three methods. From formula (20), $|R_{S/S_{D2}}(\tau)| - |R_{S/S_{D1}}(\tau)|$ and $|R_{S/S_{P2}}(\tau)| - |R_{S/S_{P1}}(\tau)|$ can suppress the noise of each component, which contribute to PAR. And SCPC owns the worst PAR due to its combination of in-phase correlation output and quadra-phase correlation output, which cause channel noise more severe. PARs of these four methods are shown in Figure 8.

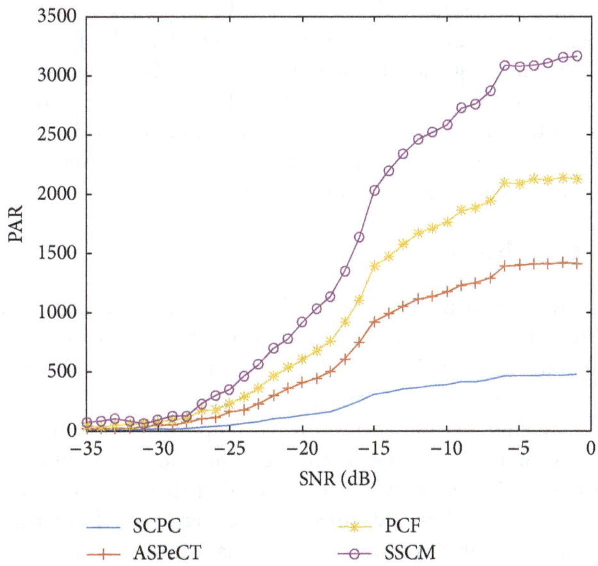

FIGURE 8: PARs of four methods.

With a fixed false alarm probability $P_{fa} = 10^{-5}$, the performance of the SSCM and PCF acquisition method is also shown in the following for comparison.

The detection probabilities of both proposed technique and PCF method are drawn in Figure 9 with different T_p and M, under the assumption of fixed false alarm probability. And the detection probabilities increase with C/N_0, T_p and M growth. And the acquisition performance of SSCM method increases 1 to 2 dB compared with the PCF method. The

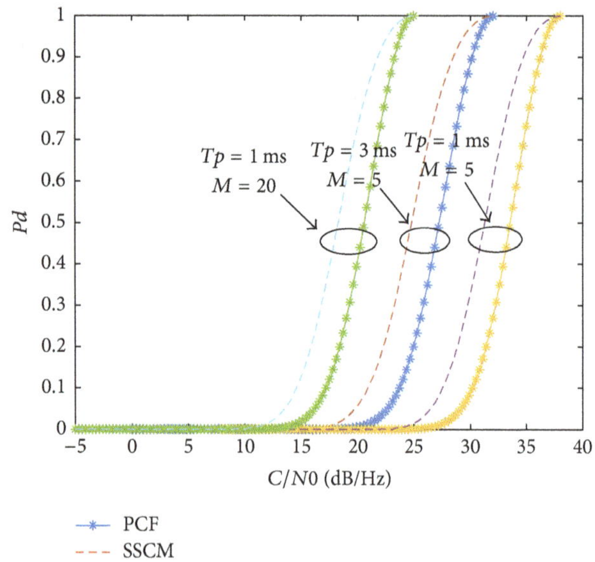

FIGURE 9: Detection probability for CBOC, $P_{fa} = 10^{-5}$.

reason for this phenomenon is that the proposed method uses data channel and pilot channel completely to get the correlation peak. Moreover, the proposed method has no threat with acquisition ambiguity problems. So the proposed scheme is robust compared with PCF scheme. Another characteristic of the proposed unambiguous methods is that it has a narrower peak. It can provide sufficiently small code delay to insure the acquisition precision.

6. Conclusion

In this paper, the principle and characteristics of CBOC(6, 1, 1/11) are studied. And a new acquisition method for CBOC(6, 1, 1/11) modulation is proposed, which can remove the ambiguity threat. Eight correlators are used in this method and result in complexity of acquisition structure increasing. However, the acquisition technique can remove side peaks completely, meaning that the ambiguity problem can be restrained. Moreover, SSCM method is suitable for working in a serious condition. And the detection probability of CBOC can be guaranteed, which devote to average acquisition time.

Conflicts of Interest

The authors declare that they have no conflicts of interest.

Acknowledgments

This work was supported by the National Natural Science Foundation of China (no. 61501309), the Program for Liaoning Innovative Research Team in University (no. LT2011005), New Century Program for Excellent Talents of Ministry of Education of China (no. NCET-11-1013), Project of Science and Technology Department of Liaoning Province (no. 20121038), and Project of Education Department of Liaoning Province (no. L2013085).

References

[1] S. Tom, "Perspective on GNSS compatibility and interoperability," in *Second meeting of the International Committee on Global Navigation Satellite System*, pp. 1025–1056, Bangalore, India, 2007.

[2] G. W. Hein, J.-A. Avila-Rodriguez, S. Wallner et al., "MBOC: the new optimized spreading modulation recommended for GALILEO L1 OS and GPS L1C," in *Proceedings of the 2006 IEEE/ION Position, Location, and Navigation Symposium*, pp. 883–892, USA, April 2006.

[3] N. Martin, V. Leblond, G. Guillotel, and V. Heiries, "BOC(x, y) signal acquisition techniques and performances," in *Proceeding of U.S. Institute of Navigation GPS/GNSS Conference*, pp. 188–198, Portland, 2003.

[4] O. Julien, C. Macabiau, M. E. Cannon, and G. Lachapelle, "ASPeCT: unambiguous sine-BOC(n,n) acquisition/tracking technique for navigation applications," *IEEE Transactions on Aerospace and Electronic Systems*, vol. 43, no. 1, pp. 150–162, 2007.

[5] J. Ren, G. Yang, W. Jia, and M. Yao, "Unambiguous tracking method based on combined correlation functions for sine/cosine-BOC CBOC and ALTBOC modulated signals," *Radioengineering*, vol. 23, no. 1, pp. 244–253, 2014.

[6] Z. Yao, "A new unambiguous tracking technique for sine-BOC(2n, n) signals," in *Proceedings of the International Technical meeting of the Satellite Division, ION GNSS 21st*, pp. 16–19, Savannah, GA, USA, 2008.

[7] Z. Yao, M. Lu, and Z. Feng, "Unambiguous technique for multiplexed binary offset carrier modulated signals tracking," *IEEE Signal Processing Letters*, vol. 16, no. 7, pp. 608–611, 2009.

[8] V. Heiries, J.-À. Àvila-Rodriguez, M. Irsigler, G. W. Hein, E. Rebeyrol, and D. Roviras, "Acquisition performance analysis of composite signals for the LI OS optimized signal," in *Proceedings of the 18th International Technical Meeting of the Satellite Division of The Institute of Navigation, ION GNSS 2005*, pp. 877–889, usa, September 2005.

[9] Y. Lee, D. Chong, I. Song, S. Y. Kim, G.-I. Jee, and S. Yoon, "Cancellation of correlation side-peaks for unambiguous BOC signal tracking," *IEEE Communications Letters*, vol. 16, no. 5, pp. 569–572, 2012.

[10] Z. Yao, M. Lu, and Z. Feng, "Unambiguous sine-phased binary offset carrier modulated signal acquisition technique," *IEEE Transactions on Wireless Communications*, vol. 9, no. 2, pp. 577–580, 2010.

[11] F. Shen, G. Xu, and D. Xu, "Unambiguous acquisition technique for cosine-phased binary offset carrier signal," *IEEE Communications Letters*, vol. 18, no. 10, pp. 1751–1754, 2014.

[12] Y. Zhou, X. Hu, T. Ke, and Z. Tang, "Ambiguity mitigating technique for cosine-phased binary offset carrier signal," *IEEE Transactions on Wireless Communications*, vol. 11, no. 6, pp. 1981–1984, 2012.

[13] J. A. Avila-Rodriguez, "Revised combined GALILEO/GPS frequency and signal performance analysis," in *Proceedings of ION GNSS*, pp. 13–16, Long Beach , Calif, USA, 2005.

[14] M. Flissi, K. Rouabah, D. Chikouche, A. Mayouf, and S. Atia, "Performance of new BOC-AW-modulated signals for GNSS system," *EURASIP Journal on Wireless Communications and Networking*, vol. 2013, no. 1, article no. 124, 2013.

[15] D. W. Lim, D. J. Cho, H. H. Choi, and S. J. Lee, "A simple and efficient code discriminator for a MBOC signal tracking," *IEEE Communications Letters*, vol. 17, no. 6, pp. 1088–1091, 2013.

[16] J. A. Avila-Rodriguez, S. Wallner, G. W. Hein, E. Rebeyrol, and O. Julien, "CBOC: an implementation of MBOC. CNES-ESA," in *Proceedings of the 1st Workshop on GALILEO Signals and Signal Processing*, Toulouse, France, 2006.

[17] J. W. Betz, "Binary offset carrier modulations for radionavigation," *NAVIGATION: Journal of the Institute of Navigation*, vol. 48, no. 4, pp. 227–246, 2001.

[18] European Union, European GNSS (Galileo) Open Service, Signal in Space Interface Control Document, SIS ICD, Issue 1.2, November, 2015.

[19] F. Bastide, O. Julien, C. Macabiau, and B. Roturier, "Analysis of L5/E5 acquisition, tracking and data demodulation thresholds," in *Proceedings of the U.S. Institute of Navigation GPS Conference*, pp. 2196–2207, Portland, Ore, USA, 2002.

[20] J. Proakis and M. Salehi, *Digital Communications*, McGraw-Hill, 2001.

[21] H. Van Trees, *Detection, Estimation And Modulation Theory, Chapter 5*, Wiley-Interscience, 2001.

6

A Detection Algorithm for the BOC Signal based on Quadrature Channel Correlation

Bo Qian, Guolei Zheng, and Yongxin Feng

School of Information Science and Engineering, Shenyang Ligong University, Shenyang 110159, China

Correspondence should be addressed to Yongxin Feng; fengyongxin@263.net

Academic Editor: Iickho Song

In order to solve the problem of detecting a BOC signal, which uses a long-period pseudo random sequence, an algorithm is presented based on quadrature channel correlation. The quadrature channel correlation method eliminates the autocorrelation component of the carrier wave, allowing for the extraction of the absolute autocorrelation peaks of the BOC sequence. If the same lag difference and height difference exist for the adjacent peaks, the BOC signal can be detected effectively using a statistical analysis of the multiple autocorrelation peaks. The simulation results show that the interference of the carrier wave component is eliminated and the autocorrelation peaks of the BOC sequence are obtained effectively without demodulation. The BOC signal can be detected effectively when the SNR is greater than −12 dB. The detection ability can be improved further by increasing the number of sampling points. The higher the ratio of the square wave subcarrier speed to the pseudo random sequence speed is, the greater the detection ability is with a lower SNR. The algorithm presented in this paper is superior to the algorithm based on the spectral correlation.

1. Introduction

The Binary Offset Carrier (BOC) signal is used in the global navigation satellite system (GNSS) and is characterized by multiple peaks in its autocorrelation function and spectrum splitting [1–3]. By using a square wave to modulate again, the synchronization precision of the BOC signal is improved and the interference of the same-frequency signals is decreased [4]. On the other hand, there are multiple side-peaks around the main peak of autocorrelation function of the BOC sequence, thus causing the ambiguity problem. To deal with the problem, several unambiguous techniques have been proposed [5, 6]. A novel cancellation technique of correlation side-peaks is proposed, by employing a combination of the subcorrelations making up the BOC autocorrelation [6].

The pseudo random sequence of the BOC signal has the characteristics of pseudo randomness and infinite periods in a short time, which is used in secret communications. Therefore, it is difficult to detect a BOC signal under non-cooperative conditions. In addition, by utilizing the direct sequence spread spectrum (DSSS), the BOC signal can be transmitted under a negative signal to noise ratio (SNR) and because the anti-interception ability is strong, it is more difficult to detect the signal.

To date, new methods of BOC signal recognition and parameter estimation have been proposed [7–11]. The detection methods are based on spectral correlation [7–9] and the methods for parameter estimation are based on autocorrelation [10, 11]. The basis of the spectral correlation methods is based on the cyclostationary characteristic of the BOC signal, so that the parameters of the carrier, square wave, and pseudo random sequence can be estimated. However, when the pseudo random sequence has an infinite period in a short time, the methods based on spectral correlation cannot work effectively.

The autocorrelation methods are based on the characteristics of the multiple autocorrelation peaks of the BOC signal. Based on demodulating the BOC signal, the parameters can be estimated effectively based on how the BOC signal correlates with the multiple autocorrelation peaks. Considering that the BOC signal is transmitted under a negative SNR in secret communications, demodulation is not easily achieved; therefore, it is difficult to estimate the parameters in a real-life environment.

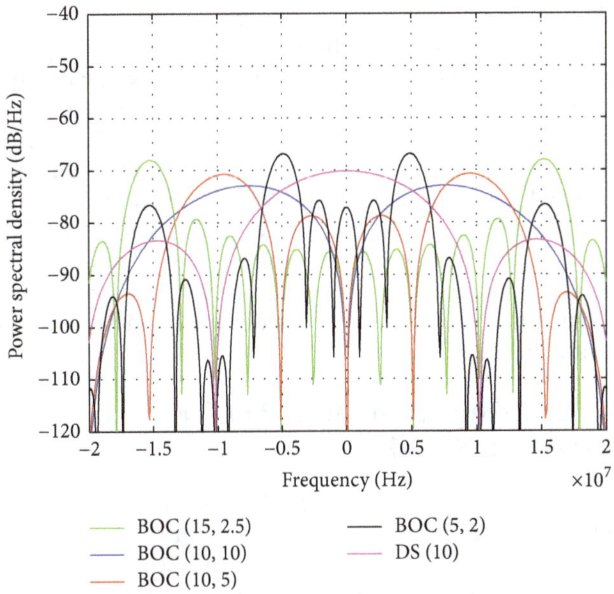

FIGURE 1: Power spectral density of the BOC signals.

In this paper, an algorithm for detecting the BOC signal is presented, using a long-period pseudo random sequence. The autocorrelation component of the carrier wave in the BOC signal is eliminated based on quadrature channel correlation. By detecting the autocorrelation peaks, the BOC signal can be detected.

The outline of this paper is as follows. In Section 2, we study the characteristics of the BOC signal. Section 3 describes the analysis of the characteristics of the multiple autocorrelation peaks for the BOC signal and the algorithm for detecting the BOC signal. Section 4 provides simulation results demonstrating the performance of the algorithm. Finally, Section 5 presents our conclusions and final comments.

2. Characteristics of the BOC Signal

The BOC signal $Y(t)$, modulated by BPSK, is given by

$$Y(t) = A \cdot D(t) P(t) S_c(t) \cos(2\pi f t + \varphi), \quad (1)$$

where A is the carrier amplitude, $D(t)$ is the baseband data, $P(t)$ is the pseudo random sequence, $S_c(t)$ is the square wave, f is the carrier frequency, and φ is the phase. The frequency of $P(t)$ is f_c, and the frequency of $S_c(t)$ is f_s.

Firstly, the spread spectrum sequence is obtained by XOR baseband data with the pseudo random sequence. Then, the spread spectrum sequence is XORed again with a square wave to generate the BOC sequence. Finally, the BOC signal is generated by modulating the BOC sequence to the main carrier. The BOC signal is denoted as BOC (N_s, N_c), where N_s means the ratio of f_s to the reference frequency f_{base}, and N_c means the ratio of f_c to the reference frequency f_{base}. In GNSS systems, the reference frequency $f_{\text{base}} = 1.023\,\text{MHz}$.

The normalized power spectral density (PSD) of the BOC signal can be expressed as [12]

$$G_{\text{BOC}(f_s,f_c)}(f) = \begin{cases} f_c \left[\dfrac{\sin(\pi f/2f_s)\sin(\pi f/f_c)}{\pi f \cos(\pi f/2f_s)}\right]^2, & n \text{ is an odd number} \\ f_c \left[\dfrac{\sin(\pi f/2f_s)\cos(\pi f/f_c)}{\pi f \cos(\pi f/2f_s)}\right]^2, & n \text{ is an even number,} \end{cases} \quad (2)$$

where

$$n = \frac{2f_s}{f_c}. \quad (3)$$

The distribution of the normalized power spectral density for the BOC signals is shown in Figure 1, where DS (10) is the normalized power spectral density of the DSSS signals, in which the frequency of the pseudo random sequence is ten times as much as f_{base}.

As shown in Figure 1, the main lobe energy of the BOC signal is split into two lobes located at $\pm f_s$ from the central frequency. The main lobe energy of the DSSS signal is concentrated in the central frequency. Therefore, the BOC signal and the DSSS signal can be transmitted on the same frequency at the same time without interfering with each other and the bandwidth efficiency is greatly improved. Because of the wider bandwidth, the BOC signal has a greater antijamming capability than the DSSS signal. Furthermore, it is difficult to estimate the carrier frequency of the BOC signal because the two lobes are not located in the carrier frequency.

The autocorrelation function of the BOC sequence can be expressed as [13]

$$R_{\text{BOC}}(\tau) = \begin{cases} 1, & \tau = 0 \\ (-1)^{l-1}\dfrac{[n-(l-1)]}{n} + (-1)^l [4f_s - f_c(2l-1)] \cdot \left(\tau - \dfrac{l-1}{2f_s}\right), & \dfrac{(l-1)}{2f_s} < \tau \le \dfrac{l}{2f_s} \\ & 0 < l \le n-1,\ l \text{ is integer} \\ (-1)^l \dfrac{[n-|l|]}{n} + (-1)^{|l|-1} [4f_s - f_c(2|l|-1)] \cdot \left(\tau + \dfrac{|l|}{2f_s}\right), & \dfrac{|l|}{2f_s} < \tau \le \dfrac{(|l|-1)}{2f_s} \\ & -n+1 \le l < 0,\ l \text{ is integer} \\ -\dfrac{1}{N} & T_c < |\tau| \le (N-1)T_c, \end{cases} \quad (4)$$

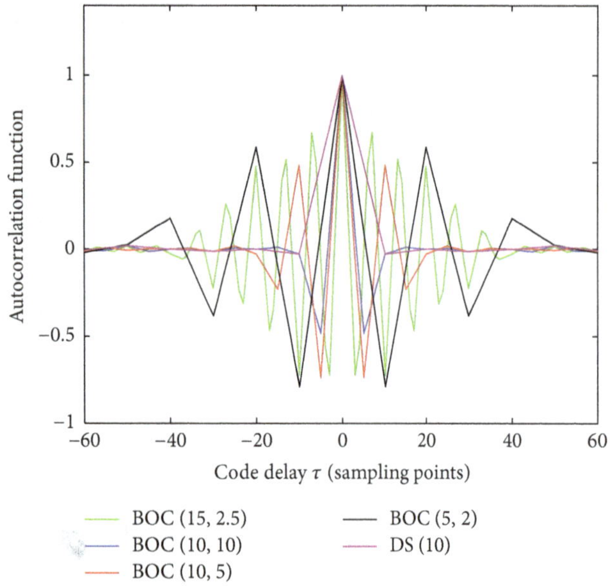

FIGURE 2: Autocorrelation function of the BOC sequence.

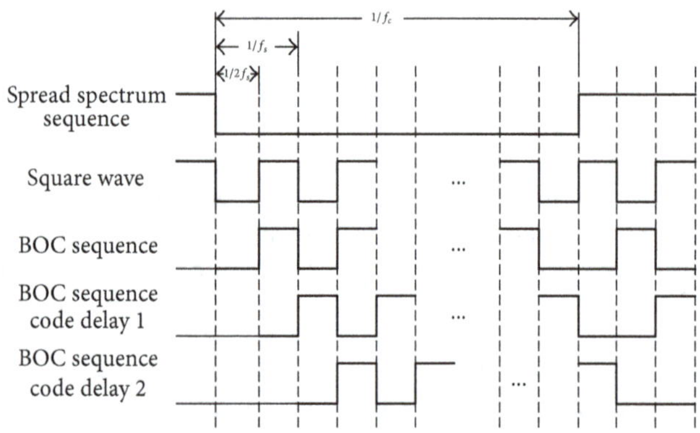

FIGURE 3: The generation process of the autocorrelation peaks.

where N is the chip number of the pseudo random sequence in one period and T_c is the period of the pseudo random sequence.

The autocorrelation functions of the BOC sequence are shown in Figure 2.

Figure 2 shows that the autocorrelation function of the BOC sequence contains multiple positive and negative peaks. The absolute values of the multiple peaks decrease gradually as the code delay increases. The sum of the multiple peaks is $2n-1$. The width of the autocorrelation function's main peak is narrower so that the BOC sequence has a better synchronous precision. The BOC sequence is highly correlated with itself. The autocorrelation function of the DSSS sequence contains only one peak and the main peak of the DSSS sequence is wider than the peak of the BOC sequence.

As shown in Figures 1 and 2, the BOC signal has two characteristics splitting spectrum peaks and multiple autocorrelation peaks, which enhance the antijamming ability and improve the precision of the acquisition [14, 15]. In addition, the BOC signal can be transmitted with other signals, greatly improving the bandwidth efficiency [16]. The advantages of the BOC signal make it highly suitable for secret communications. However, because the main lobe energy of the BOC signal is not concentrated in the carrier frequency and the BOC signal should be transmitted under the condition of a negative SNR [17, 18], it will be difficult to estimate the carrier frequency of the BOC signal. In addition, the BOC sequence, which is commonly used in secret communications, has a longer period or is aperiodic in a short time, which increases the difficulty of detecting and estimating the BOC signal.

3. The Recognition Algorithm for the BOC Signal

3.1. Multiple Autocorrelation Peaks Analysis. The multiple autocorrelation peaks are a unique feature of the BOC sequence and can be used to detect the BOC signal. This feature is needed to analyze the relevance between the multiple autocorrelation peaks. Figure 3 shows the generation process of an unambiguous autocorrelation function for the BOC sequence.

The spread spectrum sequence, which is obtained by the XOR baseband data with the pseudo random sequence is XORed again with a square wave to generate the BOC sequence (Figure 3). Then, the BOC sequence is shifted as the code delay τ increases. When τ equals half a period of a square wave, it represents the BOC sequence code delay 1. Similarly, when τ equals one period of a square wave, it represents the BOC sequence code delay 2 (Figure 3).

When τ is 0, the normalized autocorrelation result of the BOC sequence is 1. The result of the normalized autocorrelation function decreases gradually with the increase in the code delay τ from 0 to $1/2f_s$. This represents the main peak of the autocorrelation function of the BOC sequence when $\tau = 0$ as shown in Figure 2. When τ is equal to $1/2f_s$, the phase of the square wave component in the BOC sequence 1 is inverted. As a result, the autocorrelation result of the square wave component is −1. This represents the first negative peak when $\tau = 1/2f_s$ as shown in Figure 2. With an increase in the code delay τ, when τ is $1/f_s$, the phase of the square wave component in the BOC sequence 2 is the same as in the BOC sequence. As a result, the autocorrelation result of the square wave component is equal to 1. This represents the second positive peak when $\tau = 1/f_s$ as shown in Figure 2.

By parity of reasoning, because the autocorrelation result of the square wave component is changed repeatedly from 1 to −1, there are multiple positive and negative peaks in the autocorrelation function of the BOC sequence. The peaks occur at the moment when τ is the integral multiple of $1/2f_s$ and τ is less than $1/f_c$. The number of peaks is related to f_c and f_s and is equal to $2n-1$ with $|\tau| \leq 1/f_c$ by lead and lag. When τ exceeds one chip of the pseudo random sequence, the autocorrelation result of the pseudo random sequence is $-1/N$, as shown in (3). When the period of the BOC sequence is longer or aperiodic, the autocorrelation result is nearly equal to 0.

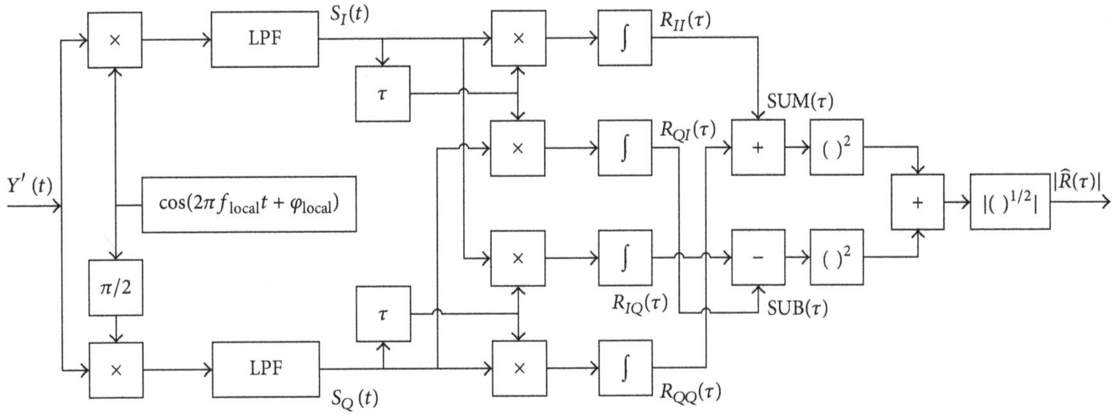

FIGURE 4: The overall process of the quadrature channel correlation.

By analyzing the generation process of the autocorrelation function for the BOC sequence, the values of the autocorrelation peaks can be expressed by

$$h(l) = \frac{(-1)^l (n-|l|)}{n}, \quad l = 0, \pm 1, \ldots, \pm n-1, \tag{5}$$

where $h(l)$ is the value of the l-th peaks and l is the sequence number of the peaks.

When l is equal to 0, $h(0)$ is the value of the main peak. l is positive when $\tau > 0$; otherwise it is negative. The values of the peaks decrease gradually as $|l|$ increases. Further, it is seen from (4) that the height difference between the l-th peak and the $(l+1)$-th peak can be represented as

$$|\Delta h| = \left| \frac{(n-|l|)}{n} - \frac{(n-|l+1|)}{n} \right| = \frac{1}{n}, \tag{6}$$

where Δh is the height difference.

Because the peaks always occur at the moment when τ is the integral multiple of $1/2f_s$, the lag difference between the l-th peak and the $(l+1)$-th peak is $1/2f_s$. It can be represented as

$$\Delta\tau = \frac{1}{2f_s}, \tag{7}$$

where $\Delta\tau$ is the lag difference.

As shown in (5) and (6), the heights and lags of the multiple autocorrelation peaks are related to $1/2f_s$. The reason is mainly due to the second modulation by the square wave. Therefore, we can detect the BOC signal by detecting multiple autocorrelation peaks of the BOC sequence. If the relevant rules shown in (5) and (6) are satisfied, the BOC signal will be detected.

3.2. Quadrature Channel Correlation. Because the BOC signal is transmitted under the condition of a negative SNR and there is a carrier wave component in the BOC signal, it is difficult to obtain multiple autocorrelation peaks of the BOC sequence.

An improved method based on quadrature channel correlation is presented to obtain multiple autocorrelation peaks. The overall process of the quadrature channel correlation is shown Figure 4.

The received BOC signal can be expressed as

$$Y'(t) = A \cdot P'(t) S_c(t) \cos(2\pi f t + \varphi) + n(t), \tag{8}$$

where $P'(t)$ is the spread spectrum sequence $P'(t) = D(t)P(t)$ and $n(t)$ is the zero-mean gauss white noise with a two-sided power spectral density N_0.

Firstly, $Y'(t)$ is multiplied by $\cos(2\pi f_{\text{local}}t + \varphi_{\text{local}})$ and $\sin(2\pi f_{\text{local}}t + \varphi_{\text{local}})$, respectively, where f_{local} is the frequency of the local oscillator and φ_{local} is the phase of the local oscillator.

Next, the high-frequency components are filtered by a low-pass filter. We obtain

$$S_I(t) = \frac{A}{2} \cdot P'(t) S_c(t) \cos[2\pi\Delta f t + \Delta\varphi] + n_I(t) \tag{9}$$

$$S_Q(t) = -\frac{A}{2} \cdot P'(t) S_c(t) \sin[2\pi\Delta f t + \Delta\varphi] + n_Q(t), \tag{10}$$

where $\Delta f = f - f_{\text{local}}$, $\Delta\varphi = \varphi - \varphi_{\text{local}}$, and $n_I(t)$ and $n_Q(t)$ are the noise components in the I and Q channels after filtering.

Subsequently, the autocorrelation functions of $S_I(t)$ and $S_Q(t)$ are calculated with the code delay τ. We take $S_I(t)$ as an example to study the autocorrelation and cross-correlation functions. The autocorrelation function of $S_I(t)$ can be expressed as

$$R_{II}(\tau) = R_{S_I S_I}(\tau) + R_{S_I n_I}(\tau) + R_{n_I n_I}(\tau), \tag{11}$$

where $R_{S_I S_I}(\tau)$ is the autocorrelation component of the BOC signal, $R_{S_I n_I}(\tau)$ is the cross-correlation component between the BOC signal and the noise, and $R_{n_I n_I}(\tau)$ is the autocorrelation component of the noise.

Because there is no correlation between the noise and the BOC signal, $R_{S_I n_I}(\tau)$ is nearly equal to 0 with enough received data. Then, formula (10) can be expressed as

$$R_{II}(\tau) = R_{S_I S_I}(\tau) + R_{n_I n_I}(\tau) \tag{12}$$

based on the assumption,

$$R_{n_I n_I}(\tau) \approx N_0 \delta(\tau). \tag{13}$$

The autocorrelation function $R_{S_I S_I}(\tau)$ of the BOC signal in I channel can be expressed as

$$R_{S_I S_I}(\tau) = \frac{1}{2T}\lim_{T\to\infty}\int_{-T}^{T} S_I(t) S_I(t+\tau)\,dt = \frac{1}{2T} \cdot \lim_{T\to\infty}\int_{-T}^{T} \frac{A^2}{8} R_{PS}(\tau)\cdot\cos\left[2\pi\Delta f(2t+\tau)+2\Delta\varphi\right]dt + \frac{A^2}{8}\cdot\cos(2\pi\Delta f\tau)\cdot R_{PS}(\tau), \quad (14)$$

where

$$R_{PS}(\tau) = P'(t) S_c(t) P'(t+\tau) S_c(t+\tau) \quad (15)$$

is the autocorrelation function of the BOC sequence.

Then, $R_{II}(\tau)$ can be expressed as

$$R_{II}(\tau) = \frac{1}{2T} \cdot \lim_{T\to\infty}\int_{-T}^{T} \frac{A^2}{8} R_{PS}(\tau)\cdot\cos\left[2\pi\Delta f(2t+\tau)+2\Delta\varphi\right]dt + \frac{A^2}{8}\cos(2\pi\Delta f\tau)\cdot R_{PS}(\tau) + R_{n_I n_I}(\tau). \quad (16)$$

Similarly, the autocorrelation function of the Q channel and the cross-correlation function of the two channels are represented as

$$R_{QQ}(\tau) = -\frac{1}{2T} \cdot \lim_{T\to\infty}\int_{-T}^{T} \frac{A^2}{8} R_{PS}(\tau)\cdot\cos\left[2\pi\Delta f(2t+\tau)+2\Delta\varphi\right]dt + \frac{A^2}{8}\cos(2\pi\Delta f\tau)\cdot R_{PS}(\tau) + R_{n_Q n_Q}(\tau) \quad (17)$$

$$R_{IQ}(\tau) = -\frac{1}{2T} \cdot \lim_{T\to\infty}\int_{-T}^{T} \frac{A^2}{8} R_{PS}(\tau)\cdot\sin\left[2\pi\Delta f(2t+\tau)+2\Delta\varphi\right]dt - \frac{A^2}{8}\sin(2\pi\Delta f\tau)\cdot R_{PS}(\tau) + R_{n_I n_Q}(\tau) \quad (18)$$

$$R_{QI}(\tau) = -\frac{1}{2T} \cdot \lim_{T\to\infty}\int_{-T}^{T} \frac{A^2}{8} R_{PS}(\tau)\cdot\sin\left[2\pi\Delta f(2t+\tau)+2\Delta\varphi\right]dt + \frac{A^2}{8}\sin(2\pi\Delta f\tau)\cdot R_{PS}(\tau) + R_{n_Q n_I}(\tau). \quad (19)$$

According to (15)~(18), we obtain

$$\mathrm{SUM}(\tau) = R_{II} + R_{QQ} = \begin{cases} \frac{A^2}{4}\cos(2\pi\Delta f\tau)\cdot R_{PS}(\tau), & \tau \neq 0 \\ \frac{A^2}{4} + N_0, & \tau = 0 \end{cases} \quad (20)$$

$$\mathrm{SUB}(\tau) = R_{QI} - R_{IQ} = \begin{cases} \frac{A^2}{4}\sin(2\pi\Delta f\tau)\cdot R_{PS}(\tau), & \tau \neq 0 \\ 0, & \tau = 0, \end{cases} \quad (21)$$

$$\widehat{R}^2(\tau) = \mathrm{SUM}^2(\tau) + \mathrm{SUB}^2(\tau) = \begin{cases} \frac{A^4}{16} R_{PS}^2(\tau), & \tau \neq 0 \\ \frac{A^4}{16} + \frac{A^2}{2} N_0 + N_0^2, & \tau = 0. \end{cases} \quad (22)$$

Then, the absolute value of the normalized multiple autocorrelation peaks is obtained as shown in the following:

$$\left|\widehat{R}(\tau)\right| = \sqrt{\widehat{R}^2(\tau)} \quad (23)$$

By using the quadrature channel correlation, the autocorrelation of the carrier component in the BOC signal has been eliminated and the multiple autocorrelation peaks of the BOC sequence are extracted.

3.3. Detection of the Autocorrelation Peaks. In order to detect the BOC modulation signal, it is necessary to detect the autocorrelation peaks. By analyzing multiple autocorrelation peaks of the BOC sequence as described in Section 2, we know that the heights and lags of the multiple autocorrelation peaks for the BOC sequence are related to $1/2f_s$ and that the sum of the multiple autocorrelation peaks is related to f_c and f_s.

The absolute values of the normalized multiple autocorrelation peaks for the BOC sequence with different f_c and f_s values are shown in Figure 5. The sampling frequency is 400 times larger than for f_{base}.

The frequency of the square wave should be larger than the frequency of the pseudo random sequence to generate the BOC signal. In Figure 5, the number of peaks for the BOC (10, 10) signal is two with a gradual increase in τ from 0 to $1/f_c$. The main peak appears when $\tau = 0$ and the secondary peak appears at the 20th sample point. Because one period of the square wave or the pseudo random sequence is sampled 40 times, when τ equals 20 sample points, the normalized autocorrelation result of the square wave is equal to −1. At the same time, the normalized autocorrelation result of the spread spectrum sequence is equal to 0.5 and the height of the secondary peak is equal to 0.5. When τ exceeds 40 sample points, the autocorrelation result of the spread spectrum sequence is nearly equal to 0 and there is no peak.

For the BOC (10, 5) signal, the frequency of the square wave is twice as much as the frequency of the pseudo random

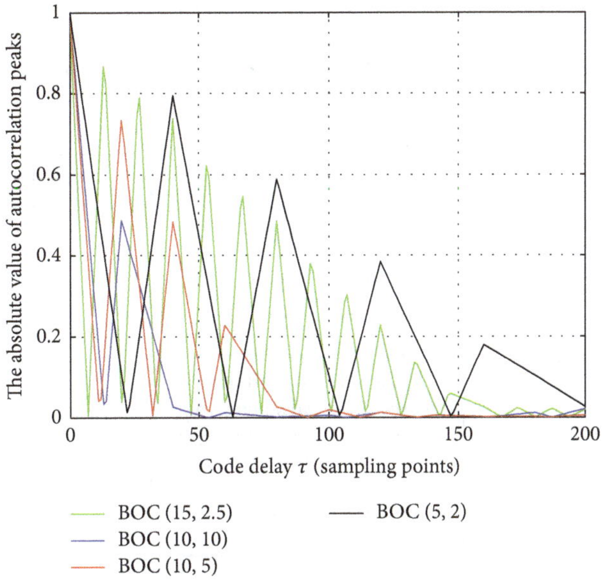

FIGURE 5: The absolute values of the autocorrelation peaks for the BOC sequence.

sequence. One period of the square wave is sampled 40 times and one chip of the pseudo random sequence is sampled 80 times. When τ equals 20 sample points, the normalized autocorrelation result of the square wave is equal to −1. At the same time, the normalized autocorrelation result of the square wave is equal to 0.75. Therefore, the absolute value of the normalized autocorrelation result for the BOC sequence is equal to 0.75. This represents the secondary peak. When τ equals 40 sample points, it represents the third peak and the height of the peak is equal to 0.5. The last peak occurs at 60 sample points and the height of the peak is equal to 0.25. There is no peak when τ exceeds 80 sample points. Similarly, for the BOC (5, 2) signal, the secondary peak occurs at 40 sample points and the height of the peak is equal to 0.8. For the BOC (15, 2.5) signal, the secondary peak occurs at 13 sample points and the height of the peak is equal to 0.87.

By analyzing the multiple autocorrelation peaks of the BOC sequence described in Section 2.1 and Figure 5, the secondary peak occurs at the moment that τ equals half a period of the square wave. Therefore, when f_c is equal to f_s, the height of the secondary peak is equal to 0.5. When f_s is larger than f_c, the height of the secondary peak is larger than 0.5, as shown in Figure 5. With a gradual increase in τ, the heights of the peaks decrease gradually. Because f_s is not less than f_c, the height of the secondary peak is not less than 0.5. Therefore, the initial threshold for detecting the secondary peaks can be equal to 0.5. The secondary peak is determined by the absolute value of the normalized autocorrelation result, which is the largest peak except for the main peak and it is not less than 0.5. If there is no peak, it is not a BOC signal. Otherwise, according to the height and lag difference, Δh and $\Delta\tau$ can be estimated. We can determine the next peak with Δh and $\Delta\tau$ until the height is nearly equal to 0. If there are multiple peaks, this means it is a BOC signal.

As shown in (21), there is an autocorrelation component of the noise in $y(\tau)$ with $\tau = 0$. When the SNR is low, the autocorrelation component of the noise is large, which will lead to greater errors for Δh, and this has a large impact on the ability to detect the BOC signal. Therefore, the absolute values of the normalized multiple autocorrelation peaks need to be adjusted. Based on the least squares fitting method, the autocorrelation results that are close to the main peak and are descending continuously are determined and the height of the main peak is adjusted.

The flow of the algorithm is shown in Figure 6.

4. Simulation Results

We assume the following parameters for the simulations: f_c = 2.046 MHz, f_s = 5.115 MHz, f_s = 20.46 MHz, and f_{local} = 10.23 MHz. The sampling frequency is 204.6 MHz and the number of sampled points is 25000.

Figure 7 shows the absolute values of the autocorrelation peaks for the BOC (5, 2) signal with SNR = 10 dB. It is evident that the autocorrelation component of the carrier wave is eliminated by using the quadrature channel correlation. The multiple autocorrelation peaks are easy to detect. For the absolute values of the normalized multiple autocorrelation peaks (without the elimination of the autocorrelation component of the carrier wave), there are more peaks that are caused by the carrier autocorrelation around the peaks of the BOC sequence. This can interfere with the detection of the peaks. From Figure 7, we can see that the multiple autocorrelation peaks of the BOC sequence are obtained by the quadrature channel correlation and the BOC signal does not require demodulation.

Figure 8 shows the absolute values of the autocorrelation peaks for the BOC (5, 2) signal with different SNR values. Because there is a noise autocorrelation component that is

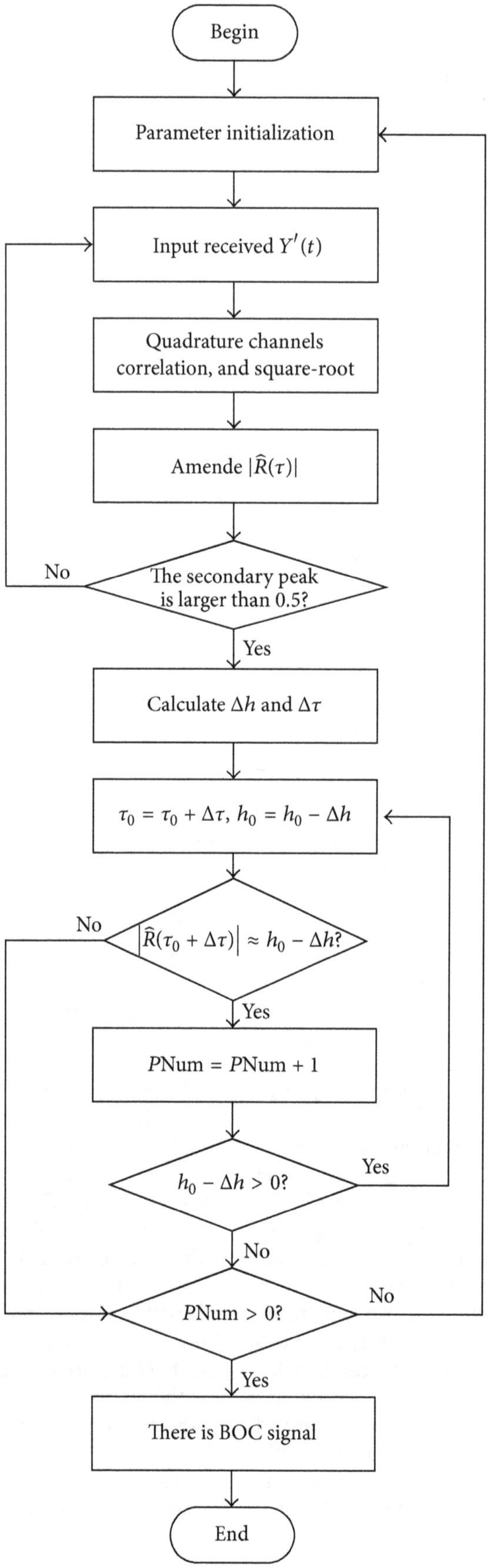

FIGURE 6: Flowchart of the proposed algorithm.

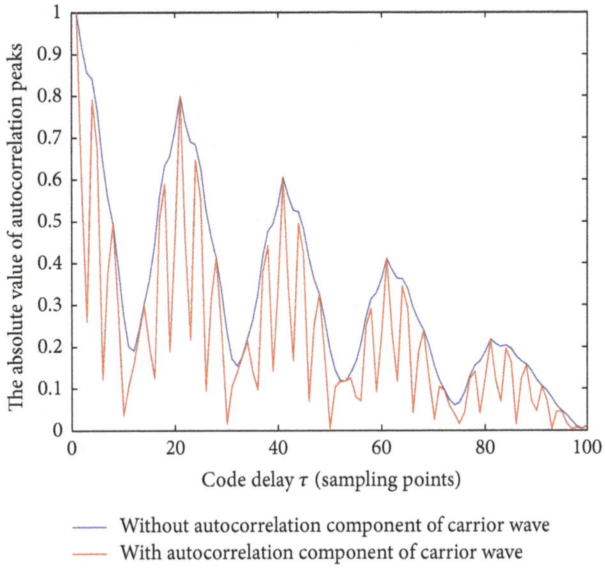

FIGURE 7: The absolute value of the autocorrelation peaks with SNR = 10 dB.

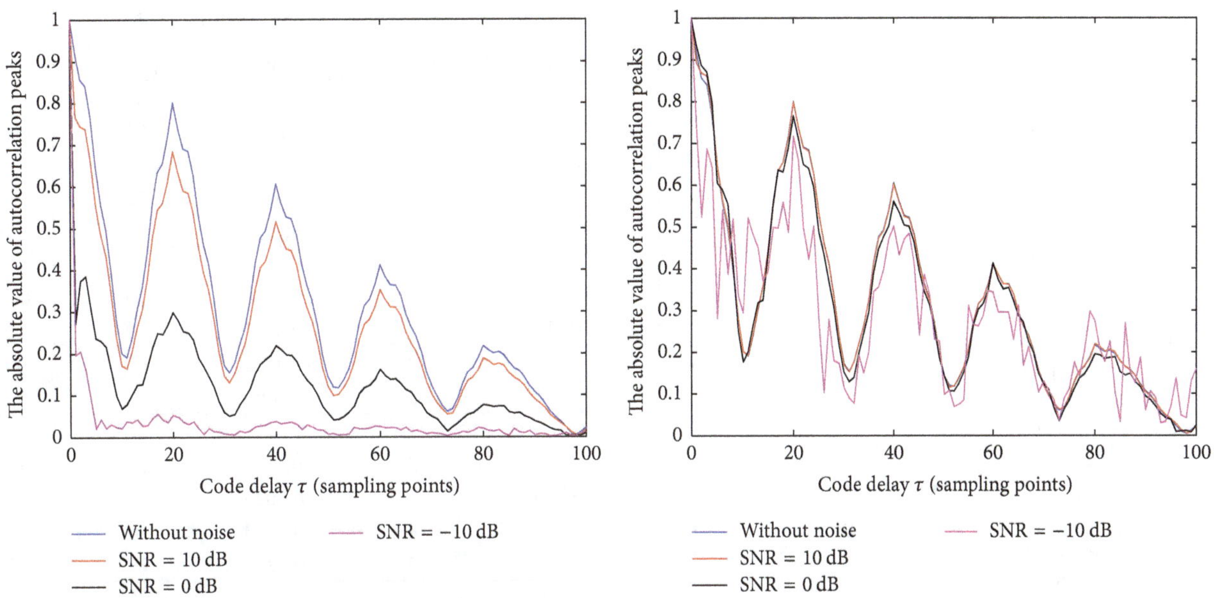

(a) Autocorrelation results without adjusted values

(b) Autocorrelation results with adjusted values

FIGURE 8: The absolute value of the autocorrelation peaks with different SNR.

included in the main peaks as shown in (a), the main peaks increase as the SNR gradually decreases. Therefore, the other normalized autocorrelation peaks are lower with a lower SNR such as the secondary peak and the third peak. The noise autocorrelation component in the main peak is eliminated after the adjustment. As a result, the adjusted result for the multiple peaks is similar to the result under a noise-free condition.

Table 1 shows the results for the detection of the multiple peaks. It is evident that the heights of the peaks decrease gradually as τ increases. For SNR = 10 dB, there are the approximate height differences Δh and lag differences $\Delta\tau$ between the adjacent peaks. For lower values of the SNR, the error gradually increases.

Figure 9 shows the performance of the proposed algorithm for different data lengths. When the data length is 25000, the BOC (5, 2) signal can be detected effectively when the SNR is greater than −12 dB. When the data length is 100000, the detection probability of the BOC (5, 2) signal increases to 1 dB. The detection performance for the BOC (5, 2) signal can be improved with an increase in the data length because the autocorrelation component of the noise is smaller with a greater data length. Therefore, the detection ability of the proposed algorithm is higher at a lower SNR. However,

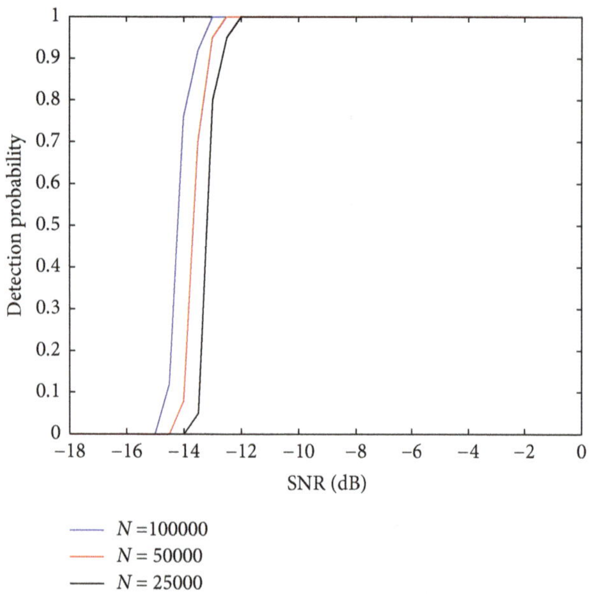

FIGURE 9: Detection probability with different data lengths.

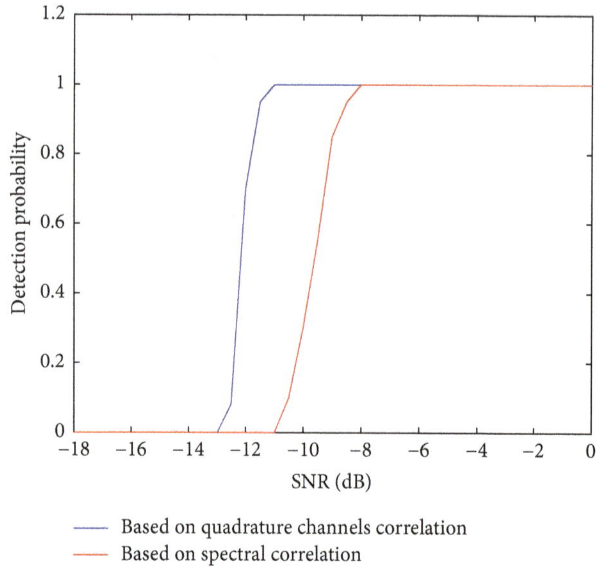

FIGURE 11: Detection probability for different algorithms.

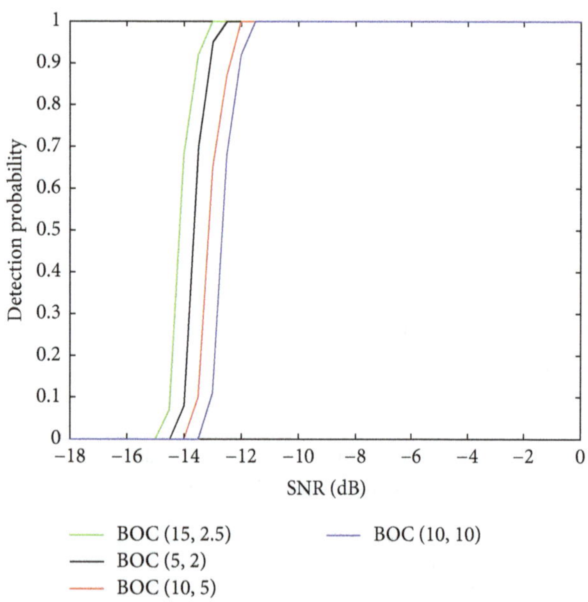

FIGURE 10: Detection probability for different values of f_c and f_s.

TABLE 1: The results of the peak detection.

SNR/dB	Number of peaks	Height	τ/sampled points	Δh
	1	1	0	0
	2	0.7992	20	0.2008
10 dB	3	0.6021	40	0.1971
	4	0.4122	60	0.1899
	5	0.2197	80	0.1925
	1	1	0	0
	2	0.7655	20	0.2345
0 dB	3	0.5613	40	0.2042
	4	0.4122	60	0.1491
	5	0.195	80	0.2172
	1	1	0	0
	2	0.7173	20	0.2827
−10 dB	3	0.5025	40	0.2148
	4	0.3473	59	0.1552
	5	0.2984	79	0.0489

when the SNR is lower than −15 dB, the BOC signal cannot be detected when the data length is 100000.

Figure 10 shows the performance of the proposed algorithm for different values of f_c and f_s with a data length of 100000. When the ratio n of f_s to f_c is larger, the probability of detecting the BOC signal is higher for the same SNR because there are more peaks with a larger n and the height of the secondary peak is larger. Therefore, it is easy to detect the peaks and detect the BOC signal.

Figure 11 shows the performance of the proposed algorithm based on the quadrature channel correlation and the spectral correlation. It is evident that the probability of detecting the BOC signal is higher for the quadrature channel correlation than for the spectral correlation algorithm. Under the same conditions, the detection probability for the BOC signal is improved by about 2 dB.

5. Conclusions

(1) The autocorrelation component of the carrier wave is eliminated by using the quadrature channel correlation. If the adjacent autocorrelation peaks have the same lag differences and height differences, the BOC signal is detected effectively by detecting the absolute value of the multiple autocorrelation peaks.

(2) The ability to detect the BOC signal is related to the data length and the ratio of f_s to f_c. Larger values for the data length and the ratio of f_s to f_c result in a higher probability to

detect the BOC signal. The algorithm presented in this paper is superior to the algorithm based on the spectral correlation.

Conflicts of Interest

The authors declare that there are no conflicts of interest regarding the publication of this paper.

Acknowledgments

This work is supported by the Science and Technology Foundation from Liaoning Education Department (no. L2015462), Shenyang Ligong University Liaoning Province Information Network and Countermeasure Technology Key Laboratory Open Foundation (2015), and Shenyang Ligong University Postdoctoral Science Foundation.

References

[1] M. Foucras, J. Leclère, C. Botteron et al., "Study on the cross-correlation of GNSS signals and typical approximations," *GPS Solutions*, vol. 21, no. 2, pp. 293–306, 2017.

[2] B. Tripathi, R. Tulsian, and R. C. Jain, "Performance Comparison of Sine and Cosine Phased BOC Signals Used for Radio-navigation," in *Proceedings of the IFAC*, vol. 45, pp. 223–228, 2012.

[3] K. Chae, S. R. Lee, H. Liu, and S. Yoon, "An unambiguous correlation function for generic sine-phased binary offset carrier signal tracking," *Computers and Electrical Engineering*, vol. 49, pp. 161–172, 2016.

[4] Z. Yanling, L. Xuejiao, and W. Xiaoqing, "Analysis of satellite navigation signal alternate binary offset carrier modulation," *Journal of Hubei University*, no. 4, pp. 334–339, 2015.

[5] Z. Yao, X. Cui, M. Lu, Z. Feng, and J. Yang, "Pseudo-correlation-function-based unambiguous tracking technique for sine-BOC signals," *IEEE Transactions on Aerospace and Electronic Systems*, vol. 46, no. 4, pp. 1782–1796, 2010.

[6] Y. Lee, D. Chong, I. Song, S. Y. Kim, G.-I. Jee, and S. Yoon, "Cancellation of correlation side-peaks for unambiguous BOC signal tracking," *IEEE Communications Letters*, vol. 16, no. 5, pp. 569–572, 2012.

[7] A. F. Khalaf, M. I. Owis, and I. A. Yassine, "Image features of spectral correlation function for arrhythmia classification," in *Proceedings of the 37th Annual International Conference of the IEEE Engineering in Medicine and Biology Society, EMBC '15*, pp. 5199–5202, IEEE, Milan, Italy, August 2015.

[8] T. Zhang, D. He, S. Chen, and S. Zhou, "Spectral correlation-based parameter estimation of BOC modulation signal," *Journal of Huazhong University of Science and Technology*, vol. 41, no. 9, pp. 11–16, 2013.

[9] S. Zhu, D. Zhao, and F. Wang, "Hybrid prediction and fractal hyperspectral image compression," *Mathematical Problems in Engineering*, vol. 2015, Article ID 950357, 10 pages, 2015.

[10] T. Former, "Phase clustering based modulation classification algorithm for PSK signal over wireless environment," *Mobile Information Systems*, vol. 2016, Article ID 2398464, 11 pages, 2016.

[11] R. Chen, J. Wang, R. Lin, and X. Zhao, "Spectrum sensing based on nonparametric autocorrelation in wireless communication systems under alpha stable noise," *Mobile Information Systems*, vol. 2016, Article ID 6753830, 6 pages, 2016.

[12] F. Guo, Z. Yao, and M. Lu, "BS-ACEBOC: a generalized low-complexity dual-frequency constant-envelope multiplexing modulation for GNSS," *GPS Solutions*, vol. 21, no. 2, pp. 561–575, 2017.

[13] M. S. Yarlykov, "Correlation functions of BOC and AltBOC signals as the inverse Fourier transforms of energy spectra," *Journal of Communications Technology and Electronics*, vol. 61, no. 8, pp. 857–876, 2016.

[14] M. S. Yarlykov and S. M. Yarlykova, "Correlation functions of complete CBOC signals derived as the inverse Fourier transform of energy spectra," *Journal of Communications Technology and Electronics*, vol. 62, no. 2, pp. 128–140, 2017.

[15] B. Yin, G. Wang, and Y. Qi, "Unambiguous sine-phased binary offset carrier modulated signal tracking technique," *Optik - International Journal for Light and Electron Optics*, vol. 132, pp. 284–290, 2017.

[16] P. Weichuan, W. Xue, and H. Chengyan, "Research on correlation peaks detection algorithm for multiplex signal of navigation systems," *Journal of Time and Frequency*, vol. 38, no. 3, pp. 163–170, 2015.

[17] F. Liu and Y. Feng, "A new acquisition algorithm with elimination side peak for all BOC signals," *Mathematical Problems in Engineering*, vol. 2015, Article ID 140345, 9 pages, 2015.

[18] K. Chae, S. R. Lee, H. Liu et al., "A novel unambiguous composite binary offset carrier (6, 1, 1/11) tracking based on partial correlations," *Computers and Electrical Engineering*, vol. 50, pp. 54–66, 2016.

7

A Vessel Positioning Algorithm based on Satellite Automatic Identification System

Shexiang Ma, Jie Wang, Xin Meng, and Junfeng Wang

School of Electrical and Electronic Engineering, Tianjin University of Technology, Tianjin, China

Correspondence should be addressed to Shexiang Ma; masx_tjut@126.com

Academic Editor: Caner Özdemir

Vessels can obtain high precision positioning by using the global navigation satellite system (GNSS), but when the ship borne GNSS receiver fails, the existence of an alternative positioning system is important for the navigation safety of vessel. In this paper, a localization method based on the signals transmitted by satellite-based automatic identification system (AIS) is proposed for vessel in GNSS-denied environments. In the proposed method, the positioning model is a modification on the basis of time difference and frequency difference of arrival measurements by introducing an additional measurement, and the measurement is obtained through the interactive multiple model algorithm. The performance of the proposed strategy is evaluated through simulations, and the results validate the feasibility and reliability of vessel localization based on satellite-based AIS.

1. Introduction

Automatic identification system (AIS) is a self-reporting system designed to protect maritime security of vessel and improve maritime efficiency [1]. It plays an important role in ship collision avoidance and maritime supervision through a series of static and dynamic vessel information automatically broadcast, and the information includes latitude, longitude, course, and velocity [2]. The geographical location reported in AIS is derived by the shipboard GNSS receiver and typically with the high accuracy [3]. However, there is a problem that followed with the GNSS being widely used in navigation of maritime. GNSS is vulnerable to accidental interference [4]; the ship will not be able to locate once the GNSS signal is deliberately disturbed or the GNSS receiver fails. So it is necessary to develop a spare navigation system for the ship.

AIS is a self-organized time division multiple access (TDMA) system, which not only can be self-reporting but also can receive AIS information [5]. Although AIS ignored the role of the satellite in its original design; it has been proven feasible to receive AIS signals by satellite [6, 7]. In the satellite-based AIS, the relative speed of satellite and ship is high, and the two are far apart; therefore, there are challenges for the correct detection of the AIS signal such as the problem of time delay, the high Doppler offset, and low signal-to-noise ratio (SNR) [8]. With the current level of AIS signal detection technology, the correct detection of AIS signal can be guaranteed with the improvement of synchronization algorithm, and the influence of high Doppler offset on carrier recovery is gradually decreasing [9]. The satellite-based AIS is already operational but focuses on the stage of "vessel transmitting, satellite receiving." In view of the fact that a large number of AIS signals are likely to reach the satellite at the same time in this stage but satellite can still detect ship signals [8], the signals can definitely be received by the ship if the satellite can send information in the AIS operating frequency band according to the AIS protocol, because the possibility of AIS signal conflict is relatively low in case of the ship reception thanks to the characteristics of signal transmission. As the technology of satellite-based AIS advances, the potentiality of AIS for navigation becomes a concern and there is the investigation on ship localization using AIS signals received by satellite [10]. In this paper, we assume that, in advanced satellite-based AIS, vessels can receive AIS signals transmitted from satellite in addition to "vessel transmitting, satellite receiving," and the information of satellite motion state is broadcast by the downlink AIS signal. On the basis of this vision, a ship localization method using AIS signals transmitted from satellite is proposed.

Among the various measurements for positioning tasks, the time of arrival (TOA), the time difference of arrival (TDOA), and the frequency difference of arrival (FDOA) are very representative choices because of the potentials in attaining high localization accuracy [11, 12]. There is a lot of research on the application of TDOA to improve the positioning accuracy of the stationary target and to locate the moving target by using frequency measurements [13–15]. In addition, the positioning methods combining two kinds of measurements such as TDOA/FDOA and TDOA/DOA are also widely discussed [16, 17]. Except for reducing the number of signals required, the combination of time and frequency measurements is attractive for improvement of positioning accuracy [18, 19]. In these methods, however, only the information extracted from the received signals is used for positioning. Taking into account the fact that AIS can obtain the ship velocity and heading by connecting external sensors, in this paper, a modification positioning model based on TDOA/FDOA is proposed by introducing an additional measurement based on the interactive multiple model (IMM) algorithm [20, 21]. The method of TDOA/FDOA and the IMM algorithm are used separately for locating or tracking the target in general; given the characteristic of AIS signal carrying information, they are combined together in the study. Besides, for the purpose of making the estimated result more suitable for the vessel status, a new probability updating method for IMM is designed in this work.

The solution of TDOA/FDOA measurement equation is complicated because of the high nonlinearity [22]. Taylor-series technique can linearize the equations but positioning result is easy to be affected by initial value setting [23]. The method of grid searching achieves the accuracy improvement with the sacrifice of computation [24]. In this paper, the localization results are obtained by Gauss-Newton iteration under the least squares criterion, and the solution of grid rough searching is chosen as the starting value. The feasibility of the positioning method based on the advanced satellite-based AIS signals is investigated through the experiment. The location error distributions of the TDOA/FDOA joint location model and the proposed localization model are analyzed in this study.

2. TDOA/FDOA Localization Based on Least Squares Estimation

In satellite-based AIS, satellites are located at a low orbit from 600 km to 1000 km above the ground. The downlink AIS signals will include the Doppler frequency shift because of the relative satellite-ship velocities, and the frequency shift is up to a maximum of ±4 kHz. In order to achieve localization with the limited number of AIS signals and improve the positioning accuracy as much as possible, the work of ship positioning is carried out on the basis of TDOA/FDOA in this paper.

2.1. Principles of TDOA/FDOA. Assuming that t_i is the time cost by the ith AIS signal transmitted from satellite to ship, the TDOA between the adjacent signals received by the ship can be expressed as

$$\begin{aligned}\Delta t_i &= t_{i+1} - t_i = \frac{|\mathbf{L}_{\mathbf{s(i+1)}} - \mathbf{L_b}| - |\mathbf{L_{si}} - \mathbf{L_b}|}{c} + \Delta n_t \\ &= \frac{1}{c}\Big[\left((x_{s(i+1)} - x)^2 + (y_{s(i+1)} - y)^2 \right. \\ &\left. + (z_{s(i+1)} - z)^2\right)^{1/2} - \left((x_{si} - x)^2 + (y_{si} - y)^2 \right. \\ &\left. + (z_{si} - z)^2\right)^{1/2}\Big] + \Delta n_t = r_{ti}(x, y, z) + \Delta n_t.\end{aligned} \tag{1}$$

The FDOA between the adjacent AIS signals can be expressed as

$$\begin{aligned}\Delta f_{ri} &= \frac{f_e}{c}\left[\frac{\mathbf{V}^{\mathbf{T}}_{\mathbf{s(i+1)}} \cdot (\mathbf{L}_{\mathbf{s(i+1)}} - \mathbf{L_b})}{|\mathbf{L}_{\mathbf{s(i+1)}} - \mathbf{L_b}|} - \frac{\mathbf{V}^{\mathbf{T}}_{\mathbf{si}} \cdot (\mathbf{L_{si}} - \mathbf{L_b})}{|\mathbf{L_{si}} - \mathbf{L_b}|}\right] + \Delta n_f \\ &= \frac{f_e}{c}\left[\frac{V_{sx_{(i+1)}}(x_{s(i+1)} - x) + V_{sy_{(i+1)}}(y_{s(i+1)} - y) + V_{sx_{(i+1)}}(z_{s(i+1)} - z)}{\left((x_{s(i+1)} - x)^2 + (y_{s(i+1)} - y)^2 + (z_{s(i+1)} - z)^2\right)^{1/2}} - \frac{V_{sxi}(x_{si} - x) + V_{syi}(y_{si} - y) + V_{sxi}(z_{si} - z)}{\left((x_{si} - x)^2 + (y_{si} - y)^2 + (z_{si} - z)^2\right)^{1/2}}\right] \\ &+ \Delta n_f = r_{fi}(x, y, z) + \Delta n_f,\end{aligned} \tag{2}$$

where f_e is the carrier frequency of the AIS signal and c is the signal propagation velocity. $\mathbf{L_b} = [x, y, z]^T$ is the vessel position vector in the ECEF reference and $\mathbf{V_{si}} = [V_{sxi}, V_{syi}, V_{szi}]^T$ and $\mathbf{L_{si}} = [x_{si}, y_{si}, z_{si}]^T$ are velocity vector and position vector of the satellite when transmitting the ith AIS signal, respectively. Δn_t is the difference of noise between the two time measurements and Δn_f is the difference of noise between the two frequency measurements.

It is assumed that the number of signals received by the ship in the visual time of the satellite is $N + 1$; the localization equation matrix based on (1) and (2) can be written as

$$\begin{bmatrix}\Delta \mathrm{T} \\ \Delta \mathrm{F}\end{bmatrix} = \begin{bmatrix}\mathbf{r_t}(x, y, z) \\ \mathbf{r_f}(x, y, z)\end{bmatrix} + \mathbf{n} \tag{3}$$

with

$$\begin{aligned}\mathbf{r_t}(x,y,z) &= [r_{t1}(x,y,z), r_{t2}(x,y,z), \ldots, r_{tN}(x,y,z)]^T \\ \mathbf{r_f}(x,y,z) &= [r_{f1}(x,y,z), r_{f2}(x,y,z), \ldots, r_{fN}(x,y,z)]^T,\end{aligned} \tag{4}$$

where $\mathbf{\Delta T} = [\Delta t_1, \Delta t_2, \ldots, \Delta t_N]^T$ is the TDOA measurement vector obtained by synchronization technique and $\mathbf{\Delta F} = [\Delta f_{r1}, \Delta f_{r2}, \ldots, \Delta f_{rN}]^T$ is the FDOA measurement vector. $\mathbf{n}$ is the measurement noise matrix.

2.2. Calculation Based on Least Squares Estimation. On the basis of least squares criterion, the estimator associated with (3) needs to minimize the differences between the measurements and predictions; the equation to be minimized can be written as

$$C(x,y,z) = [\mathbf{Y} - \mathbf{q}(x,y,z)]^T \mathbf{N}^{-1} [\mathbf{Y} - \mathbf{q}(x,y,z)], \tag{5}$$

where $\mathbf{N}$ is the noise covariance matrix, $\mathbf{Y} = [\mathbf{\Delta T} \ \ \mathbf{\Delta F}]^T$ is the measurement vector of TDOA/FDOA, and $\mathbf{q} = [\mathbf{r_t} \ \ \mathbf{r_f}]^T$.

The estimated position value $(\hat{x}, \hat{y}, \hat{z}) = \operatorname{argmin}_{(x,y,z)}\{C(x,y,z)\}$ can be obtained by Gauss-Newton iteration

$$\begin{aligned}\left[x_{j+1}, y_{j+1}, z_{j+1}\right]^T &= \left[x_j, y_j, z_j\right]^T \\ &+ \left[\mathbf{J}^T(x_j,y_j,z_j)\mathbf{N}^{-1}\mathbf{J}(x_j,y_j,z_j)\right]^{-1} \\ &\times \mathbf{J}^T(x_j,y_j,z_j)\mathbf{N}^{-1}\left[\mathbf{Y} - \mathbf{q}(x_j,y_j,z_j)\right],\end{aligned} \tag{6}$$

where matrix $\mathbf{J}$ is

$$\mathbf{J} = \begin{bmatrix}\mathbf{J_t} \\ \mathbf{J_f}\end{bmatrix} \tag{7}$$

with

$$\mathbf{J_t} = \begin{bmatrix} \frac{r_{t1}(x,y,z)}{\partial x} & \frac{r_{t1}(x,y,z)}{\partial y} & \frac{r_{t1}(x,y,z)}{\partial z} \\ \vdots & \vdots & \vdots \\ \frac{r_{tN}(x,y,z)}{\partial x} & \frac{r_{tN}(x,y,z)}{\partial y} & \frac{r_{tN}(x,y,z)}{\partial z} \end{bmatrix},$$
$$\mathbf{J_f} = \begin{bmatrix} \frac{r_{f1}(x,y,z)}{\partial x} & \frac{r_{f1}(x,y,z)}{\partial y} & \frac{r_{f1}(x,y,z)}{\partial z} \\ \vdots & \vdots & \vdots \\ \frac{r_{fN}(x,y,z)}{\partial x} & \frac{r_{fN}(x,y,z)}{\partial y} & \frac{r_{fN}(x,y,z)}{\partial z} \end{bmatrix}. \tag{8}$$

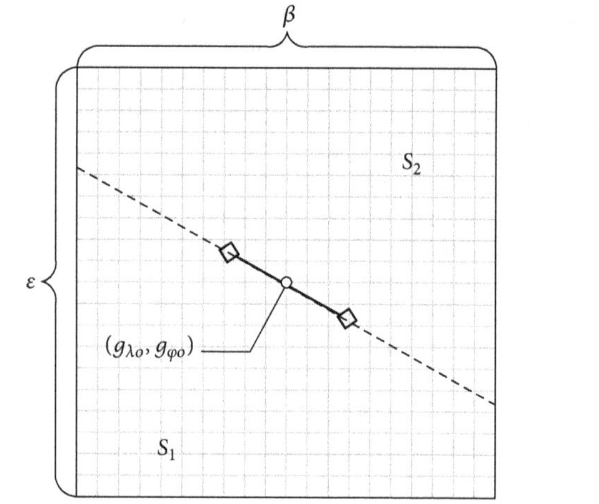

FIGURE 1: Method of determining the initialization vector.

The initial position in (6) needs to be defined in advance; the method for determining the initial position (shown in Figure 1) is as follows: (1) setting up a grid with the units of 1°. The grid is centered on the midpoint $(g_{\lambda o}, g_{\varphi o})$ of the satellite ground trajectory (the track generated during AIS signals transmission) and the range of grid geodetic coordinates (g_λ, g_φ) is $\{g_{\lambda o} - \beta/2 \le g_\lambda \le g_{\lambda o} + \beta/2,\ g_{\varphi o} - \varepsilon/2 \le g_\varphi \le g_{\varphi o} + \varepsilon/2\}$, where β and ε are the maximum visible longitude and latitude of satellite, respectively. (2) Connecting the start and end points of the satellite ground trajectory and dividing the grid into S_1 and S_2 (two parts) by extending the connecting line. (3) Searching within each part of grid and selecting two points with $(m_{\lambda i}, m_{\varphi i}) = \arg\max_{\{S_i\}}\{E = 1/(f_r - \hat{f})^2\}$ $(i = 1, 2)$. In the cost function E, f_r is the measured frequency of the AIS signal and $\hat{f} = f_c\{1 - (\mathbf{V}_{\mathbf{si}}^{\mathbf{T}}(\mathbf{L_{si}} - \mathbf{L_g})/c|\mathbf{L_{si}} - \mathbf{L_g}|)\}$ is the estimated frequency of received signal at the grid point, where $\mathbf{L_g} = [x_g, y_g, z_g]^{\mathrm{T}}$ is the position vector of grid point in ECEF coordinate. In this paper, we select "nearest point" to eliminate the false image which may occur in grid searching, that is, taking the point with the shortest distance from the origin of the ship as the optimal position (λ_0, φ_0). The transformation of vessel location from the geodetic coordinate to ECEF coordinate is defined as follows:

$$\begin{aligned} x &= R_N \cos\lambda \cos\varphi \\ y &= R_N \sin\lambda \cos\varphi \\ z &= R_N\left(1 - e^2\right)\sin\varphi, \end{aligned} \tag{9}$$

where λ and φ are the longitude and latitude coordinates of vessel and (x, y, z) are the ECEF coordinates of vessel. $R_N = \alpha/\sqrt{1 - e^2\sin^2\varphi}$ is radius of curvature in prime vertical, where $e^2 = 0.00669437999013$ and $\alpha = 6378.137$ km are the square of the first eccentricity and equatorial radius of the earth defined by WGS-84, respectively.

3. Localization Method Combining TDOA/FDOA with the Track Forecast

AIS equipped on vessel can obtain a series of dynamic information, such as speed, heading, and turning rate by connecting external sensors. Assuming that the motion state of ship remains constant during the two adjacent positioning points, it is possible with IMM algorithm to estimate the current ship position by using current dynamic information. Considering that the position of the ship at a moment is related to those of the same ship in the previous moment, in the proposed positioning method, the prediction with IMM is taken as a new measurement and added to the TDOA/FDOA joint positioning measurement.

3.1. The Motion Model of Vessel. The ship sailing at sea is a slow maneuvering target, with the consideration of the fact that balance between the model accuracy and computational cost, the constant velocity (CV) model, and the constant turn (CT) model are adopted in this paper.

The ship state at time k is defined as

$$\mathbf{X}(k) = [x(k), v_x(k), a_x(k), y(k), v_y(k), a_y(k), z(k), v_z(k), a_z(k)]^T, \quad (10)$$

where $x(k)$, $y(k)$, $z(k)$ are vessel position in ECEF coordinate and $v_x(k)$, $v_y(k)$, $v_z(k)$ and $a_x(k)$, $a_y(k)$, $a_z(k)$ are vessel velocity and acceleration in ECEF coordinate, respectively.

The CV model equation is shown as follows:

$$\mathbf{X}(k) = \mathbf{\Phi}_{CV}\mathbf{X}(k-1) + \mathbf{\Gamma}_{CV}W(k-1), \quad (11)$$

where the state transition matrix $\mathbf{\Phi}_{CV} = \begin{bmatrix} \Phi_V & 0 & 0 \\ 0 & \Phi_V & 0 \\ 0 & 0 & \Phi_V \end{bmatrix}$ with $\mathbf{\Phi}_V = \begin{bmatrix} 1 & T_S & 0 \\ 0 & 1 & 0 \\ 0 & 0 & 0 \end{bmatrix}$ (T_S is time interval between adjacent positioning points). The control input matrix $\mathbf{\Gamma}_{CV} = \begin{bmatrix} \Gamma_V & 0 & 0 \\ 0 & \Gamma_V & 0 \\ 0 & 0 & \Gamma_V \end{bmatrix}$ with $\mathbf{\Gamma}_V = [T_S^2/2 \;\; T_S \;\; 0]^T$ and $W(k-1)$ is process noise.

The CT model equation is shown as follows:

$$\mathbf{X}(k) = \mathbf{\Phi}_{CT}\mathbf{X}(k-1) + \mathbf{\Gamma}_{CT}\mathbf{W}(k-1), \quad (12)$$

where the control input matrix $\mathbf{\Gamma}_{CT} = \begin{bmatrix} \Gamma_T & 0 & 0 \\ 0 & \Gamma_T & 0 \\ 0 & 0 & \Gamma_T \end{bmatrix}$ with $\mathbf{\Gamma}_T = [T_S^3/6 \;\; T_S^2/2 \;\; T_S]^T$. The state transition matrix $\mathbf{\Phi}_{CT} = \begin{bmatrix} \Phi_T & 0 & 0 \\ 0 & \Phi_T & 0 \\ 0 & 0 & \Phi_T \end{bmatrix}$ with $\mathbf{\Phi}_T = \begin{bmatrix} 1 & \sin(\omega T_S)/\omega & [1-\cos(\omega T_S)]/\omega^2 \\ 0 & \cos(\omega T_S) & \sin(\omega T_S)/\omega \\ 0 & -\omega\sin(\omega T_S) & \cos(\omega T_S) \end{bmatrix}$ and ω is the steering rate of vessel.

3.2. Vessel Position Predicting Based on IMM Algorithm. A complete cycle of the IMM consists of four operations, namely, input mixing, model filtering, model probability update, and combination. Taking the recorded position at time $k-1$ and the dynamic information at time k (motion state is supposed to be unchanged during times $k-1$ and k) as the initial vessel state of each model, the forecasting process of ship position with IMM algorithm is as follows.

Step 1 (input mixing). p_{ij} (i, j = CV, CT) is defined as the Markov transition probability from model i to model j. The mixing probability is computed as follows:

$$\mu_{ij}(k-1 \mid k-1) = \frac{\mathbf{p_{ij}} \cdot \mu_i(k-1)}{m_j(k)} \quad (13)$$

with

$$m_j(k) = \sum_{i=CV,CT} \mathbf{p_{ij}} \cdot \mu_i(k-1), \quad (14)$$

where $\mu_i(k-1)$ is probability of mode i at time $k-1$

The mixed state estimate for model j is given by

$$\widehat{\mathbf{X}}_{\mathbf{0j}}(k-1 \mid k-1) = \sum_{i=CV,CT} \widehat{\mathbf{X}}_{\mathbf{i}}(k-1 \mid k-1)\,\mu_{ij}(k-1 \mid k-1). \quad (15)$$

The predicted covariance corresponding to the above mixed state estimate is given by

$$\begin{aligned}\mathbf{P}_{\mathbf{0j}}(k-1 \mid k-1) &= \sum_{i=CV,CT} \mu_{ij}(k-1 \mid k-1) \\ &\cdot \{\mathbf{P}_{\mathbf{i}}(k-1 \mid k-1) \\ &+ [\widehat{\mathbf{X}}_{\mathbf{i}}(k-1 \mid k-1) - \widehat{\mathbf{X}}_{\mathbf{0j}}(k-1 \mid k-1)] \\ &\times [\widehat{\mathbf{X}}_{\mathbf{i}}(k-1 \mid k-1) - \widehat{\mathbf{X}}_{\mathbf{0j}}(k-1 \mid k-1)]^T\}.\end{aligned} \quad (16)$$

Step 2 (model filtering). The extended Kalman filter (EKF) algorithm is adopted in this stage. The mixed state estimate and the predicted covariance corresponding to CV and CT in (15) and (16) are updated by the following two stages.

Time Update

$$\begin{aligned}\widehat{\mathbf{X}}(k \mid k-1) &= \mathbf{\Phi}\widehat{\mathbf{X}}(k-1) \\ \mathbf{P}(k \mid k-1) &= \mathbf{\Phi}\mathbf{P}(k-1)\mathbf{\Phi}^T + \mathbf{\Gamma}\mathbf{\Lambda}\mathbf{\Gamma}^T,\end{aligned} \quad (17)$$

where $\mathbf{P}$ is the covariance of the vessel location prediction, $\mathbf{\Phi}$ is the state transition matrix, $\mathbf{\Gamma}$ is the control input matrix, and $\mathbf{\Lambda}$ is the system noise variance matrix.

Measurement Update

$$\begin{aligned}\mathbf{K}(k) &= \mathbf{P}(k \mid k-1)\mathbf{J}^T(\widehat{\mathbf{X}}(k \mid k-1)) \\ &\times [\mathbf{J}(\widehat{\mathbf{X}}(k \mid k-1))\mathbf{P}(k \mid k-1)\mathbf{J}^T(\widehat{\mathbf{X}}(k \mid k-1)) \\ &+ \mathbf{R}]^{-1} \\ \widehat{\mathbf{X}}(k \mid k) &= \widehat{\mathbf{X}}(k \mid k-1) + \mathbf{K}(k)[\mathbf{Z}(k) \\ &- \mathbf{q}(\widehat{\mathbf{X}}(k \mid k-1))] \\ \mathbf{P}(k \mid k) &= [I - \mathbf{K}(k)\mathbf{J}(\widehat{\mathbf{X}}(k \mid k-1))]\mathbf{P}(k \mid k-1),\end{aligned} \quad (18)$$

where $\mathbf{K}$ corresponds to the gain matrix, $\mathbf{R}$ is the observation noise variance matrix, $\mathbf{I}$ is the unit matrix, and $\mathbf{J}$ is the Jacobian matrix of the measure function $\mathbf{q}(\cdot)$, as shown in (7).

Step 3 (model probability update). In the IMM algorithm, the updating of the model probability is carried out by calculating the likelihood function matching the model. However, this method is greatly influenced by the presetting model transition probability. In this work, an optimization method is developed to improve the positioning accuracy and can be described as follows.

(1) The positions estimated using EKF based on the CV model and CT model are denoted by Pos_{CV} and Pos_{CT}, respectively. Let $\text{Pos} = \mu_{\text{CV}}\text{Pos}_{\text{CV}} + \mu_{\text{CT}}\text{Pos}_{\text{CT}}$ and calculate the measurement vector $\mathbf{\Upsilon}_p$ at position Pos according to (1) and (2), where μ_{CV} and μ_{CT} ($\mu_{\text{CV}} + \mu_{\text{CT}} = 1$) correspond to the model probabilities of CV and CT models, respectively.

(2) Assuming $\mu_{\text{CV}} = \eta$ (in this paper, $\eta = 0.05$), according to $\mathbf{\Upsilon}_p$ determined by the aforementioned method, the corresponding error of measurement vector denoted by σ is calculated by $\sigma = \sqrt{(\mathbf{\Upsilon} - \mathbf{\Upsilon}_p)^T \cdot (\mathbf{\Upsilon} - \mathbf{\Upsilon}_p)}$.

(3) Define $\mu_{\text{CV}} = \mu_{\text{CV}} + \eta$ and repeat the calculation of σ recursively until $\mu_{\text{CV}} = 1 - \eta$. Select μ_{CV} ($\mu_{\text{CT}} = 1 - \mu_{\text{CV}}$) corresponding to the minimum error as output probability.

Step 4 (combination). According to the updated model probability in Step 3 and predicted state and covariance in Step 2, the combined state and covariance are represented through the following two equations:

$$\widehat{\mathbf{X}}(k \mid k) = \sum_{i=\text{CV,CT}} \mu_i \widehat{\mathbf{X}}_i(k \mid k)$$

$$\widehat{\mathbf{P}}(k \mid k) = \sum_{i=\text{CV,CT}} \mu_i \left\{ \mathbf{P}_i(k \mid k) + \left[\widehat{\mathbf{X}}_i(k \mid k) - \widehat{\mathbf{X}}(k \mid k)\right]\left[\widehat{\mathbf{X}}_i(k \mid k) - \widehat{\mathbf{X}}(k \mid k)\right]^T \right\}. \quad (19)$$

In the process of position predicting with IMM algorithm, it is desirable to have a record of ship trajectory used for initialization. Once there are no records available, the method for initializing vessel state is changed as follows: On the basis of the grid search previously described, set up a new grid centered on (λ_0, φ_0) (the optimal position obtained in grid search); the range of the grid is $2^\circ \times 2^\circ$ with the unit of 0.5°. According to the cost function E introduced before, search within the new grid and select a point $(\lambda_0{}', \varphi_0{}')$ corresponding to the maximum of E. The direction of (λ_0, φ_0) pointing to $(\lambda_0{}', \varphi_0{}')$ is taken as the vessel heading and (λ_0, φ_0) is regarded as the initial position.

3.3. The Proposed Localization Model Design. The predicted ship position with IMM is defined as $\widehat{\mathbf{L}} = [\hat{x}, \hat{y}, \hat{z}]^T$. The localization model combining this prediction with TDOA/FDOA measurements can be expressed as

$$\begin{bmatrix} \Delta\mathbf{T} \\ \Delta\mathbf{F} \\ \widehat{\mathbf{L}} \end{bmatrix} = \begin{bmatrix} \mathbf{r}_{\mathbf{t}}(x, y, z) \\ \mathbf{r}_{\mathbf{f}}(x, y, z) \\ [x, y, z]^T \end{bmatrix} + \mathbf{n}_c. \quad (20)$$

Let $\mathbf{\Upsilon}_c = \left[\Delta\mathbf{T}\ \Delta\mathbf{F}\ \widehat{\mathbf{L}}\right]^T$ and $\mathbf{q}_c = \left[\mathbf{r}_{\mathbf{t}}(x, y, z)\ \mathbf{r}_{\mathbf{f}}(x, y, z)\ [x, y, z]^T\right]^T$; on the basis of least squares criterion, the estimated position of the proposed model is achieved by applying the Gauss-Newton algorithm as follows:

$$\begin{aligned} &\left[x_{j+1}, y_{j+1}, z_{j+1}\right]^T \\ &\quad = \left[x_j, y_j, z_j\right]^T \\ &\quad + \left[\mathbf{J}_c{}^T\left(x_j, y_j, z_j\right)\mathbf{N}_c{}^{-1}\mathbf{J}_c\left(x_j, y_j, z_j\right)\right]^{-1} \\ &\quad \times \mathbf{J}_c{}^T\left(x_j, y_j, z_j\right)\mathbf{N}_c{}^{-1}\left[\mathbf{\Upsilon}_c - \mathbf{q}_c\left(x_j, y_j, z_j\right)\right], \end{aligned} \quad (21)$$

where $\mathbf{J}_c = \left[\begin{smallmatrix} \mathbf{J} \\ \mathbf{I}_{3\times3} \end{smallmatrix}\right]$, $\mathbf{N}_c = \left[\begin{smallmatrix} \mathbf{N} & \mathbf{0} \\ \mathbf{0} & \mathbf{N}_M \end{smallmatrix}\right]$, and $\mathbf{N}_M$ is the covariance matrix of ship predicted position obtained in IMM algorithm.

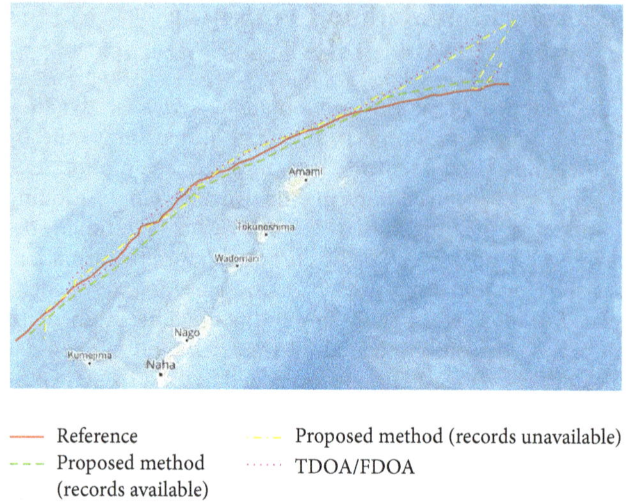

FIGURE 2: Reference and estimated vessel trajectories.

4. Simulation Research

Due to the limitation of the experimental conditions, the satellite ephemeris information is generated by Satellite Tool Kit. The orbital height of the satellite is set to 1000 km and the inclination angle is 50°. Simulations have been conducted to evaluate the estimation performance by using 3 signals for positioning (the change of ship position during the reception of the signals is negligible) and the time intervals separating signals are 20 s and 60 s. The measurement vectors of TDOA and FDOA are calculated by (1) and (2). It is supposed that, in the simulation, the time measurements are affected by an additive Gaussian noise constituting independent samples with zero mean and variance $\text{var}_t = 30\,\mu s^2$ and the frequency measurements noise is subjected to the Gaussian distribution of zero mean and variance $\text{var}_f = 400\,\text{Hz}^2$.

Figure 2 shows the estimated vessel trajectories with TDOA/FDOA positioning model and the proposed positioning model. The reference trajectory of vessel is carried out by AIS message reported by an ocean-going ship from Xiamen towards Long Beach in 24 hours and the average speed of ship is 19 knots. Figure 3 shows examples of reference position and estimated position for records availability scenario. The

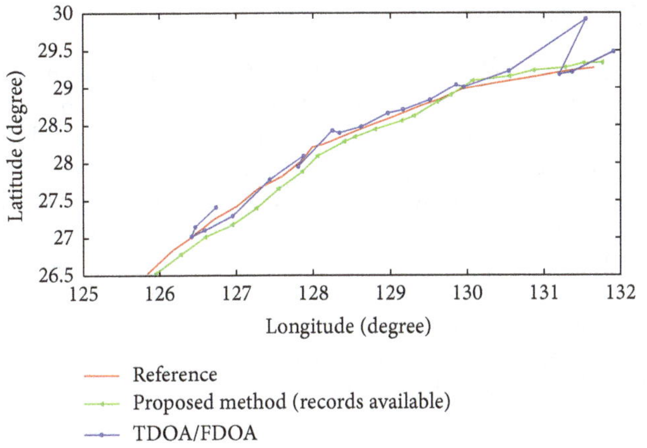

FIGURE 3: Examples of estimated vessel positions for records availability scenario.

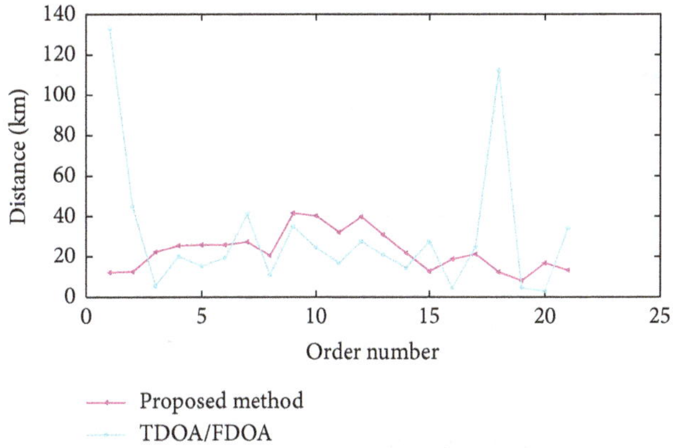

FIGURE 5: Distance between reference and estimated positions for records availability scenario.

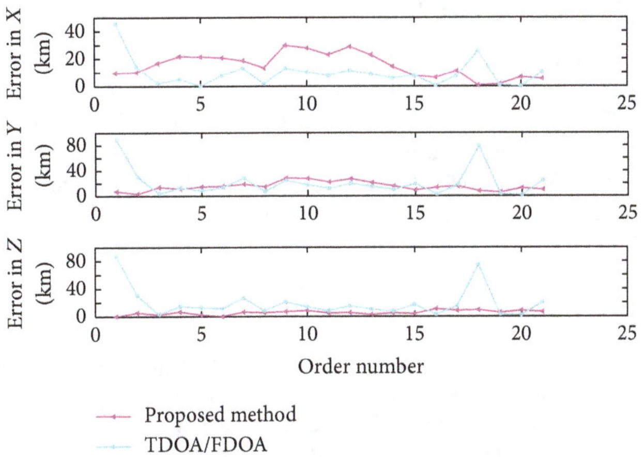

FIGURE 4: Positioning error in X, Y, Z direction for records availability scenario.

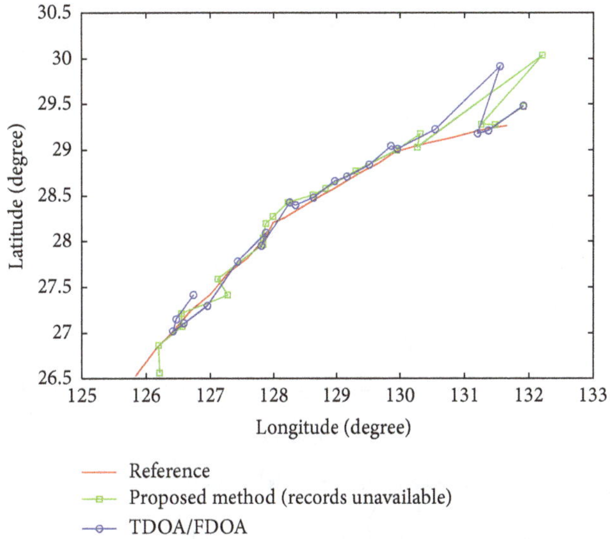

FIGURE 6: Examples of estimated vessel positions for records unavailability scenario.

number of positioning results presented is 21 and the average time separating them is approximately 1 hour. From Figures 2 and 3, it can be seen that, compared with TDOA/FDOA positioning model, the estimated trajectory with the proposed method is more in line with the reference trajectory in the case where a record of previous trajectory is used for initialization.

In Figure 3, it is assumed that the reference positions from low latitude to high latitude correspond to numbers 1 to 21 in order. Figure 4 shows the positioning error of these 21 positions in X, Y, Z direction for records availability scenario and Figure 5 shows the distance between the reference and estimated positions in the same scenario. In Figure 5, the average distance error of TDOA/FDOA positioning model is 30.4947 km, the maximum error occurs at the number 1 position, with the distance of 132.9219 km, and the minimum error is 2.8647 km at the number 20 position. Besides, with the condition of records availability, the average distance error of the proposed method is 22.9933 km and the maximum and the minimum error are 41.6187 km corresponding the number 9 position and 8.0532 km corresponding the number 19 position, respectively. As can be observed from Figures 3 and 5, the stability and accuracy of the proposed model (for records availability scenario) are better than those of the TDOA/FDOA positioning model on the whole. Although there is a case where the performance of TDOA/FDOA positioning is superior to the proposed method in individual positions, the large error values of the TDOA/FDOA method estimated at some positions cannot be ignored; because the TDOA/FDOA method is affected by the relative satellite-ship position, it is unstable in overall positioning accuracy.

The examples of reference position and estimated position for records unavailability scenario are shown in Figure 6. In Figures 3 and 6, the reference positions are the same, but it is obvious that the proposed method has a better performance in the case of records availability. Figure 7 shows the positioning error of estimated positions in X, Y, Z direction for records availability scenario.

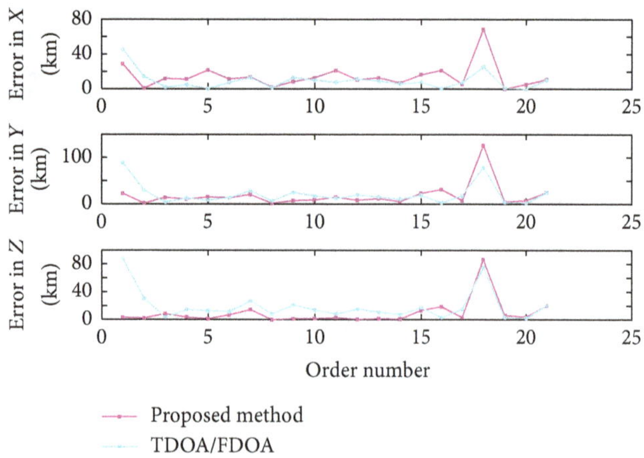

FIGURE 7: Positioning error in X, Y, Z direction for records unavailability scenario.

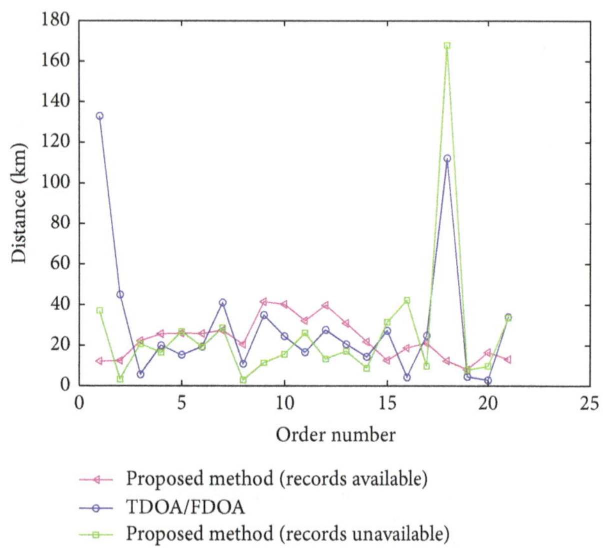

FIGURE 8: Distance between reference positions and estimated positions.

Figure 8 shows the distance between the estimated position and the reference position in different cases. For the scenario of records unavailability, the average distance error of proposed method is 26.2660 km, the maximum error occurs at the number 18 position, with the distance of 168.1150 km, and the minimum error is 2.9289 km at the number 8 position. According to Figures 5 and 8, it can be seen that, for the same reference positions, the average error with different positioning methods in descending order is as follows: 30.4947 km corresponding to TDOA/FDOA positioning method, 26.2660 km corresponding to the proposed method for records unavailability scenario, and 22.9933 km corresponding to the proposed method for records availability scenario. Overall, the performance of the proposed method in this paper is better than TDOA/FDOA positioning method, and, compared with the scenario of records unavailability, the accuracy of the proposed method is higher in the case where records are available.

5. Conclusion

A ship positioning method using AIS signals transmitted from satellite is presented in this paper. In the proposed positioning model, an additional measurement obtained by IMM algorithm is added to the TDOA/FDOA measurements. Besides, a probability update method applied in IMM algorithm is designed in this work. The feasibility of the proposed positioning method is verified by simulations. Regardless of algorithm complexity, the performance of the proposed method is better than TDOA/FDOA positioning model, especially in the case where the record of previous ship track is used for initialization.

Conflicts of Interest

The authors declare that they have no conflicts of interest.

Acknowledgments

The research presented in this paper is supported by the National Natural Science Foundation of China (Grant nos. 61371108 and 61601326), the Tianjin Research Program of Application Foundation and Advanced Technology under Grant 15JCQNJC01800, and the High School Science and Technology Developing Foundation of Tianjin under Grant 20140706.

References

[1] E. Alincourt, C. Ray, P.-M. Ricordel, D. Dare-Emzivat, and A. Boudraa, "Methodology for AIS signature identification through magnitude and temporal characterization," in *Proceedings of the OCEANS 2016*, pp. 1–6, Shanghai, China, April 2016.

[2] G. Siegert, P. Banys, C. S. Martinez, and F. Heymann, "EKF based trajectory tracking and integrity monitoring of AIS data," in *Proceedings of the IEEE/ION Position, Location and Navigation Symposium (PLANS '16)*, pp. 887–897, April 2016.

[3] F. Papi, D. Tarchi, M. Vespe, F. Oliveri, and G. Aulicino, "Radiolocation and tracking of automatic identification system signals," in *Proceedings of the 2014 IEEE Workshop on Statistical Signal Processing (SSP '14)*, vol. 9, No. 5, pp. 504–507, July 2014.

[4] X. Y. Wang and S. F. Zhang, "Evaluation of multipath signal loss for AIS signals transmitted on the sea surface," *Ocean Engineering*, vol. 146, pp. 9–20, 2017.

[5] P. Xia, T. Zhang, L. Ding, F. Yang, L. Qian, and H. Li, "Practical concern analysis on the detection probability for satellite-based AIS," in *Proceedings of the 22nd Asia-Pacific Conference on Communications (APCC '16)*, pp. 388–393, August 2016.

[6] K. Reiten, R. Schlanbusch, R. Kristiansen, F. Vedal, P. J. Nicklasson, and P. C. Berntsen, "Link and doppler analysis for space-based AIS reception," in *Proceedings of the 3rd International Conference on Recent Advances in Space Technologies (RAST '07)*, pp. 556–561, June 2007.

[7] A. N. Skauen, "Quantifying the tracking capability of space-based AIS systems," *Advances in Space Research*, vol. 57, no. 2, pp. 527–542, 2016.

[8] M. A. Cervera and A. Ginesi, "On the performance analysis of a satellite-based AIS system," in *Proceedings of the 10th International Workshop on Signal Processing for Space Communications (SPSC '08)*, pp. 1–8, October 2008.

[9] P. Burzigotti, A. Ginesi, and G. Colavolpe, "Advanced receiver design for satellite-based automatic identification system signal detection," *International Journal of Satellite Communications and Networking*, vol. 30, no. 2, pp. 52–63, 2012.

[10] R. Prevost, M. Coulon, P. Paimblanc, J. Lemaitre, J.-P. Millerioux, and J.-Y. Tourneret, "Ship localization using ais signals received by satellites," in *Proceedings of the 2013 21st European Signal Processing Conference (EUSIPCO '13)*, pp. 1–5, September 2013.

[11] T. G. Dvorkind and S. Gannot, "Time difference of arrival estimation of speech source in a noisy and reverberant environment," *Signal Processing*, vol. 85, no. 1, pp. 177–204, 2005.

[12] Y. Liu, F. Guo, L. Yang, and W. Jiang, "Source localization using a moving receiver and noisy TOA measurements," *Signal Processing*, vol. 119, article 5882, pp. 185–189, 2016.

[13] N. Wu, W. Yuan, H. Wang, and J. Kuang, "TOA-based passive localization of multiple targets with inaccurate receivers based on belief propagation on factor graph," *Digital Signal Processing*, vol. 49, pp. 14–23, 2016.

[14] J. M. Gambi, J. Clares, and M. L. García Del Pino, "FDOA post-Newtonian equations for the location of passive emitters placed in the vicinity of the Earth," *Aerospace Science and Technology*, vol. 46, article 3377, pp. 137–145, 2015.

[15] X. Qu and L. Xie, "An efficient convex constrained weighted least squares source localization algorithm based on TDOA measurements," *Signal Processing*, vol. 119, pp. 142–152, 2016.

[16] A. N. Bishop, B. Fidan, K. Doğançay, B. D. O. Anderson, and P. N. Pathirana, "Exploiting geometry for improved hybrid AOA/TDOA-based localization," *Signal Processing*, vol. 88, no. 7, pp. 1775–1791, 2008.

[17] G.-H. Zhu, D.-Z. Feng, H. Xie, and Y. Zhou, "An approximately efficient bi-iterative method for source position and velocity estimation using TDOA and FDOA measurements," *Signal Processing*, vol. 125, pp. 110–121, 2016.

[18] Y. H. Kim, D. G. Kim, and J. W. Han, "Analysis of sensor-emitter geometry for emitter localisation using TDOA and FDOA measurements," *IET Radar, Sonar & Navigation*, vol. 11, pp. 341–349, 2017.

[19] Y. Wang and Y. Wu, "An efficient semidefinite relaxation algorithm for moving source localization using TDOA and FDOA measurements," *IEEE Communications Letters*, vol. 21, no. 1, pp. 80–83, 2017.

[20] L. Hong-Tao and K. Feng-Ju, "Tracking UUV based on interacting multiple model unscented particle filter with multi-sensor information fusion," *Optik—International Journal for Light and Electron Optics*, vol. 126, no. 24, pp. 5067–5073, 2015.

[21] R. Tou and J. Zhang, "IMM approach to state estimation for systems with delayed measurements," *IET Signal Processing*, vol. 10, no. 7, pp. 752–757, 2016.

[22] S. Kay, "A fast and accurate single frequency estimator," *IEEE Transactions on Signal Processing*, vol. 37, no. 12, pp. 1987–1990, 1989.

[23] M. Morelli and U. Mengali, "Joint frequency and timing recovery for MSK-type modulation," *IEEE Transactions on Communications*, vol. 47, no. 6, pp. 938–946, 1999.

[24] D. X. Zhong, X. P. Deng, and Y. Y. Zhou, "Positioning accuracy analysis of satellite time difference based on WGS-84 ellipsoid model," *Electronic Counter Technology*, vol. 17, no. 5, pp. 19–21, 2002.

8

Speaker Recognition using Wavelet Cepstral Coefficient, I-Vector, and Cosine Distance Scoring and its Application for Forensics

Lei Lei and She Kun

Laboratory of Cyberspace, School of Information and Software Engineering, University of Electronic Science and Technology of China, Chengdu 610054, China

Correspondence should be addressed to She Kun; kun@uestc.edu.cn

Academic Editor: Mariko Nakano-Miyatake

An important application of speaker recognition is forensics. However, the accuracy of speaker recognition in forensic cases often drops off rapidly because of the ill effect of ambient noise, variable channel, different duration of speech data, and so on. Therefore, finding a robust speaker recognition model is very important for forensics. This paper builds a new speaker recognition model based on wavelet cepstral coefficient (WCC), i-vector, and cosine distance scoring (CDS). This model firstly uses the WCC to transform the speech into spectral feature vecors and then uses those spectral feature vectors to train the i-vectors that represent the speeches having different durations. CDS is used to compare the i-vectors to give out the evidence. Moreover, linear discriminant analysis (LDA) and the within-class covariance normalization (WCNN) are added to the CDS algorithm to deal with the channel variability problem. Finally, the likelihood ratio estimates the strength of the evidence. We use the TIMIT database to evaluate the performance of the proposed model. The experimental results show that the proposed model can effectively solve the troubles of forensic scenario, but the time cost of the method is high.

1. Introduction

With the increasing use of computer technology, more and more complex and tedious works can be finished by computer. Automatic speaker recognition (or speaker recognition for short) technique refers to recognizing persons from their voice using the computer software. An important application of speaker recognition is forensics. This technique is usually used for investigation and evidence reporting [1]. In the investigation task, the speaker recognition is used to compare the questioned speech with the speeches of known criminals in police's database to produce a small list of potential suspects. In the evidence reporting task, police has found the suspect and the speaker recognition is used to give out the evidence (is defined as the similarity between the questioned speech and the suspect's speech samples [2]) supporting the fact that the suspect is the author of the criminal speech.

Technologically, the forensic speaker recognition model is very similar to the common speaker recognition model [3]. It firstly uses a feature extractor to transform the digital speech signal into the feature vectors that represent unique information for a particular speaker irrespective of the speech content and then uses a learning algorithm to give out the evidence based on those feature vectors. Finally, it is required to report the likelihood ratio that estimates the strength of the evidence after it gives out the evidence [4]. The forensic context is very challenging for speaker recognition. The ambient noise cannot be controlled, the duration of speech data can vary from a few seconds to several hours, and the speeches are often obtained from different channels such as phone channel and microphone channel. It is important for forensics to find a robust speaker recognition model that is not sensitive to those factors such as noise, duration, and channel.

The speech is usually transformed into the short-term feature vector because this feature vector is simple [5]. In the speaker recognition model used for forensics, Mel-frequency cepstral coefficient (MFCC) method has been widely used to extract the short-term feature vector [6]. This method calculates only the cepstral coefficients, so the extracted feature vector just represents the static information. Fused MFCC (FMCC) [7] method is an extension of MFCC. This method calculates not only the cepstral coefficients but also the delta derivatives, so the extracted feature vector can represent both the static information and dynamic information. Both of the two methods use discrete Fourier transform (DFT) to obtain the frequency spectrum. Because the DFT is a global signal analysis approach, if a local frequency part of the speech is destroyed by noise, the ill effect of noise will be transmitted to whole feature vector. Wavelet cepstral coefficient (WCC) [8] is another method used to extract the short-term feature vector. This method uses discrete wavelet transform (DWT) to obtain the frequency spectrum. DWT is a local signal analysis approach, so the whole feature vector cannot be interfered strongly by the noise that just destroys the local frequency part. Unfortunately, this method is not used for the forensic speaker recognition. This paper tries to employ this method to extract the short-term feature vector, because it can be robust against the noise.

In the forensic context, the duration of speech can vary from a few seconds to several hours. To compare those speeches having different duration, it is necessary to represent them in a uniform way. I-vector [9] is a newly proposed feature vector and has been employed by the speaker recognition model used for forensics [10]. The main advantage of i-vector is that it can use a signal vector to represent a speech, so it can uniformly represent those speeches having different duration. In its extraction method [11], the speech signal is firstly transformed into short-term feature vectors and then those short-term feature vectors are used to train i-vector.

Based on the feature vector extracted by MFCC or Fused MFCC, Gaussian mixture model (GMM) is the conventional learning algorithm [12]. In [13], it uses the probability distribution estimated during the training phrase to determine whether the suspect is the author of the questioned speech. However, if the dimension of the input vector is very high, the curse of dimension will destroy the algorithm [14]. Cosine distance scoring (CDS) is another type of learning algorithm used for speaker recognition and [1, 15] use it to give out the evidence in the forensic speaker recognition. Because it can deal with the curse of dimension using a cosine kernel, CDS is suitable for i-vector which is high-dimensional vector compared with those short-term feature vectors. Moreover, it does not cost time to estimate the separating hyperplane, so its time cost is low. Two same speeches will become very different, if they are obtained from different channels. This is called channel variability problem. Usually, linear discriminant analysis and the within-class covariance normalization are added to the CDS to deal with this problem [16, 17].

In forensic context, the quality of speech is strongly interfered by noise, variable channel, and different duration. This is very important for forensics to find a speaker recognition model that is unsusceptible to those factors. Based on those above works of the speaker recognition used for forensics, this paper combines the WCC, i-vector approach, and the CDS learning algorithm to build a new speaker recognition model that can be robust against the noise, uniformly represent the speech having different duration, and deal with the channel variability problem. We use TIMIT to evaluate the performance of our system. The experimental results show that the proposed model can solve the trouble of forensic scenario and improve the accuracy of recognition. However, the time cost of the model is high compared with other conventional speaker recognition models.

The rest of the paper is organized as follows. Section 2 briefly describes the conventional speaker recognition model. In Section 3, we describe the i-vector-based forensic speaker recognition model. The proposed model is described in Section 4. In Section 5, we report the result of our experiment. Finally, we give out a conclusion in the last section.

2. Conventional Speaker Recognition Model

2.1. Short-Term Feature Extraction Method. In conventional speaker recognition model, the short-term feature vectors are usually extracted by MFCC and Fused MFCC methods [18]. In MFCC method, a speech signal is firstly divided into 20 ms-long frames with a 10 ms overlap for smoothing the frequency changes over those frames. After the segmentation, the frames whose energy is less than a silence threshold (= 0.0001 in this paper) are discarded as well. For each frame, the cepstral coefficient can be calculated as follows:

(i) Take DFT of the frame to obtain the frequency spectrum.

(ii) Map the power of the spectrum onto Mel scale using the Mel filter bank.

(iii) Calculate the logarithm value of the power spectrum mapped on the Mel scale.

(iv) Take DCT of logarithmic power spectrum to obtain the cepstral coefficient.

Usually, only the lower 12–14 cepstral coefficients are used to form the short-term feature vector [19]. This feature vector extracted by MFCC contains only the cepstral coefficients, so it just represents the static information of the frame. Fused MFCC (FMCC) is an extension of MFCC. It is very similar to the MFCC, except that this method calculates not only the cepstral coefficients but also the delta derivatives [20]. Assume that the feature vector extracted by MFCC is denoted as $[cc_1, \ldots, cc_{13}, cc_{14}]$. The delta derivatives are calculated by

$$d_i = \frac{\sum_{p=1}^{2} p\left(cc_{i-p} + cc_{i+p}\right)}{2\sum_{p=1}^{2} p^2},$$
$$dd_i = \frac{\sum_{p=1}^{2} p\left(d_{i-p} + d_{i+p}\right)}{2\sum_{p=1}^{2} p^2}. \tag{1}$$

The feature vector extracted by FMFCC is denoted as $cc_1, \ldots, cc_{13}, cc_{14}, d_1, \ldots, d_{13}, d_{14}, dd_1, \ldots, dd_{13}, dd_{14}$. Compared with the cepstral coefficients vector extracted by MFCC, the feature vector extracted by Fused MFCC can represent the changes over the multiple frames. The detailed algorithm of MFCC and Fused MFCC can be found in [20–22].

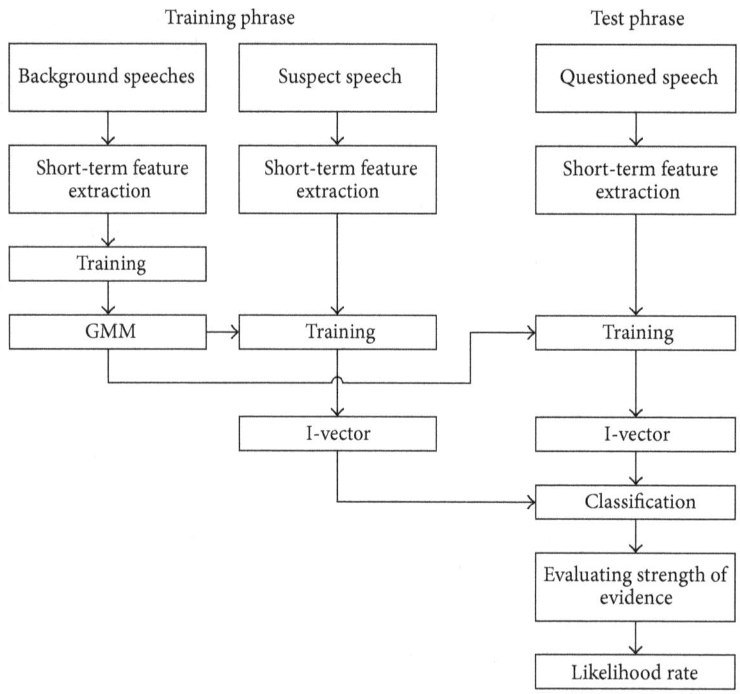

FIGURE 1: Forensic speaker recognition system.

2.2. Speaker Classification. Gaussian mixture model (GMM) is the conventional learning algorithm used for speaker classification in the conventional speaker model. This model tries to estimate the probability distribution governing the known speech data of a particular speaker and then uses the probability distribution to test an unknown speech and determine whether the unknown speech is spoken by the speaker. It is linear combination of Gaussian functions. The experiment in [23] shows that, with increasing of the number of Gaussian functions which GMM contains, the recognition accuracy of GMM will become very high. However, the large number of Gaussian functions also reduces excessive computational complexity.

Neural network (NN) is another type of learning algorithm used for speaker classification. This algorithm simulates the human brain to learn the knowledge of the known speech data of a particular speaker by iteratively adjusting the weights that connect between two neurons in the adjacent two layers and uses the knowledge to determine whether the speaker was the author of an unknown speech. Probabilistic neural network (PNN) is a special case of NN, where the sigmoid activation function is replaced by an exponential function. It uses the NN structure to directly implement the Bayesian decision and does not cost time to estimate the probability distribution compared with GMM, so the time cost of this algorithm is very low.

3. I-Vector-Based Speaker Recognition Model Used for Forensics

The i-vector-based speaker model used for forensics is shown in Figure 1.

A forensic speaker recognition model can be decomposed into training phrase and test phrase. The first two columns denote the training phrase and the last column denotes the test phrase. In Figure 1, background speeches usually contain thousands of speeches spoken by a huge number of people [5]. Suspect speech and questioned speech are collected from the suspect and the criminal scene. All the tree types of speeches have different length and are full of noise. Furthermore, they are usually obtained from different channels.

Firstly, the model transforms the three types of speeches into short-term feature vectors. The short-term feature vectors extracted from the background speeches are used to train the background model. This model represents the speaker- and channel-independent information and is implemented by a GMM. Once the GMM model is created, the short-term feature vectors extracted from the suspect speech and the questioned speech are used to train i-vector. One i-vector is trained using only the short-term feature vectors extracted from a speech. After the suspect speech and questioned speech are transformed into i-vector, a learning algorithm is used to compare those i-vectors to give out the evidence in forensics and then report the strength of the evidence as a likelihood ratio.

4. The Proposed Model

In the forensic context, the duration of speech can vary from few seconds to several hours and the recording condition is full of noise. Moreover, the speeches are usually obtained from different channels. To deal with those problems, this paper proposed a new speaker recognition model by employing WCC, i-vector, and CDS. The WCC is used to extract the short-term feature vectors that are used to train i-vector. I-vector is used to represent the speeches whose durations are very different in a uniformly way for comparing those speeches easily. CDS is used to give out the evidence. Section 4.1 describes the WCC, Section 4.2 describes i-vector, and Section 4.3 describes the CDS. Finally, the likelihood-ratio algorithm is described in Section 4.4.

4.1. Short-Term Feature Extraction. This paper uses the wavelet cepstral coefficient (WCC) to extract the short-term feature vectors, because it is able to effectively limit the ill effect of noise using discrete wavelet transform (DWT).

Wavelet transform is a type of signal processing tool that is used to obtain the frequency spectrum. A standard wavelet transform is defined by

$$Wf(n,m) = \frac{1}{\sqrt{m}} \int_{-\infty}^{+\infty} S(t) \cdot \psi\left(\frac{t-n}{m}\right) dt, \quad (2)$$

where $Wf(n,m)$ is a continue signal frame which has finite energy. $\psi(\cdot)$ is the mother wavelet and $Wf(n,m)$ represents the nth wavelet coefficient at level m. For analyzing the discrete digital signal, the discrete wavelet transform (DWT) is proposed. The DWT is usually implemented by the famous Mallet algorithm [24]. In the algorithm, the DWT is realized through a pair of low-pass and a high-pass wavelet filters that are reconstructed from a selected mother wavelet and its corresponding scaling function. Through those filters, the signal is decomposed into a low-frequency part and a high-frequency part. The low-frequency part can be further decomposed at the next decomposition level to obtain higher low-frequency resolution. The low- and high-passed filtering processes are implemented by

$$\begin{aligned} A_{m+1} &= A_m * h[2n], \\ D_{m+1} &= A_m * g[2n], \\ A_0[n] &= S[n], \\ & \qquad n = 1, 2, 3, \ldots, N, \end{aligned} \quad (3)$$

where N is the length of the analyzed signal. g and h represent the low-pass and high-pass conjugate mirror filters, respectively. $*$ is the convolution operation. Compared with DFT used in MFCC or Fused MFCC, DWT can decompose the signal into many small local frequency domains and obtain the local frequency spectrum. In other words, if one of frequency parts of signal is destroyed by noise, whole frequency spectrums will not be interfered strongly. This means that the frequency spectrum obtained by wavelet is robust against noise.

Recently, researchers have widely used a new type of feature extractor named wavelet cepstral coefficient method for short-term feature extraction [8]. The flow chart of WCC extraction algorithm used in this paper is shown in Figure 2.

In Figure 2, the speech is firstly decomposed into 20 ms-long short-term frames with a 10 ms overlap. After the segmentation, the silence frames are discarded using an energy threshold (= 0.0001). After the silence frame removing, we add a normalization method to the WCC approach to remove the ill effect of sound volume. The normalization method [25] is given by

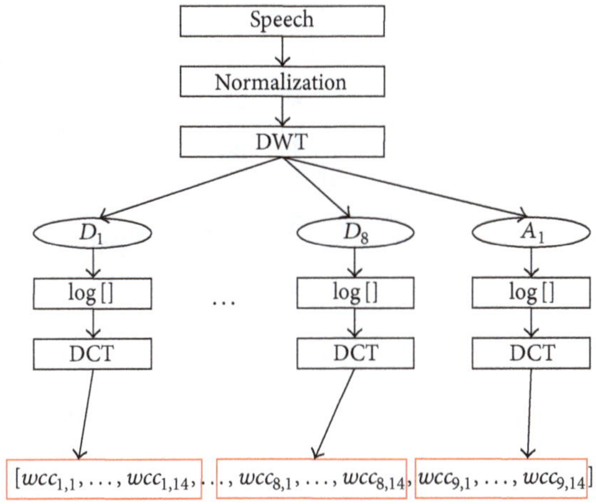

FIGURE 2: The flow chart of WCC extraction algorithm.

$$\overline{f}(n) = \frac{f(n) - \mu}{\sigma}, \quad n = 1, 2, 3, \ldots, N, \quad (4)$$

where $f(n)$ is a short-term frame that has finite energy and length. μ and σ are the mean and standard variance of the frame, respectively. N is the length of frame and $\overline{f}(n)$ is the normalized frame. The result of normalization is shown in Figure 3.

After the normalization, DWT is used to obtain the local frequency spectrum. This paper decomposes the speech into 8 levels by DWT, and therefore we obtain 9 local frequency parts such as 8 high-frequency parts denoted by $D_1, D_2, \ldots, D_8$ and one low-frequency part denoted by A_1. For each frequency part, log[] and DCT are also used to calculate the 14 cepstral coefficients. WCC is very similar to the MFCC, but the difference is that the cepstral coefficient in MFCC is calculated on a global frequency domain, but the cepstral coefficient in WCC is calculated on many local frequency domains obtained by DWT.

4.2. Training I-Vector. After the speech is transformed to short-term feature vectors, we can use those vectors to train i-vector. A background model should be trained at first. The background model represents the speaker- and channel-independent information and is implemented by a GMM. This GMM is trained by a huge set of short-term feature vectors extracted from the background speech set that contains thousands of speeches spoken by large number of speakers. For forensics, the background speech set may contain all speeches of all known criminals in police database, and two gender-dependent GMMs that generalize the characteristic of gender-dependent voice are trained using female's speeches and male's speeches, respectively [1]. Once the GMMs are trained, the suspect i-vector and questioned i-vector are trained using a particular suspect speech and questioned speech, respectively.

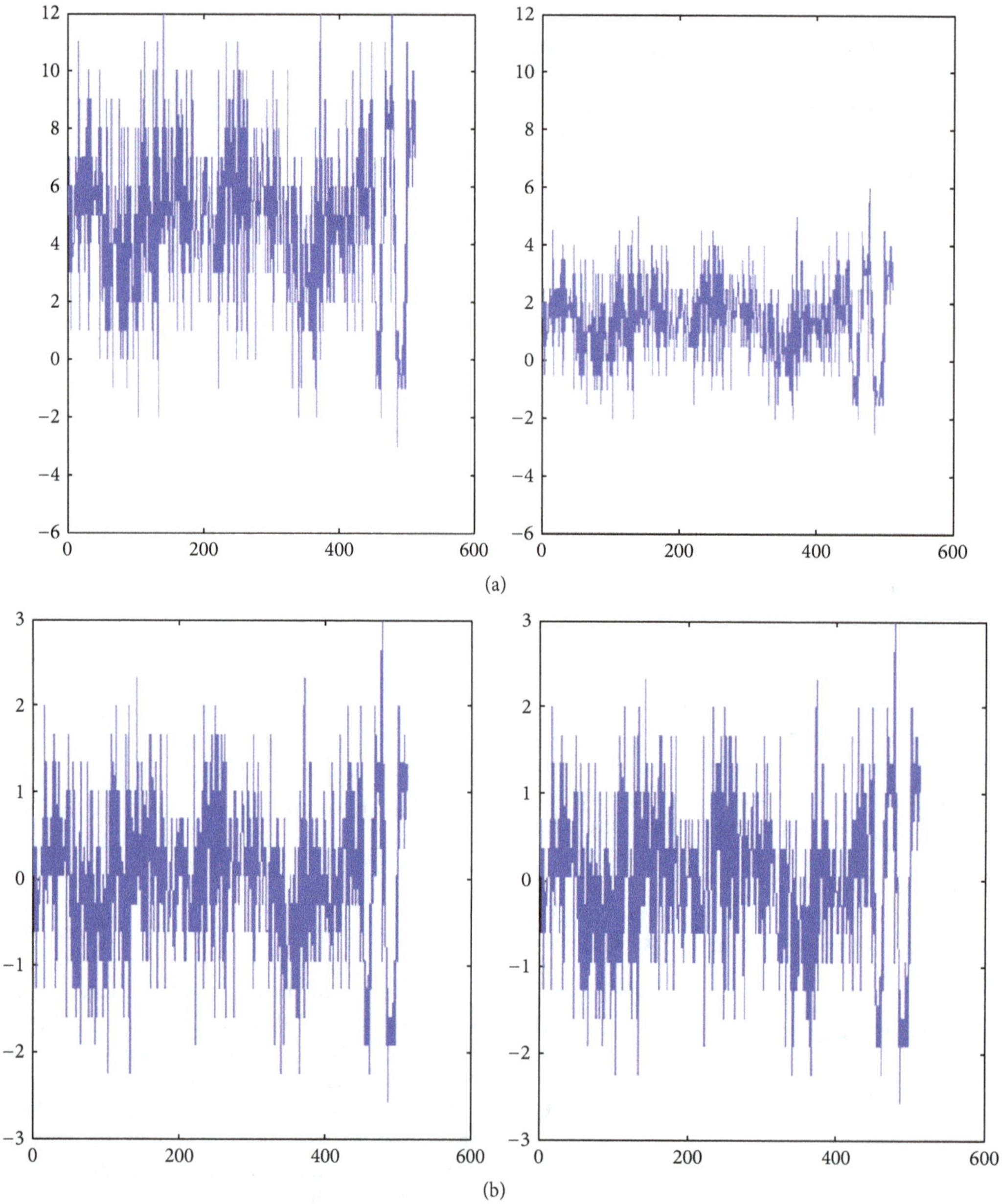

FIGURE 3: (a) Two speech signals that have different sound volume before normalization. (b) Two speech signals that have different sound volume after normalization.

Given a speech, i-vector approach assumes that the ideal speaker- and channel-dependent feature vector used to represent the speech can be modeled as

$$M = m + Zx, \tag{5}$$

where M is the ideal feature vector. m is a vector that consists of all the mean vectors of the GMM. If the mean vectors of GMM are denoted by $\mu_1^T, \mu_2^T, \ldots, \mu_I^T$, where I is the number of the mean vectors and each μ is a row vector, then m is denoted by $[\mu_1, \mu_2, \ldots, \mu_I]^T$. Z is a low rank matrix and is called the total variability matrix. x is i-vector and obeys standard normal distribution. For a given speaker and channel, m and Z are changeless, so the ideal feature vector is dependent on the value of x. In other words, x can represent the speech sample. Based on the assumption shown in (5), the i-vector training algorithm is used to iteratively estimate x and Z. x is estimated using a speech, and Z is estimated using all speeches generated from a speaker. The details of the training algorithm are described in [26].

4.3. Evidence Reporting. Cosine distance scoring (CDS) is a famous learning algorithm. It uses a cosine kernel to directly compare two input feature vectors and give out the degree of similarity between the two feature vectors. The cosine kernel is defined as

$$E = K(x_1, x_2) = \frac{x_1 x_2}{\sqrt{x_1 x_1^T}\sqrt{x_2 x_2^T}}, \tag{6}$$

where x_1, x_2 are the two feature vectors. In this paper, they are two i-vectors that represent a suspect's speech sample and a questioned sample, respectively. $K(x_1, x_2)$ is the degree of similarity between the suspect's speech and the questioned speech, so it is the evidence reported by the forensic speaker recognition model.

Two same speech samples will become very different, if they are obtained from different channels. This is called channel variability problem. In the forensic context, the speech data are usually obtained from different channels. For example, in the criminal case where a victim receives a threatening call, the questioned speech is a phone recording, but police usually record the speeches of suspect by microphone. In other words, the question speech is obtained from the phone channel, but the suspect's speeches are obtained from the microphone channel. To deal with the channel variability problem, LDA and WCCN are added to the CDS. The cosine kernel is denoted as

$$E = K(x_1, x_2) = \frac{(A^T x_1) W^{-1} (A^T x_2)}{\sqrt{(A^T x_1) W^{-1} (A^T x_1)} \sqrt{(A^T x_2) W^{-1} (A^T x_2)}}, \quad (7)$$

where A is the LDA projection matrix and W is WCCN matrix. The details of LDA and WCCN are described in [27].

4.4. Evaluating the Strength of Evidence. For evidence reporting, the speaker recognition model should report the strength of evidence as a likelihood ratio (LR). To calculate the LR, two competing hypotheses H_0 and H_1 are given. H_0 assumes that the suspected speaker is the author of the questioned speech and H_1 assumes that the suspected speaker is not the author of the questioned speech. Based on the two hypotheses, the LR [28] is defined as

$$\text{LR} = \frac{P(E \mid H_0)}{P(E \mid H_1)}, \quad (8)$$

where E is the evidence calculated by (7). Firstly, we estimate the probability distribution $P(E \mid H_0)$ and $P(E \mid H_1)$. In conventional aural speaker recognition, the degree of similarity between the questioned sample and the suspect's sample is estimated by a seven-level verbal scale shown in Table 1 [28].

To simulate this, we transform E into 7-level scale. The transform function is defined as

$$T(E) = \begin{cases} 1 & -1 \le E < -0.7 \\ \vdots & \vdots \\ 4 & -0.1 \le E < 0.1 \\ \vdots & \vdots \\ 7 & 0.7 \le E \le 1. \end{cases} \quad (9)$$

Assume that there are s ($\ge$ 2) known criminals in the police database and each one speaks m ($\ge$ 2) speeches. We iteratively select two speeches spoken by the same speaker and calculate $T(E)$. If $T(E)$ happens n_1 times, then $P(T(E) = e \mid H_0)$ is calculated as

Table 1: The verbal scales used in aural speaker recognition.

Level	Verbal equivalent
1	I am certain that the two speakers are not the same
2	I am almost certain that the two speaker are not the same
3	It is possible that the two speakers are not the same
4	I am unable to decide
5	It is possible that the two speakers are the same
6	I am almost certain that the two speakers are the same
7	I am certain that the two speakers are the same

$$P(T(E) = e \mid H_0) = \frac{n_1}{C_s^1 C_m^2} = \frac{2n_1}{sm(m-1)}, \quad (10)$$

where C_m^n denotes the number of n-combinations in a set of m elements. On the other hand, we iteratively select two speech samples spoken by different speakers and calculate $T(E)$. If $T(E) = e$ happens n_2 times, then $P(T(E) = e \mid H_1)$ is calculated as

$$P(T(E) = e \mid H_1) = \frac{n_2}{C_s^2 C_m^1 C_m^1} = \frac{2n_2}{sm^2(m-1)}. \quad (11)$$

We calculate $P(T(E) = e \mid H_0)$ and $P(T(E) = e \mid H_1)$ for each $T(E)$ to obtain the distribution of E for H_0 and H_1. Assume that there is an evidence E and $T(E) = e_1$. If we want to report its strength, we firstly search the distributions and find the value of $P(e_1 \mid H_0)$ and $P(e_1 \mid H_1)$ and then calculate the LR.

5. Results

In this section we report the outcome of our experiments. In Section 5.1, we describe the experimental dataset and procedure. In Section 5.2, we carry out an experiment to select the optimal mother wavelet for WCC method. In Section 5.3, we evaluate the performance of the proposed speaker recognition model used for investigation. In Section 5.4, we evaluate the performance of the proposed model used for evidence reporting. In Section 5.5, the time cost of the proposed model is counted.

5.1. Experimental Dataset. The results of our experiments were performed on TIMIT speech database [29]. This database contained 630 speakers (192 females and 438 males) who came from 8 different dialect regions. Each speaker supplied ten 5-second-long speeches that were sampled at 16 KHz. In forensic context, the speeches had different length and were full of noise. Moreover, the questioned speech and suspect's speeches are usually obtained from different channels. For each speaker, 3 speeches were downsampled to 8 KHz and other 7 speeches were still sampled at 16 KHz. This simulated the speech data obtained from different channels. Moreover, the 3 speeches sampled at 8 KHz were combined in a 15-second-long speech, and other 7 speeches still lasted 5 seconds. This simulates the speech data having different duration. Three types of noises such as 10 dB, 20 dB, and 30 dB were added to those speeches to simulate the speeches that were full of noise in Section 5.3. All of female speeches and all of male speeches were used to train two gender-dependent background models, respectively. The test results presented in our experiments were collected on a computer with a 2.5 GHz Intel Core i5 processor and 8 GM of memory. The experimental platform was MATLAB R2012b.

5.2. Mother Wavelet. The mother wavelet was a key issue for DWT, and good mother wavelet could improve the performance of the wavelet-based spectral speech feature such as the WCC. The goal of this experiment was to find the optimal mother wavelet for WCC. The number of the vanishing movements and the size of support were two important elements for a mother wavelet. In the theory of mother wavelet [30], if the mother wavelet had enough vanishing movements, the DWT would ignore much of unimportant information; if the mother wavelet had small enough support, the wavelet coefficients obtained by DWT would sparely and accurately represent the important information of a signal. Therefore, an optimal mother wavelet should have large number of vanishing movements and meanwhile have small support. However, we had to take tradeoff between the number of vanishing movements and the size of support, because they should satisfy the following equation:

$$L \geq 2p - 1, \tag{12}$$

where L is the size of support and p is the number of vanishing movements. In this view, Daubechies wavelets [31] were the optimal wavelets, because they had the smallest support for given vanishing movements. Moreover, these wavelets had orthogonal conjugate mirror filters which were suitable for the Mallat fast DWT algorithm.

In this experiment, we employ the normalized partial energy (NPE) [32] to evaluate the performance of Daubechies wavelets. NPE was used to quantify how well a particular transform, such as DWT, performed in capturing the important information of a signal. Assume that there was a wavelet coefficient series denoted by $\{c_1, c_2, \ldots, c_N\}$, where N is the total number of the wavelet coefficients. Form the squared magnitudes $|c_t|^2$ and order them such that

$$|c_{(1)}|^2 \geq |c_{(2)}|^2 \geq \cdots \geq |c_{(N)}|^2. \tag{13}$$

The NPE was defined by

$$\text{NPE}(n) = \frac{\sum_{u=1}^{n} |c_{(u)}|^2}{\sum_{u=1}^{N} |c_{(u)}|^2}, \quad n = 1, 2, 3, \ldots, N. \tag{14}$$

We can see that NPE(n) varied from 0 to 1 for all n. If the NPE would be close to 1 for small n, the DWT was able to capture the key information. In other words, the mother wavelet used in the DWT was optimal. The Daubechies wavelet was denoted by db$N$, where N is its number of vanishing movements. This experiment employed db 1–8 to decompose 200 speeches that were randomly selected from our dataset. For each mother wavelet, we calculated 200 NPEs and count the average value of those NPEs. Table 2 shows the average NPE of those mother wavelets when n was equal to 5, 10, and 20, respectively.

Table 2: The NPEs of the DWT using different mother wavelets.

Mother wavelets	$n = 5$	$n = 10$	$n = 20$
db1	0.51	0.73	0.91
db2	0.66	0.75	0.93
db3	0.80	0.85	0.95
db4	**0.86**	**0.92**	**0.98**
db5	0.79	0.87	0.98
db6	0.73	0.83	0.97
db7	0.69	0.76	0.95
db8	0.65	0.76	0.95

In Table 2, we could find that the NPEs of all mother wavelets reached higher than 0.9 when n was equal to 20 and db4 and db5 obtained the highest NPEs of 0.98. When n was equal to 10, only db4 obtained the NPE of 0.92 compared with other mother wavelets which obtained the NPEs of less than 0.9. Moreover, the db4 wavelet could use only 5 wavelet coefficients to obtain the NPE of 0.86, but other mother wavelets obtain the NPE of less than 0.8. Those results show that the db4 wavelet was the most suitable mother wavelet, because its DWT can capture the key information of speech. In [33], researchers suggested that the Symlet wavelets could also obtain good performance. However, the complex conjugate mirror filters of Symlet wavelets produced the complex wavelet coefficients whose imaginary parts were redundant for real signal such as speech, so we abandoned the Symlet wavelets.

5.3. The Accuracy Rate of Investigation. This experiment tested the performance of the speaker recognition model when it was used for investigation. In investigation task, the speaker recognition model was used to compare the questioned speech with all of the speeches of known criminals in police database to produce a small list of potential suspects. In our experiment, we selected 384 speakers (192 females and 192 males) to form the large criminal set. For each speaker, the 7 recordings sampled at 16 KHz are used as the known criminal's speeches (called known speeches for short) and the recordings sampled at 8 KHz are used as the questioned speech. The similarity between the criminal's speech sample and the questioned speech was defined as

$$s = \frac{1}{7}\sum_{i=1}^{7} K(y_i, x), \tag{15}$$

where y is the known speeches and x is the questioned speech. $K(\cdot,\cdot)$ is the kernel function defined in (7). For each questioned speech, we required the speaker recognition model to produce a list of top 10 potential suspects that obtained the highest similarity. If the "real criminal" is in the

Table 3: The accuracy of investigation.

Short-term feature extractor	Accuracy rate of investigation (%)
MFCCGMM	82.46
FMFCCPNN	86.81
MFCCICDS	93.75
WCCICDS	95.48

list, the speaker recognition model got one score; if the "real criminal" was not found in the list, the model got zero score. We summed the score of the speaker recognition model for the 384 questioned speeches and calculated the accuracy rate that is defined as

$$ACC = \frac{\text{score}}{384} \times 100\%. \tag{16}$$

Because the speaker recognition model is used for investigation, the likelihood ratio is not required to be calculated.

For speaker recognition, many types of models were proposed by researchers. The typical speaker recognition model is MFCCGMM [34]. This model used 14D short-term feature vectors obtained by MFCC method to directly represent the speeches and used GMM for classification. In [20], researchers proposed a model based on Fused MFCC and probabilistic neural network (PNN), which is named FMFCCPNN. This model used 52D short-term feature vectors obtained by Fused MFCC method for speech representation and used the PNN for classification. Recently, [15] proposed a speaker model based on the MFCC, i-vector, and CDS, which is named MFCCICDS. This model used i-vector for speech representation and CDS was used for classification. Moreover, i-vector was trained using 14D short-term feature vectors obtained by MFCC method. Inspired by the above model, we proposed new speaker model based on WCC, i-vector, and CDS, which was called WCCICDS. This model was similar to the above one, but we used 126D short-term feature vectors obtained by WCC to train i-vector. Moreover, the mother wavelet used in our model was db4. For comparison, we employed the above 4 models to achieve the investigation task. The accuracy rates were shown in Table 3. In this experiment, we just used the clear speeches.

In Table 3, MFCCGMM and FMCCPNN obtained low accuracy of 82.46% and 86.81%, respectively. However, the two models in [20] obtained the accuracy of higher than 89% and 92%, respectively. This was because the speeches in [4] were obtained from same channel and had the same duration, but the speeches in our experiment were obtained from different channels and their durations were different too. This shows that the CDS could deal with the channel variability problem and i-vector was able to model the different length speeches to improve the performance of speaker recognition. Compared with the two models based on i-vector and CDS, we found that MFCCICDS obtained lower accuracy than the WCCICDS did. This was because WCC used DWT to analyze the speech signal, but MFCC used DFT. DFT used the fixed window to decompose signal, but DWT used the variable window that could obtain high frequency resolution at low frequency and high time resolution at high frequency [35]. For the explodent sound that had short duration and high frequency, variable window can capture its information, but the fixed window might make the information fuzzy [22]. Therefore, the model which used WCC obtained higher accuracy.

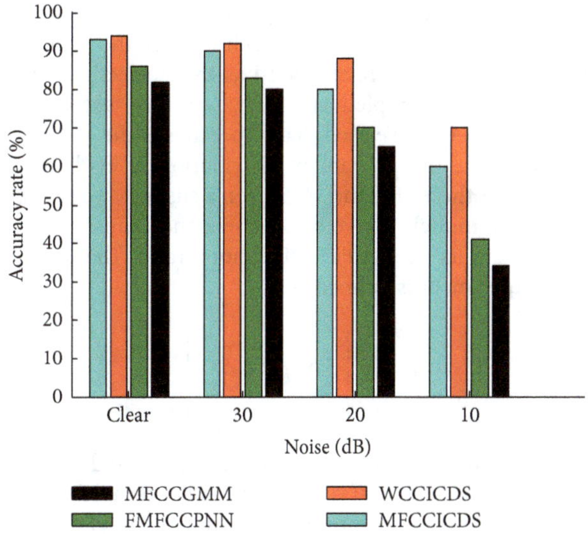

Figure 4: The accuracy rates of speaker recognition in noisy environment.

In forensic context, the ambient noise cannot be controlled. In particular, the questioned speech is usually recorded in the condition that is full of noise. We therefore carried out an experiment to evaluate the performance of our model in noisy environment. We added 30 dB, 20 dB, and 10 dB noises to the speeches used in the above experiment. All noise was generated by the MATLAB Gaussian white noise function. We repeated the above experiment and the accuracy rates were shown in Figure 4. We could find that the accuracy rates of the four models decreased by less than 3% when the noise is 30 dB. This shows that the weak noise could not interfere in those models. However, if we enhanced the noise, we found that the accuracy rates of MFCCGMM, FMFCCPNN, and MFCCICDS dropped off rapidly. When the noise increased to 10 dB, the accuracy of the three models decreased by about 48.7%, 45.2%, and 33.4%, respectively. This shows that the two models were susceptible to noise. Compared with the above two models, the proposed model, named WCCICDS, performed better. When the 10 dB noise was added to those speeches, the accuracy of the model decreased by about 24.6%. This shows that the model using WCC was more robust against noise.

5.4. The Performance of Evidence Reporting. This experiment evaluated the performance of speaker recognition model when it was used for evidence reporting. In the evidence reporting, the suspect had been found and police used the speaker recognition model to give out the evidence that is defined as the degree of similarity between the suspect speech and the questioned speech. In this case, the speaker recognition model was required to report the strength of the evidence as a likelihood ratio. In this experiment, we also used the above 384 speakers. For each speaker, the 7 speeches sampled at 16 KHz were also used as the suspect's speeches and the speech sampled at 8 KHz was also used as the questioned speech. We used all the speeches of the 384 speakers to estimate the probability distribution of evidence for the two hypotheses H_0 and H_1. Given a suspect, we required the speaker model to give out the evidence and report the evidence and its strength as a likelihood ratio. The evidence in this experiment is defined as

$$E = T\left[\frac{1}{7}\sum_{j=1}^{7} K\left(x, y_j\right)\right], \quad (17)$$

where x and y are the questioned speech and suspect's speeches, respectively. $K(x, y_j)$ was the CDS kernel defined in (7) and $T(\cdot)$ was a transform function defined in (9).

This section used Tippett plots [36] to evaluate the performance of our model. It was originally used for the forensic DNA analysis and then was used for evaluating the forensic speaker recognition model. Assume that there were m questioned speeches and M suspects that had been found by police. The speaker recognition model reported the likelihood ratio (LR) for each of questioned speeches and each suspect, so we obtained $m \times M$ LRs. The Tippett plot was defined as

$$T(t) = \frac{n}{N} \times 100\%, \quad 0.1 < t < 10, \quad (18)$$

where $T(\cdot)$ was the Tippett plot; n was the number of LRs that were greater than the threshold t. $N = m \times M$ was the total number of LRs. We varied the threshold t from 0.1 to 1 to obtain different Tippett plot. To evaluate the performance of our model, we calculated two types of Tippett plot. The first one was calculated in the assumption that the questioned speech and the suspect speech were spoken by same speaker and we call it T_1. The second one was calculated in the assumption that the questioned speech and the suspect speech were spoken by different speakers and we called it T_2. LR presented the strength of the evidence that the suspect was the criminal. For good speaker recognition model, the reported LR would be very high if the questioned speech and the suspect's speeches were spoken by same speaker. On the other hand, the LR would be very low if the questioned speech and the suspect's speeches were spoken by different speakers. In other words, the separation between the two types of Tippett plots is an indication of the performance of the model [37]. Given small t, a larger separation implied better performance than a smaller one. The results of this experiment were shown in Figure 5.

In Figure 5, the two types of Tippett plots of WCCICDS separated from each other when t is less than 0.5 and greater than 0.1, but the two Tippett plots of MFCCICDS stuck together. The two curves of MFCCICDS separated from each other until the threshold increased to about 1. This shows that the separation between the two types of Tippett plots of WCCICDS was slightly larger than MFCCICDS', so the WCCICDS performed better than the MFCCICDS. When

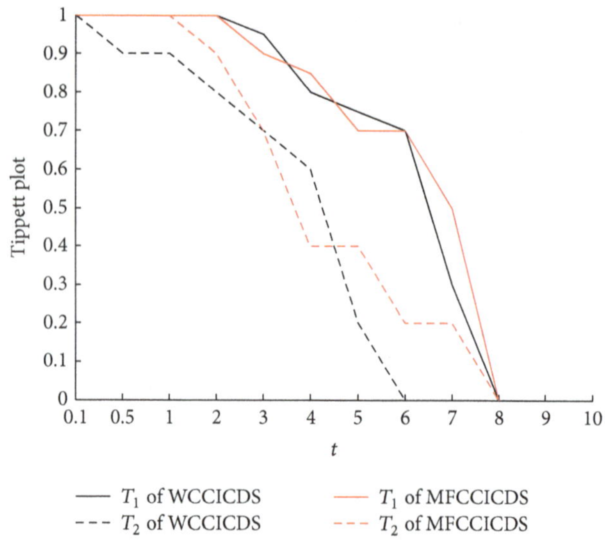

FIGURE 5: Tippett plot curve of the speaker recognition models.

the threshold t increased to 8, all plots decreased to 0. This was because of the fact that the threshold became too high and no LR was greater than it.

5.5. The Time Cost of Speaker Recognition Models. I-vector in many cases can improve the accuracy of recognition at cost of increasing the computational complexity, so in this experiment we test the time cost of the four speaker recognition models, MFCCGMM, FMCCPNN, MFCCICDS, and our WCCICDS. We used 200 5-second-long speeches to test their time cost and calculated the average time cost. The time cost of inputting those speeches was discarded. The result was shown in Table 4.

In Table 4, MFCCGMM and FMFCCPNN did not employ i-vector for speech representation, so they did not cost time to train i-vector. FMFCCPNN, MFCCICDS, and WCCICDS used the PNN and CDS for speaker classification. Because the PNN and CDS were the unsupervised learning models, the three models need not cost time during the training step of speaker classification. The time cost of short-term feature extraction of the proposed WCCICDS was the highest. This was because this model used WCC to calculate 126 cepstral coefficients from 9 local frequency domains compared with other three models that used MFCC or FMFCC to calculate 14 cepstral coefficients from a global frequency domain. Training i-vector was also a time-consuming process, so i-vector-based model costs more time than the model that did not employ i-vector. In all, WCC and i-vector could slightly improve the performance of the speaker recognition model at cost of increasing the time cost, so selection of speaker recognition model was the process that found the balance between the performance and time cost.

Parallel computation was an effective way to reduce the time cost, because many loops in the linear computation could be finished at once using a parallel algorithm. For example, we used DWT to decompose a signal at M levels. In the linear algorithm, we had to run a filtering process

TABLE 4: Average time cost of the speaker recognition models.

Model	Feature extraction (s/speech)		Speaker classification (s/speech)	
	Short-term feature extraction	I-vector training	Training	Recognition
MFCCGMM	0.45	—	1.92	0.81
FMFCCPNN	1.29	—	—	0.83
MFCCICDS	0.51	2.21	—	0.85
WCCICDS	2.91	2.19	—	0.89

whose time cost was $O(\log N)$ N times for each level, so the time complexity of DWT was $O(MN \log N)$. However, if we used a parallel algorithm to implement the DWT, we could use N independent cores to compute N filtering processes at once, and therefore the time cost reduced to $O(M \log N)$. In a further study, the parallel computation may be used to reduce the time cost of the proposed model.

6. Conclusions

In the forensic context, the speaker recognition model is usually used for investigation and evidence reporting. In the investigation, police assume that the real criminal is in a large set of known criminals and the speaker recognition model is used to produce a small list of potential suspects from a large set of known criminals. In the evidence reporting, the suspect is found, and the speaker recognition model is used to report the evidence that supports the fact the suspect was the real criminal. In this case, speaker recognition model also should report the strength of evidence as a likelihood ratio.

The forensic scene is very challenging for speaker recognition, because the ambient noise cannot be controlled and the much-change speech data are usually obtained from different channels. In this paper we propose a new speaker recognition model based on WCC, i-vector, and CDS. WCC has good performance on antinoise, because the DWT employed by WCC is a local analysis approach that can prevent the noise interfering in whole frequency domain. I-vector is a robust way to represent a speech utterance using a signal i-vector, so it can model the much-change speech data effectively. CDS employs the LDA and WCCN to compensate the channel to deal with the channel variability problem. Our experiments simulate the investigation and evidence reporting tasks. The result of our experiments shows that the proposed WCCICDS obtained high accuracy rate in the investigation task and also obtained good performance in the evidence reporting task, but its time cost was higher compared to other models. The result also shows that the parallel algorithm could effectively reduce the time cost of the model based on i-vector.

In the future, we will use the parallel algorithm to reduce the time cost of the proposed model. Moreover, we will combine audio and visual features to improve the performance of the forensic speaker recognition system.

Competing Interests

The authors declare that they have no competing interests.

Acknowledgments

The authors thank Professor Kun She for assistance in preparation of the manuscript and acknowledge support by Key Technology Support Program of Sichuan Province (Grant no. 2015GZ0102).

References

[1] M. Mandasari, M. McLaren, and D. A. V. Leeuwen, "Evaluation of I-vector speaker recognition systems for forensic application," in *Proceedings of the INTERSPEECH*, pp. 21–24, August 2011.

[2] A. Eriksson, "Aural acoustic vs. Automatic methods in forensic phonetic case work," in *Forensic Speaker Recognition*, pp. 41–69, 2012.

[3] A. Drygajlo, "From speaker recognition to forensic speaker recognition," *Biomedical Authentication*, vol. 8897, pp. 93–104, 2014.

[4] J. R. Conzalez, A. Drygajilo, D. R. Castro, and M. G. Garcia, "Robust estimation interpretation and assessment of likelihood ratios in forensic speaker," *Computer Speech & Language*, vol. 20, pp. 331–355, 2006.

[5] T. Kinnunen and H. Li, "An overview of text-independent speaker recognition: from features to supervectors," *Speech Communication*, vol. 52, no. 1, pp. 12–40, 2010.

[6] C. Champod and D. Meuwly, "Inference of identity in forensic speaker recognition," *Speech Communication*, vol. 31, no. 2, pp. 193–203, 2000.

[7] M. A. Silveria, C. P. Schroeder, J. P. C. L. da Costa et al., "Convolutive ica-based forensic speaker identification using mel frequency cepstral coefficients and gaussian mixture model," *The International Journal of Forensic Computer*, vol. 1, pp. 27–34, 2013.

[8] S. Srivastava, S. Bhardwaj, and A. Bhandari, "Wavelet packet based mel frequency cepstral features for text independent speaker identification," *Advances in Intelligent Systems and Computing*, vol. 182, pp. 237–247, 2013.

[9] P. Kenny, G. Boulianne, and P. Dumouchel, "Eigenvoice modeling with sparse training data," *IEEE Transactions on Speech and Audio Processing*, vol. 13, no. 3, pp. 345–354, 2005.

[10] D. A. Leeuwen and N. Brümmer, "The distribution of calibrated likelihood-ratios in speaker recognition," in *Proceedings of the*

14th Annual Conference of the International Speech Communication Association (INTERSPEECH '13), pp. 1619–1623, Lyon, France, August 2013.

[11] N. Dehak, R. Dehak, and P. Keney, "Support vector machines versus fast scoring in the low-dimensional total variability space for speaker verification," in *Proceedings of the 10th Annual Conference of the International Speech Communication Association (INTERSPEECH '09)*, vol. 9, pp. 1559–1562, September 2009.

[12] H. Arsikere, S. M. Lulich, and A. Alwan, "Estimating speaker height and subglottal resonances using MFCCs and GMMs," *IEEE Signal Processing Letters*, vol. 21, no. 2, pp. 159–162, 2014.

[13] T. Becker, M. Jessen, and C. Grigoras, "Forensic speaker verification using formant features and Gaussian mixture models," in *Proceedings of the 9th Annual Conference of the International Speech Communication Association (INTERSPEECH '08)*, pp. 1505–1508, September 2008.

[14] C. Bouveyron and C. Brunet-Saumard, "Model-based clustering of high-dimensional data: a review," *Computational Statistics & Data Analysis*, vol. 71, pp. 52–78, 2014.

[15] K. K. George, C. S. Kumar, K. I. Ramachandran, and A. Panda, "Cosine distance features for robust speaker verification," in *Proceedings of the 16th Annual Conference of the International Speech Communication Association (INTERSPEECH '15)*, pp. 234–238, Dresden, Germany, September 2015.

[16] M. Li, A. Tsiartas, M. Van Segbroeck, and S. S. Narayanan, "Speaker verification using simplified and supervised i-vector modeling," in *Proceedings of the 38th IEEE International Conference on Acoustics, Speech, and Signal Processing (ICASSP '13)*, pp. 7199–7203, Vancouver, Canada, May 2013.

[17] A. Kanagasundaram, D. Dean, R. Vogt, M. McLaren, S. Sridharan, and M. Mason, "Weighted LDA techniques for I-vector based speaker verification," in *Proceedings of the IEEE International Conference on Acoustics, Speech, and Signal Processing (ICASSP '12)*, pp. 4781–4784, Kyoto, Japan, March 2012.

[18] J. P. Campbell, W. Shen, W. M. Campbell, R. Schwartz, J.-F. Bonastre, and D. Matrouf, "Forensic speaker recognition," *IEEE Signal Processing Magazine*, vol. 26, no. 2, pp. 95–103, 2009.

[19] S. K. Kopparapu and M. Laxminarayana, "Choice of Mel filter bank in computing MFCC of a resampled speech," in *Proceedings of the 10th International Conference on Information Sciences, Signal Processing and Their Applications (ISSPA '10)*, pp. 121–124, Kuala Lumpur, Malaysia, May 2010.

[20] K. S. Ahmad, K. J. Thosar, A. S. Nirmal, and J. H. Pande, "A unique approach in text-independent speaker recognition using MFCC feature sets and probabilistic neural network," in *Proceedings of the 8th International Conference on Advances in Pattern Recognition*, pp. 1–6, Kolkata, India, January 2015.

[21] N. Lulla and N. Purohit, "An improved algorithm for efficient computation of MFCC," in *Proceedings of the 11th IEEE India Conference (INDICON '14)*, pp. 1–4, Pune, India, December 2014.

[22] X.-Y. Zhang, J. Bai, and W.-Z. Liang, "The speech recognition system based on bark wavelet MFCC," in *Proceedings of the 8th International Conference on Signal Processing (ICSP '06)*, pp. 16–20, Beijing, China, November 2006.

[23] D. A. Reynolds and R. C. Rose, "Robust text-independent speaker identification using Gaussian mixture speaker models," *IEEE Transactions on Speech and Audio Processing*, vol. 3, no. 1, pp. 72–83, 1995.

[24] Y. Zhao, L. Zhang, J. Hu, and T. Liao, "Mallat wavelet filter coefficient calculation," in *Proceedings of the 5th International Conference on Computational and Information Sciences (ICCIS '13)*, pp. 963–965, Chengdu, China, June 2013.

[25] K. Daqrouq, "Wavelet entropy and neural network for text-independent speaker identification," *Engineering Applications of Artificial Intelligence*, vol. 24, no. 5, pp. 796–802, 2011.

[26] Y. Lei, L. Burget, and N. Scheffer, "A noise robust I-vector extractor using vector taylor series for speaker recognition," in *Proceedings of the 38th IEEE International Conference on Acoustics, Speech, and Signal Processing (ICASSP '13)*, pp. 6788–6791, Vancouver, Canada, May 2013.

[27] M. McLaren and D. Van Leeuwen, "Improved speaker recognition when using i-vectors from multiple speech sources," in *Proceedings of the 36th IEEE International Conference on Acoustics, Speech, and Signal Processing (ICASSP '11)*, pp. 5460–5463, May 2011.

[28] A. Alexander, D. Dessimoz, F. Botti, and A. Drygajlo, "Aural and automatic forensic speaker recognition in mismatched conditions," *International Journal of Speech, Language and the Law*, vol. 12, no. 2, pp. 214–234, 2005.

[29] A. Biswas, P. K. Sahu, A. Bhowmick, and M. Chandra, "Feature extraction technique using ERB like wavelet sub-band periodic and aperiodic decomposition for TIMIT phoneme recognition," *International Journal of Speech Technology*, vol. 17, no. 4, pp. 389–399, 2014.

[30] S. Mallat, *A Wavelet Tour of Signal Processing*, Elsevier, 2012.

[31] I. Deubechies, *Ten Lectures on Wavelets*, Society for Industrial and Applied Mathematics, 1992.

[32] D. B. Percival and A. T. Walden, *Wavelet Methods for Time Series Analysis*, vol. 4, Cambridge University Press, 2006.

[33] I. S. Prabha, M. K. Rao, and B. Manjusha, "Voice verification system using symlet wavelet," *International Joural of Research in Computer Applications and Robotics*, vol. 1, pp. 24–30, 2013.

[34] B. Saha and K. Kamarauslas, "Evaluation of effectiveness of different methods in speaker recognition," *Elektronika ir Elektrochnika*, vol. 98, pp. 67–70, 2015.

[35] K. Daqrouq and T. A. Tutunji, "Speaker identification using vowels features through a combined method of formants, wavelets, and neural network classifiers," *Applied Soft Computing Journal*, vol. 27, pp. 231–239, 2015.

[36] G. A. Morrison, "Vowel inherent spectral change in forensic voice comparison," in *Vowel Inherent Spectral Change*, pp. 263–282, Springer, 2013.

[37] G. Zadora, A. Martyna, and D. Ramos, *Statistical Analysis in Forensic Science: Evidential Value of Multivariate Physicochemical Data*, John Wiley & Sons, New York, NY, USA, 2013.

9

An Experimental Study and Concept Evaluation on Tree-Interior Imaging Radar using Sinusoidal Template-Based Focusing Algorithm

Betül Yılmaz, Serhat Gökkan, and Caner Özdemir

Department of Electrical and Electronics Engineering, Mersin University, Ciftlikkoy 33343, Mersin, Turkey

Correspondence should be addressed to Serhat Gökkan; sgokkan@mersin.edu.tr

Academic Editor: Ping Feng Pai

An algorithm for detecting cavities inside a tree-body is presented with simulation and measured examples. The details of the imaging algorithm that is based on sinusoidal template focusing routine are given. First, the algorithm is tested with the simulation scenario for which perfect reconstruction of the simulated cavity structure together with tree-body is successfully formed in MATLAB programming environment. Then, the algorithm is applied to the measurement data that have been collected from a laboratory set-up. Collected backscattered measurements from the tree-body (with cavity) structure are used to generate the image of the scene by the help of our proposed algorithm. The resultant radar images of the measured data collected from the laboratory arrangement have shown the applicability of the developed algorithm for the detection of cavity structures inside tree-bodies.

1. Introduction

The detection of voids or cavities within tree-bodies at microwave band has become an emerging technology for the determination of unhealthy and weak trees [1, 2]. Every year tens of people are being killed all around the world due to falling tree incidents and related incidents [1–4]. Although most of the trees seem to be healthy and strong when looking from the outside, it is not easy to realize the existence of interior cavities that might occur due to fungus and worm attacks by using conventional methods such as acoustic imaging techniques [5], resistivity measurement methods [6], and thermographic inspection [7]. It is known by the experts that decaying wood is expected to have a different moisture content and density; therefore, this part of the tree-body should have different electrical characteristics such as dielectric constant and conductivity [8]. One of the most applied methods is the ultrasonic tomography that can sense decay structures only under special circumstances [9]. This technique uses the mechanical behavior of wood that usually depicts a strong anisotropy such that ultrasonic pulses are being used to sense this discontinuity caused by the cavity. On the other hand, ultrasonic pulses in wood are usually strongly attenuated; therefore, this method requires high signal sensitivity and dynamic range while acquiring and processing the raw data [10, 11]. Recently, using obstacle-penetrating radar technologies to detect and image cavities within a tree-body has been increasingly gaining attention thanks to recent advances in radar hardware and radar signal processing techniques [12–15]. It is eminent by the electromagnetic (EM) community that microwave imaging techniques have the advantages of providing high resolution features and requiring relatively smaller amount of power to penetrate such obstacles. Algorithms used in ground-penetrating-radar (GPR) [16–18] and through-wall-imaging-radar (TWIR) [19–21] seem to provide promising solutions to this special problem. Also, circular Radon transform could be an opportunity to focus the scattered electric field data collected from a circular measurement set-up for a different application [22].

In this work, we offer a practical but also a fast focusing/migration method that can be readily used for tree-interior imaging radar (TIIR) applications. We are presenting a conceptual study on TIIR by suggesting an imaging/focusing method that has been specially tailored to

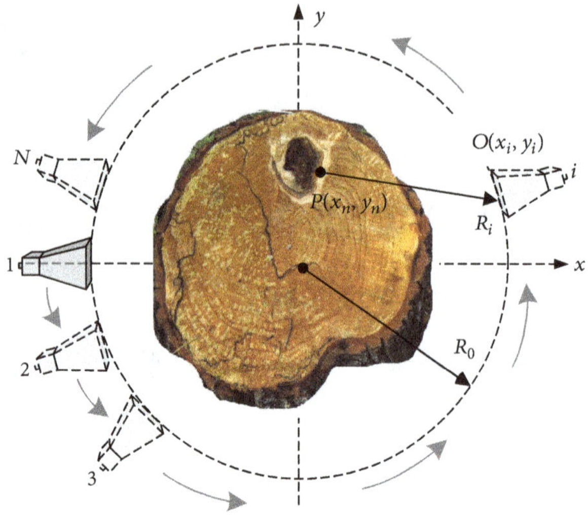

FIGURE 1: The geometry of TIIR problem.

this specific application and also demonstrating a proof-of-concept study by providing and assessing the results of related simulations and measurements from a test-bed.

The paper is organized as follows: in the next section, we present the steps and the details of our imaging algorithm for the TIIR problem. In the third section, the validity of the proposed method is first tested with the help of simulated data that are generated with point scatterer model. In the following section, a measurement set-up for TIIR experiments is constructed and both the simulation of the experimental set-up and measurements conducted with real tree-body with cavity are presented. The results after applying the proposed methods are provided. Resultant radar images of both the simulation and the measurement data validate the effectiveness and the success of our suggested algorithm. In the final section, the work is concluded by discussing the applicability and assessment of the measurement system with the proposed algorithm.

2. Imaging/Focusing Algorithm for TIPR

The geometry of our TIIR problem can be described as illustrated in Figure 1 where the tree-body is circularly scanned by a directive antenna for a total of M distinct look-aspects covering the whole azimuth angles. The radius of the scanning circle is R_0. The frequency-diverse backscattered electric field data are collected in the monostatic configuration of the radar. For any point scatterer, $P(x_n, y_n)$ was assumed to be on the perimeter of the cavity within the radius R_0; therefore, one-dimensional (1D) electric field data can be obtained by the antenna of the i_{th} position as

$$\mathbf{E}_\mathbf{i}^s(\mathbf{k}) = A_i \exp(-j2\mathbf{k}r_i). \tag{1}$$

Here, $r_i = \text{sqrt}[(x_n - x_i)^2 + (y_n - y_i)^2]$ is the distance between the scatterer and the antenna and $\mathbf{k}$ is the wavenumber vector. This result is valid for the following assumptions: first, we assume isotropic radiation of the electromagnetic (EM) wave inside the tree-body such that the EM wave travels along its propagation direction. This assumption is usually true for the decayed trees since such trees usually dry in the inside due to decaying phenomenon. It is reported that the dielectric constant of dried wood material is on the order of 1.4 to 2.9 in practice [23]. Therefore, the moisture content for the decayed trees is usually very low and does not present significant anisotropy in terms of the water content of the tree-body. This situation also helps radar signal to penetrate the tree-body better since the relative electric permittivity contrast between the air and the tree structure is not high. Secondly, the scattering structures within the tree-body are assumed to be represented with point scatterer model that is very effective and commonly used in many radar applications [24]. Therefore, near field boundary conditions for the air void and the tree-body are not imposed. Although such modelling will provide a more realistic representation, the surface modelling of air void would be quite complex and yield a very complex calculation of the EM reflection. Therefore, we have encountered a more basic and simplified EM scattering calculation model to ease the numerical load of the problem. Furthermore, such kind of study is out of this research work. Yet, modelling boundary points such as surface of cavities with perfect point scatters has proven to be an effective way of calculating the scattering from such boundaries in many radar problems [25, 26]. Thirdly, any multipath phenomenon is not accounted as all radar imaging algorithms are based on only single-reflection assumption.

Based on the above assumptions, then, (1) can be rewritten as a summation over scatterer and provides a two-dimensional (2D) data by considering different look angles as the following for a finite number of point scatterers (M) that form the tree-body:

$$\mathbf{E}^s(\mathbf{k}, \phi) = \sum_{i=1}^{M} A_i \exp(-j2\mathbf{k}\mathbf{R}_\mathbf{i}), \tag{2}$$

where $\mathbf{R}_\mathbf{i} = \text{sqrt}[(\mathbf{X} - x_i)^2 + (\mathbf{Y} - y_i)^2]$ is the distance vector between the scatterer and all the positions of radar for whole azimuth observation angles of ϕ ranging from 1 to M and $\mathbf{k}$ is the wave number for the stepped frequencies for a total of N distinct frequencies. Then, $\mathbf{X}$ and $\mathbf{Y}$ vectors correspond to Cartesian coordinates of M antenna positions, accordingly as given below:

$$\begin{aligned} \mathbf{X} &= R_0 \cos\phi, \\ \mathbf{Y} &= R_0 \sin\phi. \end{aligned} \tag{3}$$

Taking the one-dimensional (1D) inverse Fourier transform (IFT) of (2) along the spatial frequency axis, one can easily get the 2D range-angle data, $E^s(\mathbf{r}, \phi)$. In this resultant range-angle data, any point $P(x_n, y_n)$ exhibits a sinusoidal behavior for the 360° variation of azimuth angle ϕ because of circular scanning arrangement.

Our detection and imaging algorithm can be briefly summarized as follows: (i) after obtaining the 2D range-angle backscattered electric field data, $\mathbf{E}^s(\mathbf{r}, \phi)$, we select an image window in 2D (range, angle) domain that covers the whole geometry. It is important to note that the range-extent of the

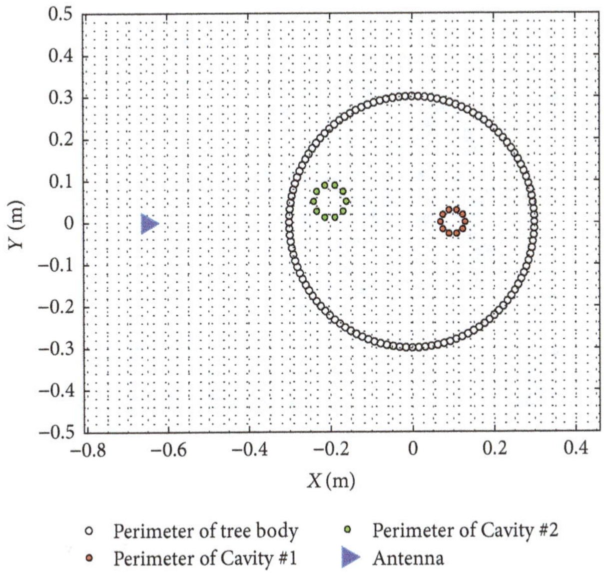

FIGURE 2: Simulation geometry for the TIIR algorithm.

image window should be larger than or at least equal to the diameter of scanning, that is, $2R_0$. (ii) Next, we iteratively pick all the points; $P(x_p, y_p)$ within the selected window. (iii) Then, we form a template vector for each look angle of ϕ_i by writing the following template equation:

$$\mathbf{R}^{\mathbf{n}}_{\text{temp}}(\phi_{\mathbf{n}}) = \text{sqrt}\left[\left(\mathbf{X} - x_p\right)^2 + \left(\mathbf{Y} - y_p\right)^2\right]. \quad (4)$$

(iv) Afterwards, we go back to 2D $E^s(\mathbf{r}, \phi)$ data and pick the data under $(\mathbf{R}^{\mathbf{n}}_{\text{temp}}, \phi_{\mathbf{n}})$ points. At this point, a precise data interpolation scheme needs to be employed to lessen the error associated with data interpolation. (v) In the final step, we add all the data under this template vector and record the result in a new 2D image matrix at (x_p, y_p) location. The new, reconstructed image data at (x_p, y_p) location becomes a pixel in the final image. The algorithm successively continues until all the points in the selected image window are passed through the algorithm. The resultant $\mathbf{I}(x, y)$ matrix is the focused image.

3. Simulation Results

The above-explained algorithm was first tested by a simulation scenario whose illustration is given in Figure 2. The simulation scene was constructed in MATLAB programming environment [27] with the assumptions that were listed in the previous section. To be able to test and evaluate the performance of this technique with ideal conditions, perimeter of the tree-body (shown as black dots) was assumed to be composed of a total of *60* perfect point scatterers with the identical reflectivity amplitudes of *"1,"* that is, independent of frequency and aspect. The radius of the tree-body is taken as *60 cm*. In the simulation, there were two cavity regions; one was centered at (10 cm, 0) while the other one was located at (−20 cm, 5 cm). Both cavities were considered to be in the form of circles with diameters of 8 cm and 6 cm for the first and second void regions, respectively. The circumference of the cavities was also assumed to be composed of perfect point scatterers with equal EM reflectivities of *"0.5"* that are independent of frequency and angle. Dielectric constant for the tree-body was taken as 1.65 and assumed to be homogeneous within the tree-body. The radar antenna was positioned at 35 cm away from the tree's surface and moved along a circular track for a total of 360 distinct look angles to collect the backscattered data. For each of the look angles, frequency-diverse backscattered electric field was gathered for the stepped frequencies ranging from 1 to 8 GHz for the total of 100 discrete frequency points. Therefore, the backscattered electric field data were collected on the 2D frequency-angle plane such that a matrix of 100 × 360 points was obtained.

We apply the steps of our focusing algorithm that was based on summation of the data beneath the sinusoidal template. In Figure 3(a), the range profile matrix for different look angles, $\mathbf{E}^s(\mathbf{r}, \phi)$, was obtained. We have several observations about the range-angle raw data in Figure 3(a) as follows: (i) the scattering from the nearest point from the circular tree-body geometry shows up as a straight line since both the scanning path and the tree-body are in the form of circles. This line occurred at a range distance of 35 cm as expected. (ii) Scatterers on the perimeter of both cavities come out as sinusoids of different oscillating amplitudes due to different round-trip distances between the antenna and the scatterers due to circular movement of the radar while collecting the data. In fact, the peak amplitudes of these oscillating sinusoids depend on their distances from the center of rotation. For our simulation example, these numbers are 4 cm and 3 cm for the first and the second cavity, respectively. The range-angle image in Figure 3(a) exactly exhibits this behavior as we have anticipated. (iii) The widths of the sinusoids represent the diameters of the cavities. From the figure, the widths of the sinusoidal behavior are 8 cm and 6 cm that are precisely the same values of real diameters for the cavity regions. (iv) The starting point of the sinusoidal behavior actually pinpoints the exact location of the cavity: for example, if the sinusoids start from the maximum value, the cavity is located at the nearest point to the antenna's location. If the sinusoid starts from the minimum value, then, the cavity is at 180° away from the antenna. (v) Center point of the scanning will show up as a straight line in the range-angle image. Therefore, if there is a small cavity at the center, its image will be a horizontal line at the half wave between the scatterings from near and far points of the tree-body. (vi) Second straight line around 95 cm in range-angle image in Figure 3(a) corresponds to scattering from far-most point of tree-body. As the radar moves around a circle, the distance from the far-most point of the tree-body stays constant at 95 cm as it was obtained in Figure 3(a). After analyzing the range-angle image in Figure 3(a) that certainly tells us many facts, we applied the steps of the algorithm such that the final focused image $\mathbf{I}(x, y)$ was obtained as given in Figure 3(b). As it is clearly seen from the figure, both the tree-body and the cavity regions were successfully imaged in the resultant focused image at their correct locations in the Cartesian coordinates. Of course, the image strength of cavities is much

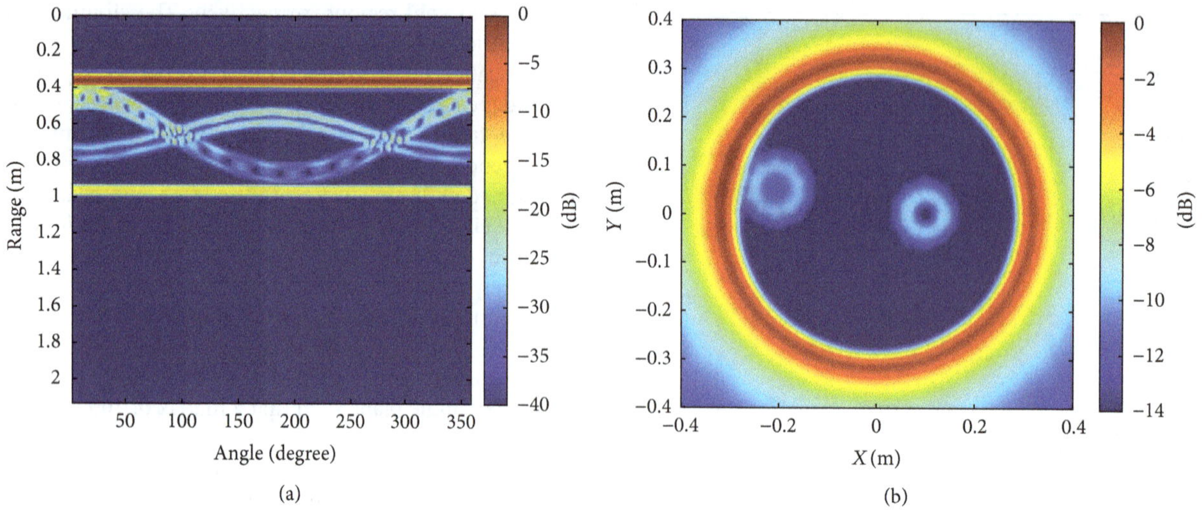

FIGURE 3: Simulation results for the geometry in Figure 2: (a) range-angle image; $\mathbf{E}^s(r, \phi)$, (b) focused image; $\mathbf{I}(x, y)$ after applying our algorithm.

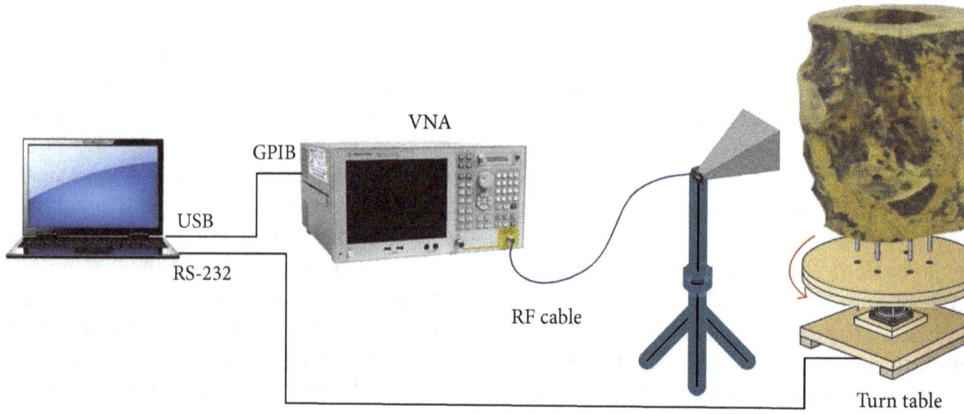

FIGURE 4: The measurement set-up for the TIIR experiment.

smaller than that of tree's surface due to decaying of EM wave with $1/(4\pi R^2)$ term.

The fidelity of the resultant image in our simulation is compared to other methods such as Radon transform [21] and ultrasonic techniques are as follows [9]: focused image can be obtained from Radon transform simulation method that uses projection data associated with cross-sections of an object but this method requires complex calculations for obtaining the final image. Therefore, its implementation is more difficult and takes much more time. Ultrasonic simulation method also can be used for imaging but the disadvantage of this method is the weakness of the ultrasonic signals to be able to penetrate the tree-body. To ensure the required minimum detectable signal level, many receivers are usually required along the perimeter of tree-body. Our method is more practical and has faster migration method to be applied for TIIR applications.

4. Measurement Set-Up and Experimental Results

To assess the effectiveness of our TIIR detection/imaging algorithm with real experimental data, we have constructed a measurement test-bed at our anechoic chamber of Mersin University's Advanced Technologies Research Center (MEATRC) facility. The geometry of the measurement set-up is shown in Figure 4. In this set-up, a Vector Network Analyzer (VNA) [Agilent ENA5071B] was utilized to generate/transmit/receive the radar signal by using an ultra-wide band (UWB) double-ridged horn antenna [Geozondas GZ0126DRH] that was used in the monostatic configuration. The antenna frequency range is 1–26 GHz. Beam width of the antenna at E plane is 37.7° at 1 GHz and 22.4° 8 GHz. A turn table with automation software was employed to get different look angle profiles of the tree-body. Also, a PC has to be used

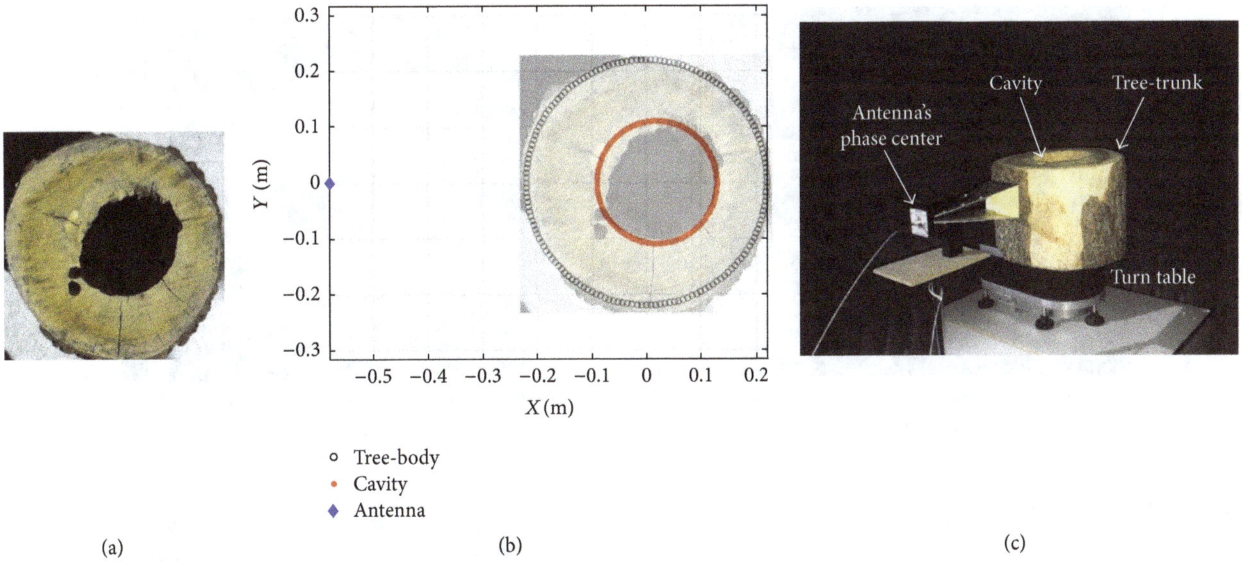

FIGURE 5: (a) Picture of the piece of pine tree trunk, (b) simulation scenario for the experiment, and (c) picture from the real TIIR experiment.

to synchronously control the VNA and the turn table with a MATLAB [27] script.

For the presented configuration in Figure 4, a piece of pine tree trunk with outer-diameter of approximately 42 cm was used as the target to be imaged. The tree has a cavity whose center is about 2 cm away from the center of the tree with a radius of about 11 cm as pictured in Figure 5(a). Before starting the experiment, we first constructed an approximate 2D simulation of the measurement that we were going to do to visualize the results out of the experiment and to compare it to real experiment that we were going to conduct later on. In Figure 5(b), the simulation for the experiment was constructed again in MATLAB. Our antenna is assumed to be 58 cm away from the center of the tree or 36 cm away from the body of the tree trunk. The geometry for the tree trunk and the cavity was constructed in accordance with the real values as illustrated in Figure 5(b).

For the simulation, the frequency is altered from 1 to 8 GHz for a total of 100 discrete frequencies while the look angle was also varied from 1° to 360° for a total of 361 distinct aspects. The tree-body's frequency-independent reflectivity amplitude was chosen to be "1" while that of cavity was selected as "0.1" to emulate the attenuation of EM wave while travelling inside the tree-body. The 2D range-angle data after running the simulation is presented in Figure 6(a) where we can observe different scattering mechanisms easily: the reflection from the tree-body showed up as a straight line around 36 cm from the antenna as expected. The front point and the back point of the cavity experienced two sinusoids as the look angle varies along the whole azimuth. Of course, the reflection magnitude for the back point of the cavity is less than that of the front point of the cavity due to attenuation of the EM wave with distance. After applying our sinusoidal template-based focusing algorithm, the TIIR image of this simulation for the experiment was gathered as plotted in Figure 6(b). By comparing this figure with the simulation geometry in Figure 5(b), one can easily see that the suggested algorithm successfully forms the TIIR image of the scenario with good fidelity.

As the final study, we have conducted a real through-the-tree-imaging-radar experiment as pictured in Figure 5(c). The experiment was performed with the measurement test-bed at our anechoic chamber of MEATRC facility. During this experiment, the tree trunk whose picture is seen in Figure 5(a) was put on the turn table. The antenna in monostatic configuration was positioned as seen in Figure 5(c) such that the phase center of the measurement set-up is about 58 cm away from the scanning axis of the tree trunk just as similar to the simulation. By using the automation software that was adopted by our MATLAB script, turn table has been turned for the full azimuth coverage of 360° at 1° increments. For each look angle measurement, the frequency of the VNA was changed from 1 to 8 GHz for a total of 100 discrete frequency points as in the case of the simulation. Therefore, a 2D scattered electric field matrix, $\mathbf{E}^s(k,\phi)$, of 100 × 360 measurement points was collected. Theoretical achievable resolution is calculated as $\Delta r = c/(2 * B * \sqrt{\epsilon_r}) = 1.66\,\text{cm}$ for a bandwidth of $B = 7\,\text{GHz}$ and $\epsilon_r \cong 1.65$ (relative electric permittivity of tree trunk). Maximum range is $R_{\max} = N * \Delta r = 166\,\text{cm}$ for number of the discrete frequency points $N = 100$.

After applying the IFT operation along the frequency axis, 2D range-aspect data, $\mathbf{E}^s(r,\phi)$, were obtained as plotted in Figure 6(c). We can notice the scattering from the tree's body as a wavy line around 36 to 40 cm since the perimeter of the tree was not a perfect circle. We also notice the scattering from cavity that experiences the same behavior of the tree's body as the look angle varies during the measurement. Although the attenuation of the EM wave inside the tree texture looks more intense than we have anticipated in the simulation, we can still sense the disturbance of the scattering energy due to existence of the cavity around 50 to 70 cm in range direction. Next,

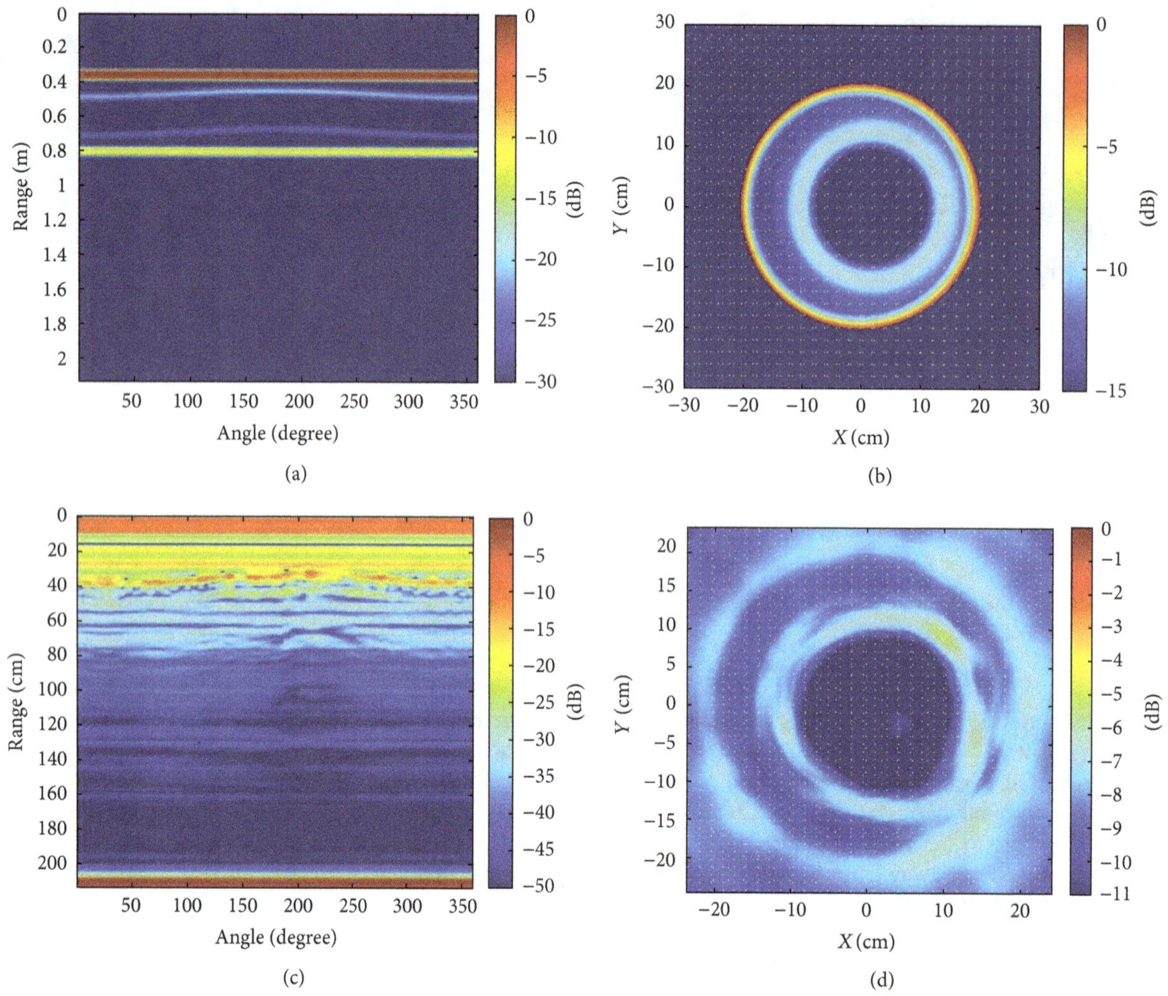

FIGURE 6: Simulation and measured results for the experiment: (a) simulated range-angle data, (b) simulated TIIR image, (c) measured range-angle data, and (d) measured TIIR image.

we have employed our sinusoidal template based focusing algorithm to migrate the data from range-aspect domain to the image domain such that we have obtained the final TIIR image of the tree trunk as depicted in Figure 6(d). Thanks to the focusing ability of our algorithm, the reflections from the tree-body and the cavity's perimeter have been amplified. Therefore, we can easily notice the borders of impedance changes from air to tree-body and tree-body to cavity and vice versa. Visually comparing the results in Figure 6(d) [measurement] and Figure 6(b) [simulation] with the real picture of tree trunk in Figure 5(a), one can clearly realize that resultant TIIR images successfully form the geometry of tree-body and the cavity around their correct locations such that the cavity inside the tree trunk is detected.

5. Conclusion

In this work, we have presented our recent work on tree-interior imaging radar research. On this scope, we have introduced a new algorithm based on summation of raw data over sinusoidal templates to be effectively used for TIIR or similar applications. The algorithm is specially developed for multiangle multifrequency monostatic radar set-up with circular scanning geometry of object under test. The details of the algorithm were given together with the assumptions. Then, the algorithm is first tested with simulation data and then a real experiment was conducted at Mersin University's MEATRC facility. Both the simulated and the measured results of TIIR images demonstrate the validity and the success of the proposed algorithm in detecting and localizing the cavity region inside the tree-body.

Algorithm can be modified to include the effect caused by change in the velocity of the EM waves inside the tree-body as a future study. This algorithm can also be applied to similar problems such as detecting and imaging breast tumors by employing a very similar geometry of data collection.

Conflicts of Interest

The authors declare that they have no conflicts of interest.

Acknowledgments

This work has been supported by the Scientific and Research Council of Turkey (TUBITAK) under 2211-D Ph.D. scholarship programme and funded by Mersin University Scientific Research Project Unit with Grant no. BAP 2015-TP3-1158. The authors are also thankful to Mersin University MEATRC management for their help and support while conducting the experiments.

References

[1] J. Fazio, *"How to Recognize and Prevent Hazard Trees," Tree City USA Bulletin No. 15*, National Arbor Day Foundation, Nebraska City, NE, USA, 1989.

[2] J. Albers and E. Hayes, *How to Detect, Assess and Correct Hazard Trees in Recreational Areas*, DNR, St. Paul, MN, USA, 1993.

[3] A. Brookes, "Preventing death and serious injury from falling trees and branches," *Australian Journal of Outdoor Education*, vol. 11, no. 2, pp. 50–59, 2007.

[4] J. Watt and D. J. Ball, "Trees and the risk of harm, report for the national tree safety group," Forestry Commission Rept., Middlesex University, Scotland, 2009.

[5] V. Bucur, *Acoustics of Wood*, CRC Pres, Boca Raton, FL, USA, 1995.

[6] S. Al Hagrey, "Electrical resistivity imaging of tree trunks," *Near Surface Geophysics*, vol. 4, no. 12, pp. 177–185, 2006.

[7] G. Catena, L. Palla, and M. Catalano, "Thermal infrared detection of cavities in trees," *European Journal of Forest Pathology*, vol. 20, no. 4, pp. 201–210, 1990.

[8] A. Franchois, "Microwave permittivity measurements of two conifers," *IEEE Transactions on Geoscience and Remote Sensing*, vol. 36, no. 5, pp. 1384–1395, 1998.

[9] V. Bucur and I. Böhnke, "Factors affecting ultrasonic measurements in solid wood," *Ultrasonics*, vol. 32, no. 5, pp. 385–390, 1994.

[10] M. Hasegawa, M. Takata, J. Matsumura, and K. Oda, "Effect of wood properties on within-tree variation in ultrasonic wave velocity in softwood," *Ultrasonics*, vol. 51, no. 3, pp. 296–302, 2011.

[11] H. Berndt, A. P. Schniewind, and G. C. Johnson, "Ultrasonic energy propagation trough wood: where, when, how much," in *Proceedings of the Proc. Of 12th Int. Symposium on NDT of Wood*, pp. 57–66, Sopron, 2000.

[12] Y. Zhou, "Microwave imaging based on wideband range profiles," *Progress in Electromagnetics Research Letters*, vol. 19, pp. 57–65, 2010.

[13] G. Lawday, J. A. Dolwin, D. Lonsdale, and J. R Barnett, "Development and use of stress wave meter, to detect the presence of decay in wood blocks," in *Proceedings of the Proc. of 12th Int. Symposium on NDT of Wood*, pp. 187–196, Sopron, September.

[14] V. Bucur, "Techniques for high resolution imaging of wood structure: A review," *Measurement Science and Technology*, vol. 14, no. 12, pp. R91–R98, 2003.

[15] B. Yilmaz, C. Ozdemir, and S. Gokkan, "A focusing algorithm for tree-penetrating radar imaging: An experimental study and concept evaluation," in *Proceedings of the 2016 17th International Radar Symposium (IRS)*, pp. 1–4, Krakow, Poland, May 2016.

[16] J. R. Butnor, J. A. Doolittle, L. Kress, S. Cohen, and K. H. Johnsen, "Use of ground-penetrating radar to study tree roots in the southeastern United States," *Tree Physiology*, vol. 21, no. 17, pp. 1269–1278, 2001.

[17] C. Özdemir, Ş. Demirci, E. Yiğit, and B. Yilmaz, "A review on migration methods in b-scan ground penetrating radar imaging," *Mathematical Problems in Engineering*, vol. 2014, Article ID 280738, 2014.

[18] C. Ozdemir, S. Demirci, E. Yigit, and A. Kavak, "A hyperbolic summation method to focus B-scan ground penetrating radar images: An experimental study with a stepped frequency system," *Microwave and Optical Technology Letters*, vol. 49, no. 3, pp. 671–676, 2007.

[19] B. Yılmaz, Ş. Demirci, E. Yiğit, and C. Özdemir, "An Experimental Study of Through-the-Wall Radar for Life sign Detection," in *Proceedings of Progress in Electromagn, Research Symposium (PIERS '13)*, pp. 1602–1604, 2013.

[20] M. G. Amin, *Through-the-Wall Radar Imaging*, Taylor and Francis, Bocaraton, FL, USA, 2010.

[21] A. Muqaibel, A. Safaai-Jazi, A. Bayram, A. M. Attiya, and S. M. Riad, "Ultrawideband through-the-wall propagation," *IEE Proceedings—Microwaves, Antennas and Propagation*, vol. 152, no. 6, pp. 581–588, 2005.

[22] N. J. Redding and G. N. Newsam, "Inverting the Circular Radon Transform," *DTSO Research Report DTSO-RR-0211*, August 2001.

[23] http://web.hep.uiuc.edu/home/serrede/P435/Lecture_Notes/Dielectric_Constants.pdf.

[24] C. Özdemi, *Inverse Synthetic Aperture Radar Imaging with Matlab Algorithms*, John Wiley Sons, Hoboken, NJ, USA, 2012.

[25] R. Bhalla, J. Moore, and H. Ling, "Global scattering center representation of complex targets using the shooting and bouncing ray technique," *IEEE Transactions on Antennas and Propagation*, vol. 45, no. 12, pp. 1850–1856, 1997.

[26] C. Ozdemir, R. Bhalla, and H. Ling, "A radiation center representation of antenna radiation patterns on a complex platform," *IEEE Transactions on Antennas and Propagation*, vol. 48, no. 6, pp. 992–1000, 2000.

[27] MATLAB, R2015a, MathWorks Inc., Natick, MA, 2015.

10

Study on Electrophysiological Signal Monitoring of Plant under Stress based on Integrated Op-Amps and Patch Electrode

Weiming Cai and Qingke Qi

School of Information Science and Engineering, Ningbo Institute of Technology, Zhejiang University, Ningbo 315100, China

Correspondence should be addressed to Weiming Cai; caiwm@zju.edu.cn

Academic Editor: Jit S. Mandeep

Electrophysiological signal in plant is a weak electrical signal, which can fluctuate with the change of environment. An amplification detection system was designed for plant electrical signal acquisition by using integrated op-amps (CA3140, AD620, and INA118), patch electrode, data acquisition card (NI USB-6008), computer, and shielded box. Plant electrical signals were also studied under pressure and flooding stress. The amplification detection system can make nondestructive acquisition for *Aquatic Scindapsus* and *Guaibcn* with high precision, high sensitivity, low power consumption, high common mode rejection ratio, and working frequency bandwidth. Stress experiments were conducted through the system; results show that electrical signals were produced in the leaf of *Aquatic Scindapsus* under the stress of pressure. Electrical signals in the up-leaf surface of *Aquatic Scindapsus* were stronger than the down-leaf surface. Electrical signals produced in the leaf of *Guaibcn* were getting stronger when suffering flooding stress. The more the flooding stress was severe, the faster the electrical signal changed, the longer the time required for returning to a stable state was, and the greater the electrical signal got at the stable state was.

1. Introduction

There are potential changes when life activities occur among living body, biological tissues and organs, and biological cell, which are called electrophysiological signal. The electrophysiological signal is a basic character as well as a behavior in living beings. It is also the general reflection of a series of physical and chemical reactions [1–4]. Some plants would react to external stimulus. For example, the leaves will close with the change of potential when mimosa is stimulated. Stahlberg [5] reviewed the history of electrical signals from the first recordings of action potentials (AP) in sensitive *Dionaea* and *Mimosa* plants at the end of the 19th century to their rediscovery in common plants in the 1950s, from the first intracellular recordings of AP in giant algal cells to the identification of the ionic mechanisms by voltage-clamp experiments. Sinyukhin and Britikov [6] who published a thesis about potential build reproductive system in journal called Nature. In recent years, there are a lot of researches about electrical signal in plant all over the world [7–9]. For instance, Ding and Lu's research group recognized plant stress factor by the fuzzy optimal wavelet packet and set up a model based on the environmental factor of electrical signal in plant [10, 11]. However, the electrical signal is very weak, although magnitudes of electrical reactions can be often about tens mV (up to 100 mV or more) under environment stimulation; if we take the general detection method to amplify and process the signal, it can be influenced by the zero drift of operational amplifier, the noise, the electromagnetic wave, and the spread of the channel; thus it is difficult for us to obtain authentic signals [12]. Chatterjee et al. [13] provide a platform for realizing a plant signal-based biosensor to classify the environmental stimuli by using raw electrical signals from plants. How to successfully detect which stimuli caused the signal is quite promising because it not only can open the possibility of remotely monitoring the environment of a large geographical area, but also can help in taking timely preventive measures for natural or man-made disasters. However, at present, it is difficult to find an instrument and equipment which is specifically applied to detect the electrophysiological signal in plant. And it is not conducive to satisfy models to guide agricultural production. It is of great significance for developing professional plant electrical signal equipment, setting up a prediction model based on the electrical signal in plant, and promoting the application of plant electrical signal in agriculture to find ways to detect electrical signal in plant.

Besides, China always suffers from serious flooding. About two-thirds of the land area suffers different degrees of flooding [14]. The waterlogging stress is a kind of water stress, and the water stress includes drought stress (moisture loss) and waterlogging stress (flood). The damage caused by flood is called waterlogging injury. Waterlogging stress is very harmful to plant growth. There is a limit to the demand for water plants; too less or too much water will be harmful to plant growth. Water deficit can cause drought, and waterlogging stress can cause the plant roots decay. At present, people have partly understood the mechanism of plants' response to the flooding stress but there is still a need for more study to be done [15, 16].

Therefore, a method for detecting the physiological signals of plants is proposed in the paper. Through studied changes in the electrophysiological signals of stress on *Aquatic Scindapsus* and *Guaibcn*, the detecting system will be tested on whether stable plant electrical signal can be collected correctly; moreover, stress experiments were conducted through the detecting system to find the rule of waterlogging stress influence on the plant electrical signals.

2. Principles and Methods

2.1. Composition of the Detection System of Electrical Physiological Signal in Plant. As shown in Figure 1, the detection system consists of electrode, amplifying circuit, data acquisition card, shield, and so on. The graphite patch electrode was used to measure the electrical signals on the surface of the plant. The amplifying circuit consists of three integrated operational amplifiers (CA3140, AD620, and INA118), two integrated voltage stabilizing circuits (CW7805, CW7905), and one data acquisition card (NI USB-6008). The shield was made of aluminum foil.

2.2. The Design Principle of Amplifying Circuit. Figures 2 and 3 are the amplifying circuit principle diagram of the electrophysiological signal in plant and PCB plate-making figure, respectively.

As shown in Figures 2 and 3, in view of the uncertainty of the size of plant resistance and the frequency interference caused by the ends of the amplifying, we should add two high input impedance circuit operational amplifiers (CA3140) and form a depth series voltage negative feedback in order to get less interference, and it can also form a differential circuit with instrumentation amplifier (AD620). Since the common mode voltage output voltage and drift which are caused by CA3140 can counteract each other, the amplifying circuit has advantages of small voltage output drift and strong common mode rejection capability.

The signal is amplified by AD620; however, if the signal is just amplified by AD620, circuit might burn out which is caused by the sudden increase of signal. So we used INA118 which contains input protection circuit to amplify the signal secondly. If the inputs overload, protection circuit can ensure the safety of the follow-up circuit by limiting the input current in 1.5 mA to 5 mA. INA118 has some advantages like high precision, low consumption, high membrane rejection ratio, wide working frequency band, and so on. It is suitable for the amplification of small signals. Its gain formula is

$$G = 49.4\frac{K}{R} + 1. \tag{1}$$

In Figure 2, the external resistor of CA3140 is 10 k ohms of the potentiometer; the first stage for amplification circuit is AD620; the external resistor of AD620 is a 200 ohms of the potentiometer. After adjusting potentiometer to 167 ohms, magnification of 300 times was obtained. The second stage for amplification circuit is INA118; the external resistor of INA118 is 2 k ohms of the potentiometer. After adjusting potentiometer to 2 k ohms, the magnification of 20 times was obtained, and the amplifying circuit can magnify the electrical signals 6000 times totally.

2.3. Design of Experiments on Plant Electrical Signal Detection under Stress. In order to examine whether the amplification detection system of electrophysiological signal in plant can gather reliable electrical signal and study the changing rules of the electrophysiological signals in plants under stress, based on the *Aquatic Scindapsus* and *Guaibcn* as the research object, the following tests were designed.

(a) Electrical Signal Detection of Aquatic Scindapsus Leaf Surface under Compressive Stress. Aquatic Scindapsus that grows well was chosen in the study. Electrical signals on the Aquatic Scindapsus surface were detected when the leaves of Aquatic Scindapsus were clamped by clip. When *Aquatic Scindapsus* is under pressure, the potential voltage change of electrical signals on the up-leaf and down-leaf surface of *Aquatic Scindapsus* was measured. Specific experimental steps are as follows.

(1) Attach graphite patch electrode A (measurement electrode) on the up-leaf or down-leaf surface of one leaf of *Aquatic Scindapsus* and graphite patch electrode B (reference electrode) posted on the *Aquatic Scindapsus* stem which is connected with the leaf.

(2) After equipping with batteries and turning on the circuit, let it stand for 40 minutes until the numerical digit of voltage becomes stable; then save the voltage data in the computer.

(3) Clamp the leaves of *Aquatic Scindapsus* by clip, record the instantaneous potential change, and observe the potential changes. Then remove the clamp and repeat parallel experiments 10 times.

(4) Remove graphite patch electrode A, attach to the *Aquatic Scindapsus* under the leaf (the same as the measuring objects before, detecting the same leaf; the only difference was that one is to detect the up-leaf surface but the other is to detect the down-leaf surface), and attach a graphite patch electrode on the *Aquatic Scindapsus* stem which is connected with the leaf.

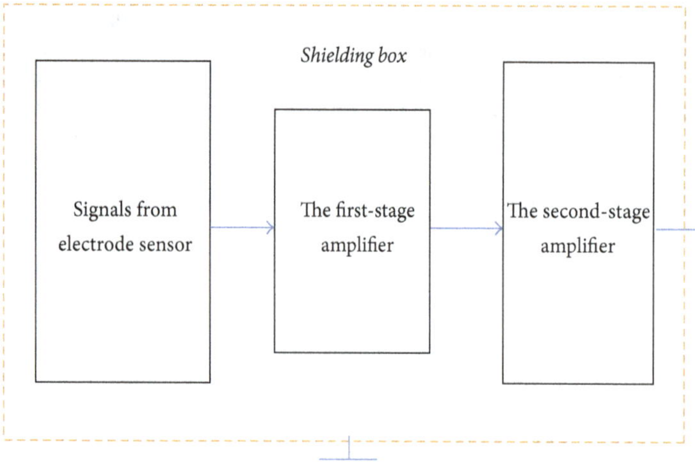

FIGURE 1: Structure diagram of the electrical signals detection system for plant.

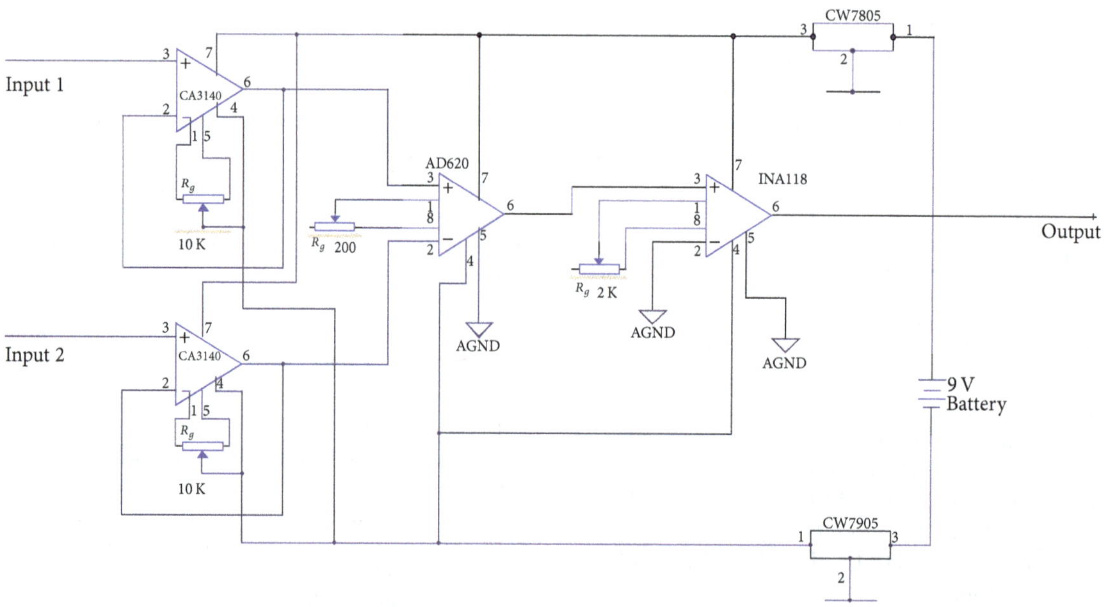

FIGURE 2: Schematic diagram of electrical signals amplifying circuit for plant.

(5) After equipping with batteries and turning on the circuit again, let it stand for 40 minutes until the numerical digit of voltage becomes stable; then save the voltage data in the computer.

(6) Clamp the leaves of *Aquatic Scindapsus* by clip, and record the instantaneous potential change as well as observing the potential changes. Then remove the clamp, to repeat parallel experiments 10 times; compare the potential difference between up-leaf surface and down-leaf surface.

(b) Electrical Signal Detection of Guaibcn under Waterlogging Stress. *Guaibcn* that grows well was chosen in the study. By watering *Guaibcn* continuously (simulating the rainy day) or applying one-time watering (simulating rainstorm), then, observe the change of the electrical signal of *Guaibcn*.

In order to study the *Guaibcn* electrical signals in rainy season affected by the persistent plum rains, the *Guaibcn* was watered persistently with 50 ml each time, then observing the change of the electrical signal of *Guaibcn*. Specific experiment steps are as follows.

(1) Attach graphite patch electrode A (measurement electrode) on the down-leaf surface of one leaf of *Guaibcn* and the graphite patch electrode B (reference electrode) posted on the *Guaibcn* stem which is connected with the leaf.

(2) After equipping with batteries and turning on the circuit, let it stand for 40 minutes until the numerical digit of voltage becomes stable, then saving the voltage data in the computer.

(3) Water the *Guaibcn*, observe the change of the electrical signal of *Guaibcn*, and record the voltage value every 30

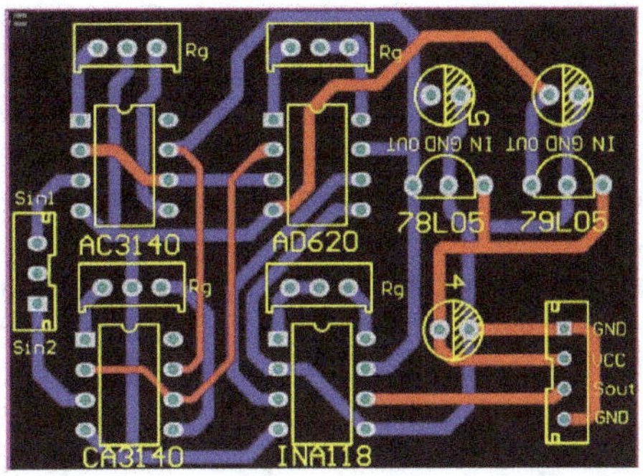

FIGURE 3: Printed circuit board diagram of electrical signals amplifying circuit for plant.

seconds. Water the *Guaibcn* by 50 ml water again after it keeps stable (the time points of watering are 6 min, 20 min, 25 min, and 30 min).

In order to observe the change of the electrical signal of *Guaibcn* when rainstorm occurs, add 200 ml water into the culture soil of the *Guaibcn* and then save the voltage data every 30 seconds in the computer. Specific experiment steps are as follows.

(1) Attach graphite patch electrode A (measurement electrode) on the down-leaf surface of one leaf of *Guaibcn* and graphite patch electrode B (reference electrode) posted on the *Guaibcn* stem which is connected with the leaf.

(2) After equipping with batteries and turning on the circuit, let it stand for 40 minutes until the numerical digit of voltage becomes stable and then save the voltage data in the computer.

(3) Add 200 ml water into the culture soil of the *Guaibcn* and then save the voltage data every 30 seconds in the computer.

(c) Statistical Analysis. There are ten repetitions in the study. Data were analyzed using an analysis of variance (ANOVA) multiple comparison (single factor). The differences between the experiment and control were tested by the method. Stress effects were considered to be significant at $P < 0.05$.

3. Results and Discussion

3.1. Aquatic Scindapsus Electrical Signal before and after the Pressure of Leaves. Figures 4 and 5 showed the comparison chart about voltage values (the real magnitude is 1/6000 times the magnitude after amplification as shown in the figures) of electrical signals in the up-leaf surface of *Aquatic Scindapsus* before and after the pressure, respectively.

In Figure 4, the maximum change of electrical signal on the leaf is 0.0033 V, and the minimum is 0.000 V, the median is 0.0017 V, the average is 0.0006 V, the standard deviation is 0.001092, and the coefficient of dispersion is 1.82. In Figure 5, the maximum number of the changes of electrical signal under leaf is 0.0047 V, the minimum is 0.0010 V, the median is 0.0029 V, the average is 0.0032 V, the standard deviation is 0.001361, and the coefficient of dispersion is 0.42. From Figures 4 and 5, we can find that the signal of up-leaf surface

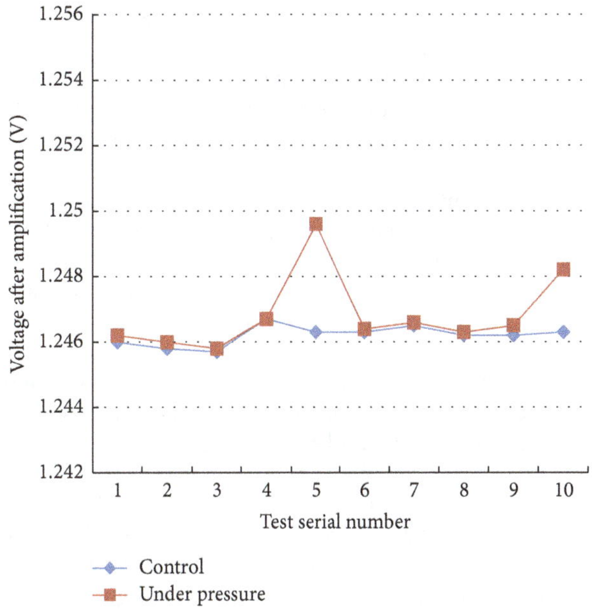

FIGURE 4: Voltage values of electrical signals in the up-leaf surface of *Aquatic Scindapsus*.

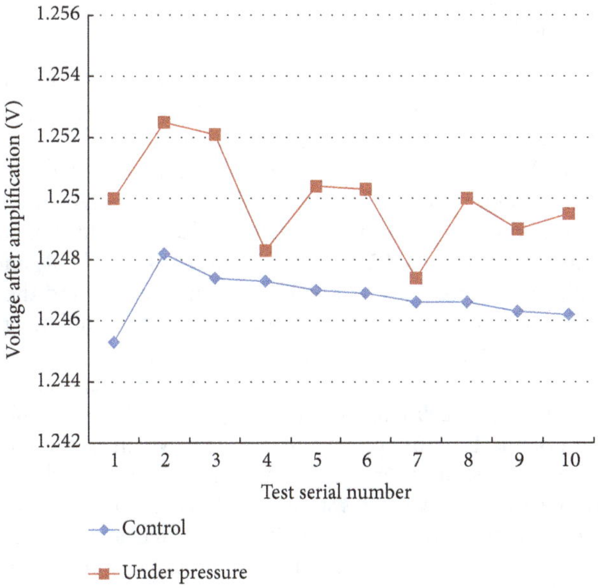

FIGURE 5: Voltage values of electrical signals in the down-leaf surface of *Aquatic Scindapsus*.

does not change obviously before and after the clamping; only two sets of data have large potential change, but the changes of down-leaf surface are obvious; there are 8 sets of data that have obvious potential rise phenomenon. As we know, many plants can efficiently produce electric signals in the form of action potentials (AP) and variation potentials (VP) and the long-distance propagation of these signals proceeds in the vascular bundles [5]. Therefore, the experimental results indicated that the changes of down-leaf surface are more obvious than the up-leaf surface, probably because the leaf vein (as the vascular bundles) is more near to the down-leaf of down-leaf surface.

Since there are two sets of data values (5 and 10) deviating from the value of the other eight sets of data, we suppose the two sets of data as the unreliable data caused by interference. After removing the two unreliable data, we find that the maximum number of the changes of electrical signal on leaf is 0.0003 V, the minimum is 0.0000 V, the median is 0.0002 V, the average is 0.00012 V, the standard deviation is $8.57E-05$, and the coefficient of dispersion is 0.623. After removing those two groups of data, the discrete coefficient decreased significantly.

By analysis and comparison of the data above, we found that the electrical signal of *Aquatic Scindapsus* will change and amplitude will increase under the action of external pressure. After the signal is amplified by amplification, the potential of the up-leaf surface increases in average by 0.00006 V and the down-leaf surface increase in average by 0.0032 V. The growth rate of down-leaf surface is greater than the up-leaf surface. The reason why the electrical signal in plant falls fast after removing the pressure might relate to the production mechanism of the electrical signal in plant and the mechanism of conduction [17].

3.2. Changes of the Electrical Signal of Guaibcn under Waterlogging Stress

3.2.1. Changes of the Electrical Signal of Guaibcn under One-Time Flooding. Figure 6 is the change of the electrical signal of *Guaibcn* under one-time water flooding.

In Figure 6, the time we started to add water is on the red arrow point (0 moment on x-axis). As shown in Figure 6, before adding water, curve is flat and fluctuates in a small scope. After we added 200 ml water, curve rose gradually and achieved the highest point about 28 minutes later, and then the curve declined and returned back to the potential before watering. After we added a lot of water, the potential rises quickly, and the rate of change was in the positive; then the rate of change declined, until becoming in the negative. When the potential almost returns to the data which is the data before we added water, the fluctuation of the rate of change was near zero. The result is similar to the avocado tree's action with irrigation provided by Gurovich and Hermosilla [18].

Flooding can result in an anaerobic situation in the plant root system that induces pH changes in the apoplast and symplast. Therefore, electrical potential increased during flooding is probably because of pH changes in the apoplast and symplast.

3.2.2. The Change of the Electrical Signal of Guaibcn under Continuous Adding of Water. Figure 7 showed the change of the electrical signal of *Guaibcn* under continuous water flooding.

In Figure 7, the time we started to add water is on the red arrow point (0 moment on x-axis). The potentials of stem and leaf rise slowly after watering the first 50 ml water. After 2 minutes, the potential rises quickly. Six minutes later, the electrical signal gradually stabilized. After watering the second 50 ml, the potentials of stem and leaf rise immediately, and three minutes later, the electrical signal gradually stabilized. After watering the third and fourth 50 ml water, the potentials of stem and leaf rise immediately. But the rise degree was less than the previous two, and the fourth was also less than the third. The electrical signal gradually stabilized two minutes later.

Several physiological mechanisms that explain electric responses to irrigation have been postulated. Action potential (AP) and variation potential (VP) lead to a physiological reaction by informing distant cells about local stimuli [19]. Additional signalling mechanisms in plants have been reported, including modifications of cytoplasmatic pH [20, 21].

The above two kinds of experiments of *Guaibcn* were carried out under the condition of flooding stress. The first experiment added water four times, and 50 ml water was added each time; the electrical signal of *Guaibcn* changes obviously after we added the first 50 ml water for 3 minutes. The second experiment added 200 ml water one time, and the electrical signal changes immediately after adding water. Through the above analysis, we found the following:

(1) The main phenomenon of the change of *Guaibcn* under water flooding was the amplitude of electrical signal increases.

(2) The change of electrical signal of *Guaibcn* relates to the amount of water, the more severe the water flooding is, the quicker the electrical signal changes, the larger the amount of change is, and the longer the time required for returning to a stable state is.

(3) The electrical signal produced under the water flooding can restore its original state as long as the water is in a certain range.

4. Conclusions

(1) Patch electrode and integrated operational amplifiers (CA3140, AD620, and INA118) were used to design an amplifier circuit with advantage of high precision, high sensitivity, low power consumption, high membrane rejection ratio, and wide working frequency band. The amplification detection system can make nondestructive acquisition for *Aquatic Scindapsus* and *Guaibcn*.

(2) According to the experiment results, the electrophysiological signal detection system can collect stable electrical signals and satisfy the need of analysis of the electrophysiological signal based on the combination of the patch electrode and amplifying circuit (consisting of CA3140, AD620, INA118, CW7805, and CW7905).

(3) According to the study about the change of electrical signal of *Aquatic Scindapsus* under compressive stress, we found that *Aquatic Scindapsus* could generate the electrical signal under compressive stress, and the change of down-leaf is greater than the up-leaf.

(4) According to the study about the change of electrical signal of *Guaibcn* under purpose of stress, we found that the electrical signal of *Guaibcn* would rise under waterlogging stress, and the change degree relates to the amount of water; the more severe the waterlogging stress is, the quicker the electrical signal changes, the larger the amount of change is, and the longer the time required for returning to a stable state.

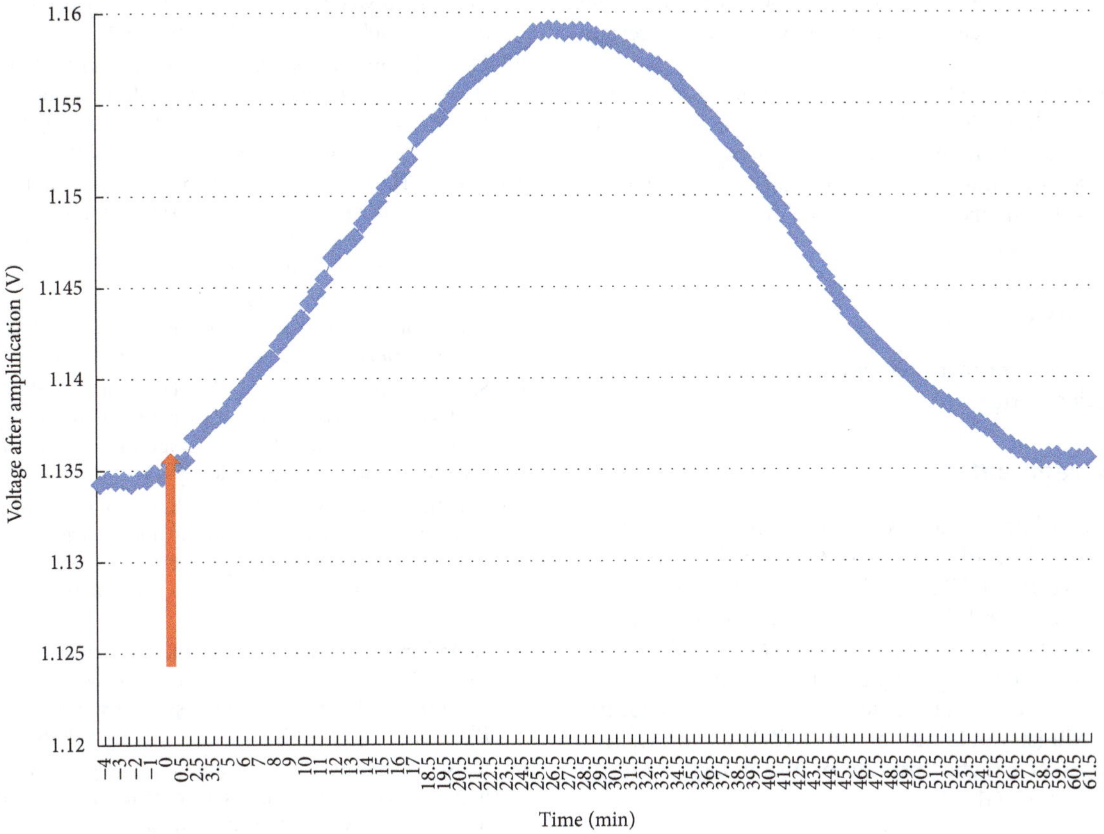

FIGURE 6: Changes of electrical signals for *Guaibcn* under one-time water flooding.

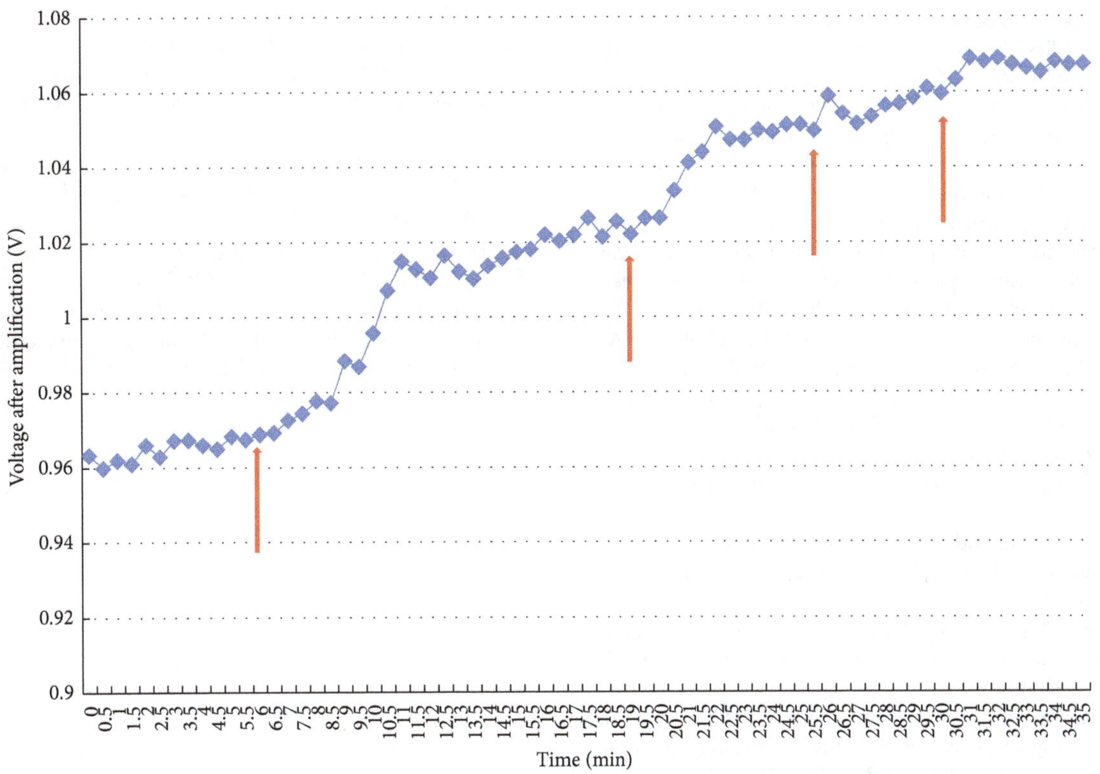

FIGURE 7: Changes of electrical signals for *Guaibcn* under continuous water flooding.

The electrical signal produced under the water flooding can restore its original state as long as the water is in a certain range.

The relationship between the plant physiological parameters and the electrical signals is very complex. The results of the study provided a method for developing electrophysiological signal monitoring device of plant and a reference for studying the change rule of the plant electrophysiological signals under stress.

Conflicts of Interest

The authors declare that they have no conflicts of interest regarding this work.

Acknowledgments

The authors acknowledge that the research was supported by the General Scientific Research Project of Zhejiang Education Department (Y201534782), Ningbo Natural Science Foundation (2014A610155), Ningbo Enriching Program of Science and Technology (2015C10014), Research Project of Public Welfare Technology Application in Zhejiang Province (2016C32106), the Humanities and Social Sciences Foundation of the Ministry of Education (16YJCZH112), and College Talent Funds of Ningbo Institute of Technology, Zhejiang University (1141257G1402).

References

[1] Y.-C. Wang, P. Li, and D.-F. Chi, "Electrophysiological and behavioral responses of Tenebrio molitor L. to fourteen kinds of plant volatiles," *Journal of Asia-Pacific Entomology*, vol. 19, no. 2, pp. 261–267, 2016.

[2] E. Michard, M. M. Wudick, M. A. Lizzio, C. O. Ramírez, C. Campos, and J. A. Feijó, "Plant glutamate receptors: electrophysiological characterization and evolutionary perspectives," *Biophysical Journal*, vol. 110, no. 3, pp. 404-405, 2016.

[3] X. Xu, X. Cai, L. Bian, Z. Luo, Z. Xin, and Z. Chen, "Electrophysiological and behavioral responses of Chrysopa phyllochroma (Neuroptera: Chrysopidae) to plant volatiles," *Environmental Entomology*, vol. 44, no. 5, pp. 1425–1433, 2015.

[4] M. R. Zimmermann, A. Mithöfer, T. Will, H. H. Felle, and A. C. U. Furch, "Herbivore-triggered electrophysiological reactions: Candidates for systemic signals in higher plants and the challenge of their identification," *Plant Physiology*, vol. 170, no. 4, pp. 2407–2419, 2016.

[5] R. Stahlberg, "Historical overview on plant neurobiology," *Plant Signaling and Behavior*, vol. 1, no. 1, pp. 6–8, 2006.

[6] A. M. Sinyukhin and E. A. Britikov, "Action potentials in the reproductive system of plants," *Nature*, vol. 215, no. 5107, pp. 1278–1280, 1967.

[7] V. Sukhov, "Electrical signals as mechanism of photosynthesis regulation in plants," *Photosynthesis Research*, vol. 130, no. 1–3, pp. 373–387, 2016.

[8] V. A. Vodeneev, L. A. Katicheva, and V. S. Sukhov, "Electrical signals in higher plants: mechanisms of generation and propagation," *Biophysics*, vol. 61, no. 3, pp. 505–512, 2016.

[9] H. Y. Weng, D. P. Ye, J. G. Zheng, H. Y. Li, X. Liang, and S. H. Chen, "Research on characteristics of plant electrical signal under high voltage pulse stimulation," *Applied Mechanics and Materials*, vol. 513–517, pp. 3122–3125, 2014.

[10] H. Lu, X. Jing, W. M. Ding, H. M. Yu, and W. L. Ling, "Plant stress recognition based on fuzzy-rule using optimal wavelet packet," *Transactions of the Chinese Society for Agricultural Machinery*, vol. 43, no. 12, pp. 217–221, 2012 (Chinese).

[11] J. X. Lu, H. M. Yu, and S. J. Chen, "Environment factor prediction models based on plant electrical signals," *Transactions of the Chinese Society for Agricultural Machinery*, vol. 44, no. 11, pp. 229–233, 2013 (Chinese).

[12] H. Wu, L. G. Tian, S. Hu, Z. L. Chen, and M. Li, "Detection system on weak electrical signal in plants," *Applied Mechanics and Materials*, vol. 427–429, pp. 2037–2040, 2013.

[13] S. K. Chatterjee, S. Das, K. Maharatna et al., "Exploring strategies for classification of external stimuli using statistical features of the plant electrical response," *Journal of the Royal Society Interface*, vol. 12, no. 104, pp. 1–13, 2015.

[14] Y. Lou, Y. Pan, C. Gao, M. Jiang, X. Lu, and Y. J. Xu, "Response of plant height, species richness and aboveground biomass to flooding gradient along vegetation zones in floodplain wetlands, Northeast China," *PLoS ONE*, vol. 11, no. 4, Article ID e0153972, 2016.

[15] S. Orchard, R. J. Standish, D. Nicol, V. V. S. R. Gupta, and M. H. Ryan, "The response of fine root endophyte (Glomus tenue) to waterlogging is dependent on host plant species and soil type," *Plant and Soil*, vol. 403, no. 1-2, pp. 305–315, 2016.

[16] B. Xu, Y. Cheng, X. Zou, and X. Zhang, "Ethanol content in plants of Brassica napus L. correlated with waterlogging tolerance index and regulated by lactate dehydrogenase and citrate synthase," *Acta Physiologiae Plantarum*, vol. 38, no. 3, article 81, pp. 1–9, 2016.

[17] E. Davies, "New functions for electrical signals in plants," *New Phytologist*, vol. 161, no. 3, pp. 607–610, 2004.

[18] L. A. Gurovich and P. Hermosilla, "Electric signalling in fruit trees in response to water applications and light-darkness conditions," *Journal of Plant Physiology*, vol. 166, no. 3, pp. 290–300, 2009.

[19] J. Fromm and S. Lautner, "Electrical signals and their physiological significance in plants," *Plant, Cell & Environment*, vol. 30, no. 3, pp. 249–257, 2007.

[20] H. H. Felle, "pH: Signal and messenger in plant cells," *Plant Biology*, vol. 3, no. 6, pp. 577–591, 2001.

[21] V. Gloser, H. Korovetska, A. I. Martín-Vertedor et al., "The dynamics of xylem sap pH under drought: a universal response in herbs?" *Plant and Soil*, vol. 409, no. 1-2, pp. 259–272, 2016.

11

A Four Quadrature Signals' Generator with Precise Phase Adjustment

Xiushan Wu,[1] Yanzhi Wang,[1] Siguang An,[1] Jianqiang Han,[1] and Ling Sun[2]

[1]*College of Mechanical and Electrical Engineering, China Jiliang University, Hangzhou 310018, China*
[2]*Jiangsu Key Laboratory of ASIC Design, Nantong University, Nantong, China*

Correspondence should be addressed to Xiushan Wu; wuxiushan@cjlu.edu.cn

Academic Editor: Ahmed M. Soliman

A four-way quadrature signals generator with precise phase modulation is presented. It consists of a phase precision regulator and a frequency divider. The phase precision regulator generates two programmable currents by controlling the conduction of the tail current sources and then changes the currents into two bias voltages which are superimposed on the clock signals to adjust the phase difference of the four quadrature signals generated by the frequency divider, making the phase difference of 90 degrees. The four quadrature signals' generator with precise phase modulation has been implemented in a 0.18 μm mixed-signal and RF 1P6M CMOS technology. The size of the chip including the pads is 675 μm $*$ 690 μm. The circuit uses a supply voltage of 1.8 V, a bias current of 7.2 μA, and the bits of phase-setting input level $n = 6$ in the design. The measured results of the four orthogonal signals' phase error can reach ±0.1°, and the phase modulation range can reach ±3.6°.

1. Introduction

The method of integrated orthogonal signal generator is RC-CR phase-shift method, and the RC-CR phase-shift network can be achieved by the input signal phase shift of 45 degrees. In the literature [1], the design of *I/Q* generator is used in the S band, its unbalance amplitude is 0.1 dB, and the unbalance phase is 0.1 degrees. But this method cannot adjust the phase of the signals. The method of RC-CR network is complex, and once integrated, it cannot be used for phase error compensation. In addition, the capacitor and resistance should not be too large; otherwise integrated circuit is also difficult to be integrated in the chip. The second method often uses quadrature voltage-controlled oscillator cross coupling method. In the literature [2], a QVCO which is low in power consumption is manufactured and used in 2.4 GHz PLL. This design reduces power consumption and improves the noise coefficient, but the unbalance of *I/Q* phase is 2.21 degrees and the phase cannot be adjustable. Then the digital quadrature signal generator is used to generate orthogonal signals [3–6], but the phase error of the signal is not adjustable and compensated. In the previous study [7], the quadrature phase error caused by the mismatch of the capacitor is very large, and this phenomenon is more serious with the increase of the frequency. The more important problem is that the implemented integrated circuit is able to generate orthogonal signals, and the phase difference of the orthogonal signal is exactly 90 degrees in the early simulation stage. However, after the chip is processed, the phase difference often deviates from 90 degrees due to the limitation of the technology of integrated circuit production. Therefore, an integrated precise adjustment circuit structure is needed to compensate quadrature signals for the phase deviation caused by the integrated circuit process.

In this paper, an integrated four quadrature signals' generator is presented. The generator cannot only produce four orthogonal signals, but also can generate a programmable current by controlling the conduction of the tail current sources. The current is converted into a bias voltage superimposed on the clock signal to adjust the phase difference of the four signals, so as to make the phase difference be 90 degrees.

2. Circuit Design

As shown in Figure 1, the structure of the quadrature signals generator is composed of a phase precision regulator

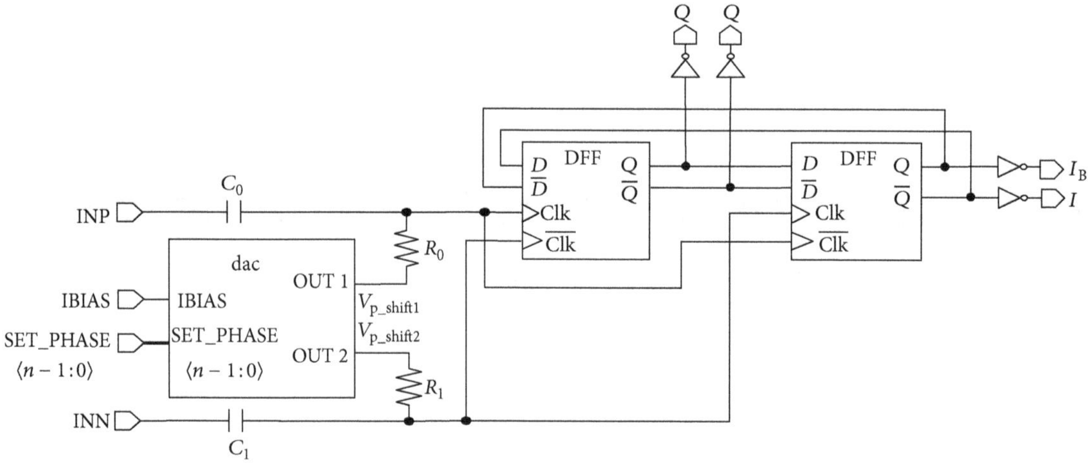

FIGURE 1: Circuit structure of the quadrature signal generator.

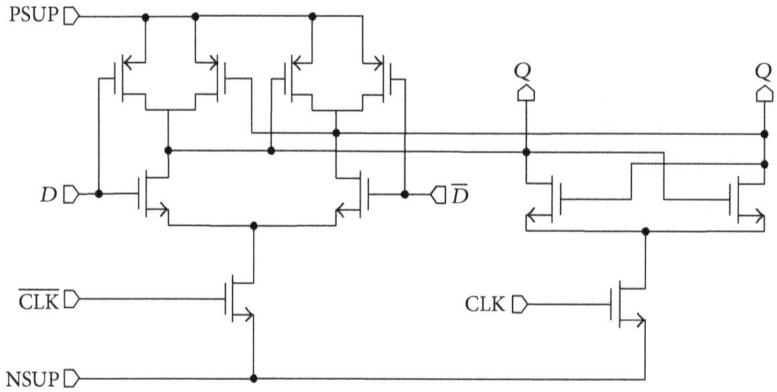

FIGURE 2: Circuit structure of D flip-flop (DFF).

unit (Ph_reg for short) and a frequency divider. The phase precision regulator unit can produce a programmable current by controlling the conduction of the tail current sources, and then the current can be converted into a bias voltage superimposed on the clock signal to precisely adjust the phase change. The frequency divider which consists of two D triggers (DFFs) is used to generate the four quadrature signals. The SET_PHASE$\langle n-1:0\rangle$ input level is represented by thick solid lines because it is an n-bit bus.

2.1. The Design of Divider. The frequency divider consists of two DFFs which are connected in the form of a two-stage ring with the differential input signal injected into the clock terminals [8]. As shown in Figure 1, the outputs of the first DFF are connected with the inputs of the second DFF, and the outputs of the second one connect back to the first one's input terminals which are in reversed polarity to achieve the extra phase shift of 180°. The clock terminals of the two DFFs are tied in reversed polarity and used to inject the differential input signal. The output signals can be taken from the data terminals of the second DFF; each output terminal's frequency is half of the input frequency.

The schematic of the DFF is shown in Figure 2. The cell of DFF contains two parts: the trigger part of the input signal is sent to the output and the storage part of the memory output logic level. The trigger part is realized by differential pairs; the lock part is realized by a cross coupling. The two parts are driven by a pair of clock signals, which are used to control the trigger circuit and the latch circuit, respectively [9].

The specific work process of the divider is as follows: when the input clock is a rising edge, the first DFF in Figure 1 is in the trigger state; that is to say, the output varies with the input. The second DFF in the lock state will remain the same state with the previous one, and its output will be sent back to the first DFF by reverse phase. When the input clock is a falling edge, the first DFF is in the lock state; the second one changes into the trigger state, and the state of its output will be locked in the first one. In this way, the time of a period of each DFF's output signal is the same as two periods of the clock signal, and the output frequency is just half of the input frequency,

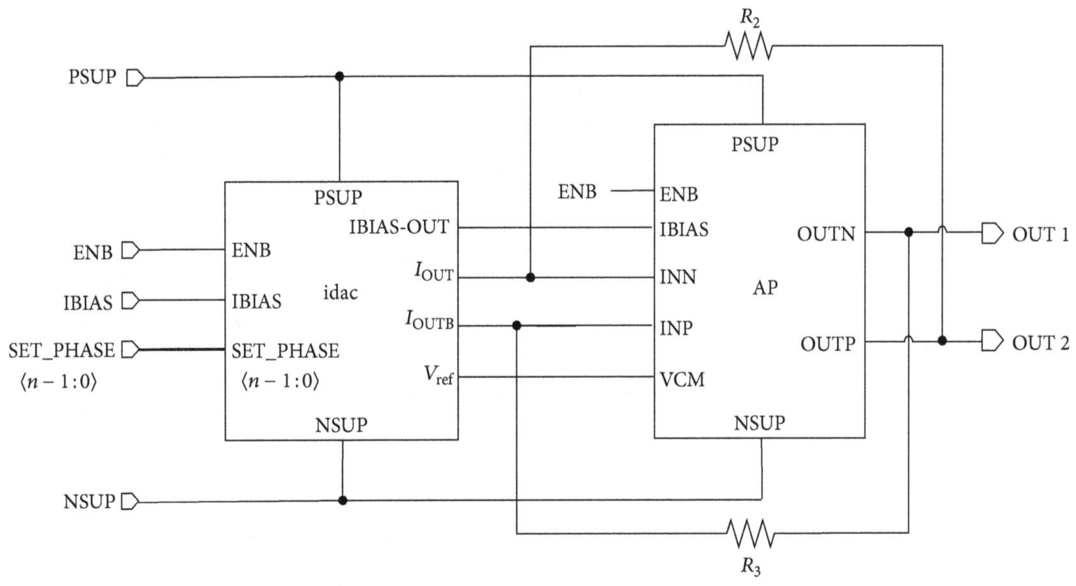

FIGURE 3: Circuit structure of the phase precision regulator unit.

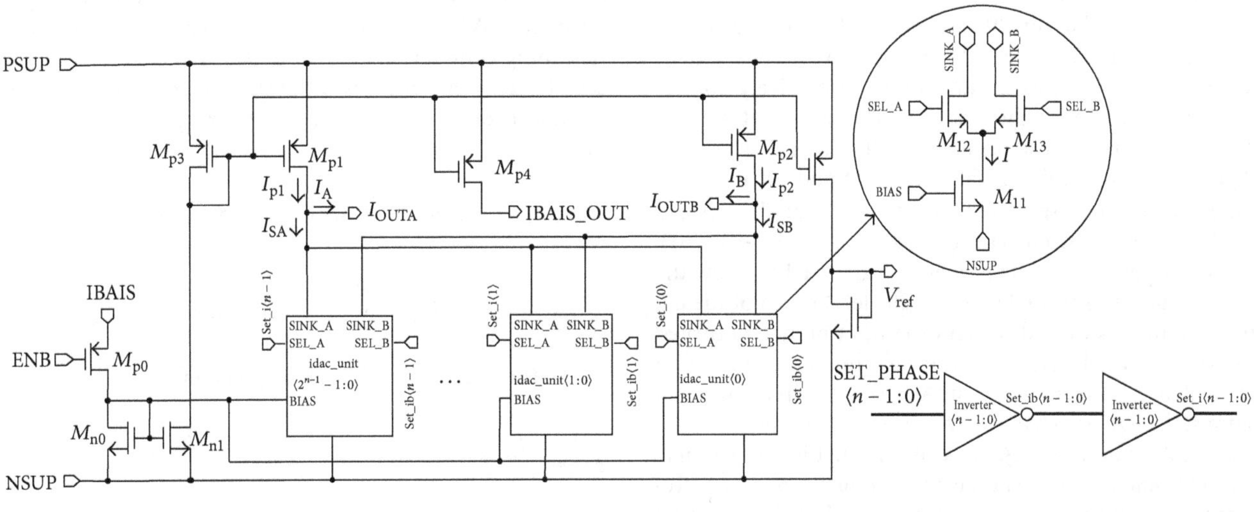

FIGURE 4: Structure of the programming current output cell.

thus achieving function of divide-by-2. The output terminals of the two DFFs are all used as the output signals, and then the four orthogonal signals are obtained.

2.2. The Design of Phase Modulator. The precise phase precision regulator unit can produce two programmable currents by controlling the conduction of the tail current sources, and then the programmable currents can be converted into two-way bias voltages by a COMS operational amplifier. The two-way bias voltages are superimposed on the clock signals to precisely adjust the phase change. Figure 3 shows the specific circuit of phase modulator unit, which consists of the programming current output cell (idac for short) and the current converting voltage cell. Two programmable currents are produced by n-bits phase-setting input level to control the conduction of the tail current sources in the programming current output cell. The current converting voltage cell is composed of a full differential CMOS amplifier (AP for short) and two resistors of R_2 and R_3.

The schematic of the idac is shown in Figure 4; the idac receives the external n-bits phase-setting input level of SET_PHASE$\langle n-1:0\rangle$ and generates two pair inverse strobe levels Set_i$\langle n-1,0\rangle$ and Set_ib$\langle n-1,0\rangle$ through the inverters. These pair levels are used to control the conductions of the tail current sources (idac_unit for short in Figure 4). The first bit phase-setting input signal (SET_PHASE$\langle 0\rangle$) generates two inverse strobe signals Set_i$\langle 0\rangle$ and Set_ib$\langle 0\rangle$ by the inverter

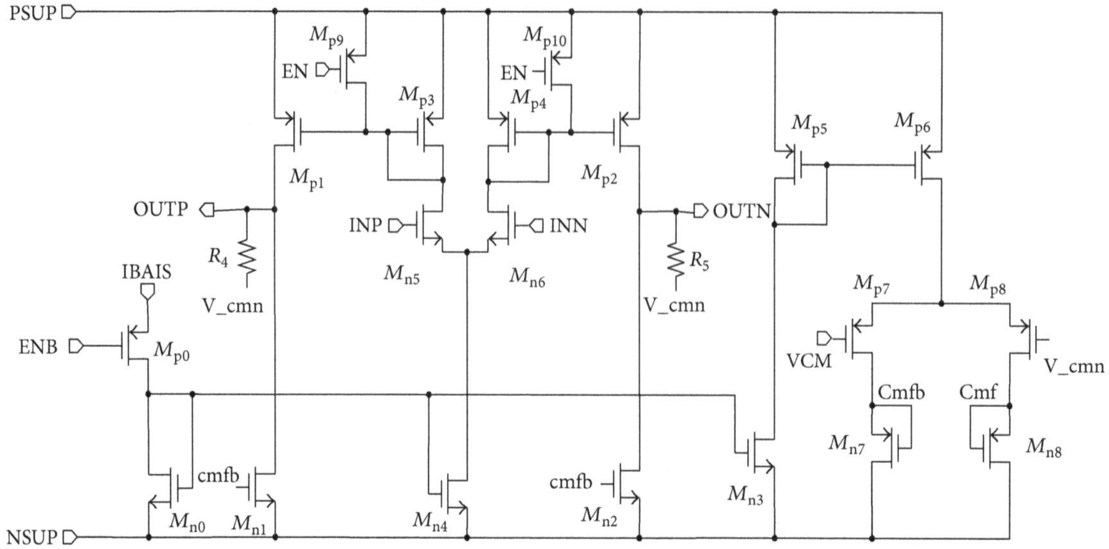

Figure 5: Structure of differential CMOS amplifier.

to, respectively, control the left and right branch of the current source (idac_unit⟨0⟩), the second bit phase-setting input signal (SET_PHASE⟨1⟩) generates two strobe signals Set_i⟨1⟩ and Set_ib⟨1⟩ by the inverter to, respectively, control two parallel left and right branches of the current sources (idac_unit⟨1 : 0⟩), and so on, and the n bit (SET_PHASE⟨$n-1$⟩) generates two strobe signals S_i⟨$n-1$⟩ and Set_ib⟨$n-1$⟩ by the inverter to, respectively, control 2^{n-1} parallel left and right branches of the current sources (idac_unit⟨$n-1$: 0⟩), and the suspension points in Figure 4 are used to show the omitted idac_units from 2 to $n-2$. The left branch of the tail current source is turned on, and the right branch is turned off correspondingly; similarly, the right branch is turned on and the left branch is turned off. As a small map is shown in Figure 4, the left branch of the ith tail current module chooses the control terminal SEL_A to receive the ith bit strobe signal, the right branch of the ith tail current module chooses control terminal SEL_B to receive the ith bit signal of the inverting signal, the ith tail current module is composed of 2^{i-1} tail current sources in parallel, i is a natural number, and $1 \le i \le n$.

Assuming every branch of conduction current of each tail current source is I, when the highest bit of is 1, the others are 0 (or the highest bit is 0, and the others are 1) for the SET_PHASE⟨$n-1$: 0⟩ in Figure 3, the current difference between I_A and I_B is minimum, and the value is I. Then I_A and I_B are converted into voltages by the current converting voltage unit, and the voltage difference is also the smallest. The minimum voltage difference is used to adjust and compensate the phase of the four output signals determining the accuracy of the circuit. When all bits of the SET_PHASE⟨$n-1$: 0⟩ are 1 (or all bits are 0), the current difference between I_A and I_B is maximum, and the value is $(1+2+4+\cdots+2^{n-1}) * I$, that is, equal to $(2^n-1) * I$, and the voltage difference is also the largest. The maximum voltage difference is used to adjust and compensate the phase of the four output signals determining the adjustment range of the circuit. When the value of I_A is increased, I_B is decreased. Assuming conduction current of M_{p1} is I_{p1}, M_{p2} is I_{p2}, and each tail current source is I, the external n-bits phase-setting input level of SET_PHASE⟨$n-1$: 0⟩ is $k_{n-1}k_{n-2}\cdots k_1k_0$, and then

$$I_{SA} = \sum_{i=0}^{n-1} k_i \cdot 2^i \cdot I \quad (k_i = 1),$$
$$I_{SB} = \sum_{i=0}^{n-1} k_i \cdot 2^i \cdot I \quad (k_i = 0). \tag{1}$$

So the outcurrent of I_A and I_B is equal to

$$I_A = I_{p1} - I_{SA} = I_{p1} - \sum_{i=0}^{n-1} k_i \cdot 2^i \cdot I \quad (k_i = 1),$$
$$I_B = I_{p2} - I_{SB} = I_{p2} - \sum_{i=0}^{n-1} k_i \cdot 2^i \cdot I \quad (k_i = 0). \tag{2}$$

Yet, the sum of the two currents is constant, and its value is $(I_{p1}+I_{p2}) - (2^n-1) * I$.

The output of the two programmable currents I_A and I_B is converted into two bias voltages through a full differential operational CMOS amplifier, the resistors R_2 and R_3, respectively. Because the two programmable currents' size and direction can be programmed with the selected external n-bits phase-setting input level of SET_PHASE⟨$n-1$: 0⟩, the two-way bias voltages of OUT1 and OUT2 shown in Figure 3 are programmed. So in this design, a classic double-end differential CMOS amplifier is used as shown in Figure 5.

3. Results and Discussion

According to the simulations' results, the design can generate two programmable currents by controlling the conduction of

Table 1: The W/L values of the CMOS in Figure 3.

Device	W/L (μm)
M_{p0}	10/1
M_{p1}	12/1.2
M_{p2}	12/1.2
M_{p3}	12/1.2
M_{p4}	12/1.2
M_{n0}	3.2/2
M_{n1}	1.6/2
M_{11}	1.6/2
M_{12}	1/2
M_{13}	1/2

Table 2: The W/L values of the differential CMOS amplifier in Figure 5.

Device	W/L (μm)
M_{p0}	10/0.25
M_{p1}	100/1
M_{p2}	100/1
M_{p3}	20/1
M_{p4}	20/1
M_{p5}	40/1
M_{p6}	40/1
M_{p7}	40/1
M_{p8}	40/1
M_{p9}	10/0.25
M_{p10}	10/0.25
M_{n0}	20/3
M_{n1}	5/1
M_{n2}	5/1
M_{n3}	50/3
M_{n4}	20/3
M_{n5}	10/3
M_{n6}	10/3
M_{n7}	3/1
M_{n8}	3/1

the tail current sources and then changes the currents into two bias voltages superimposed on the clock signals to adjust the phase difference of the four signals, making the phase difference of 90 degrees.

In the design, the values of resistances are $R_0 = R_1 = 12\,\text{k}\Omega$, $R_2 = R_3 = 200\,\text{k}\Omega$, and $R_4 = R_5 = 100\,\text{k}\Omega$, respectively. The W/L values of the CMOS in Figure 3 are listed in Table 1. The W/L values of the differential CMOS amplifier in Figure 5 are listed in Table 2. Parameter settings and part of the simulation results are listed in Table 3.

For demonstration, the presented circuit has been fabricated in SMIC's 0.18 μm CMOS process with a 4 GHz phase-locked loop together. The chip microphotograph is shown in Figure 6, and the size of the chip including the pads is 675 μm $*$ 690 μm. A LC_tank voltage-controlled

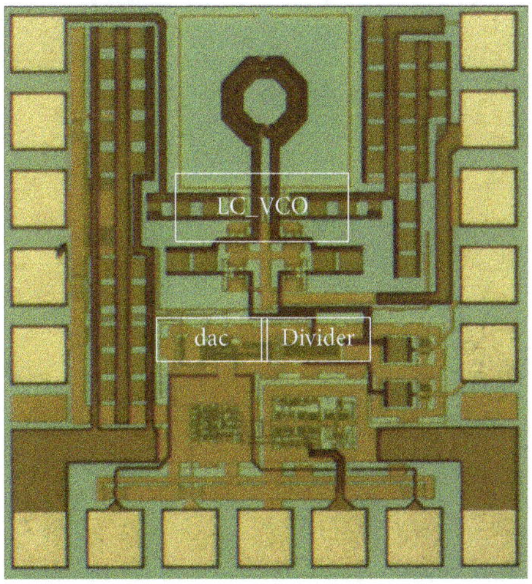

Figure 6: The microphotograph of the quadrature signal generator.

oscillator (LC_VCO) is used in the phase-locked loop. The used differential inductor has an inner diameter of 30 μm, a metal width of 8 μm, and the spacing of 1.5 μm. The value of the inductor is about 2.4 nH, and the effective quality factor is about 10 under the 4 GHz frequency. The range of the capacitor is about 0.3–0.68 pF. The tuning range of VCO is 300 MHz, from 3.85 GHz to 4.15 GHz, with the center frequency being about 4 GHz. The two output signals of the LC tank voltage-controlled oscillator are sinusoidal wave, and the signals are divided by four and used as the input clock signals of INN and INP shown in Figure 1. The circuit uses a supply voltage of 1.8 V, a bias current of 7.2 μA, and the bits of phase-setting input signal $n = 6$ in the design. There are no effective methods to measure the phase error because the present oscillography could not measure the phase error between four-channel several hundred MHz signals. The phase error is measured between every two of them based on the time domain outputs, the output four orthogonal signals' phase error precision can reach $\pm 0.1^\circ$, and the phase modulation range is $\pm 3.6^\circ$, but the phase difference in the simulation is about $\pm 2.7^\circ$. The main reason of the error is caused by different ways in the measurement and simulation; the signals' phase error cannot be directly measured because of the limitation of the instruments. The die bonding leads to error of measurement too.

4. Conclusions

In this paper, a four phase quadrature signals' generator with precise phase modulation is proposed. The design can generate two programmable currents by controlling the conduction of the tail current sources and then changes the currents into two bias voltages superimposed on the clock signals to adjust the phase difference of the four signals generated, making the phase difference of 90 degrees. It has been implemented in 0.18-μm CMOS process. The measurement result shows the proposed quadrature signal generator could achieve $\pm 0.1^\circ$ phase error, and the phase

TABLE 3: Parameter settings and results of the simulations.

SET_PHASE ⟨5 : 0⟩	IBIAS (μA)	I (μA)	I_{p1} (μA)	I_{p2} (μA)	I_A (μA)	I_B (μA)	V_{p_shift1} (mV)	V_{p_shift2} (mV)	Phase difference (°)
000000	7.2	0.72	23.2	23.2	23.14	−22.45	854.9	360.9	2.7
010000	7.2	0.72	23.2	23.2	11.55	−10.89	730.5	486.2	1.5
011000	7.2	0.72	23.2	23.2	5.75	−5.11	667.6	549.2	0.8
011011	7.2	0.72	23.2	23.2	5.03	4.38	643.9	572.9	0.3
011111	7.2	0.72	23.2	23.2	0.68	−0.04	612.4	604.1	0.1
100000	7.2	0.72	23.2	23.2	−0.04	0.68	604.5	612.4	−0.1
100100	7.2	0.72	23.2	23.2	−2.93	3.58	572.9	644.0	−0.3
101000	7.2	0.72	23.2	23.2	−5.83	6.47	541.4	675.5	−0.8
110000	7.2	0.72	23.2	23.2	−11.62	12.27	478.3	738.4	−1.6
111111	7.2	0.72	23.2	23.2	−22.46	23.13	360.9	854.9	−2.7

modulation range is ±3.6° with the bits of phase-setting input signal $n = 6$.

Competing Interests

The authors declare that there is no conflict of interests regarding the publication of this paper.

Acknowledgments

This research was financially supported by the National Natural Science Foundation of Zhejiang (LY4F040004), Zhejiang Province Instruments and science priority subjects to open fund (no. JL130102), and the National Natural Science Foundation of China (51407172 and 61376114).

References

[1] Y.-M. Hu, G.-W. Yu, and Y.-F. Zhang, "Research on characteristics of RC polyphase network for quadrature signal generator," *Journal of Radars*, vol. 2, no. 4, pp. 476–480, 2013.

[2] C.-T. Lu and H.-H. Hsieh, "A low-power quadrature VCO and its application to a 0.6-V 2.4-GHz PLL," *IEEE Transactions on Circuits and Systems*, vol. 57, no. 4, pp. 793–802, 2010.

[3] X.-J. Hao, "Quadrature signals generator based on FPGA," *Automated Measurement & Control*, vol. 27, no. 5, pp. 77–79, 2008.

[4] M.-M. Lei, Y.-M. Li, and Y.-H. Sun, "A 1.8 V 0.9 mW 4.8 GHz frequency divider in 0.18 μm CMOS process," *Microelectronics*, vol. 37, no. 2, pp. 279–281, 2007.

[5] C. Qi, L. Wang, C.-D. Ling, and X. Yang, "Design of quadrature frequency divider based on SCL," *Microcomputer & Its Applications*, vol. 31, no. 22, pp. 27–28, 2012.

[6] X.-Z. Yin, Y.-F. Yu, C.-Y. Ma et al., "Design of quadrature 2:1 frequency divider for GNSS receivers," *Optics and Precision Engineering*, vol. 20, no. 5, pp. 1015–1016, 2012.

[7] M. Bagheri, R. Bagheri, J.-F. Buckwalter, and L.-E. Larson, "Tuning-range enhancement through deterministic mode selection in RF quadrature oscillators," *IEEE Transactions on Microwave Theory and Techniques*, vol. 63, no. 11, pp. 3713–3726, 2015.

[8] N. Kim, J. Yun, and J.-S. Rieh, "A 120 GHz voltage controlled oscillator integrated with 1/128 frequency divider chain in 65 nm CMOS technology," *Journal of Semiconductor Technology and Science*, vol. 14, no. 1, pp. 131–137, 2014.

[9] X.-Y. Gui, Z.-M. Chen, and M.-M. Green, "Analysis of nonlinearities in injection-locked frequency dividers," *IEEE Transactions on Microwave Theory and Techniques*, vol. 63, no. 3, pp. 945–953, 2015.

12

Robust Self-Contained Pedestrian Navigation by Fusing the IMU and Compass Measurements via UFIR Filtering

Meng Hou,[1] **Yuan Xu**,[2] **and Xiao Liu**[1]

[1]*School of Electrical Engineering and Automation, Qilu University of Technology, Jinan 250353, China*
[2]*The School of Electrical Engineering, University of Jinan, Jinan 250022, China*

Correspondence should be addressed to Meng Hou; houmeng@qlu.edu.cn

Academic Editor: Choon Ki Ahn

In order to overcome the poor observability of yaw measurement for foot-mounted inertial measurement unit (IMU), an integrated IMU+Compass scheme for self-contained pedestrian navigation is presented. In this mode, the compass measurement is used to provide the accurate yaw to improve the accuracy of the attitude transformation matrix for the foot-mounted IMU solution. And then, when the person is in a stance phase during walk, a unbiased finite impulse response (UFIR) filter based on the self-contained pedestrian navigation scheme is investigated, which just needs the state vector size M^U and the filtering horizon size N^U, while ignoring the noise statistics compared with the Kalman filter (KF). Finally, a real test has been done to verify the performance of the proposed self-contained pedestrian navigation using the IMU and compass measurements via UFIR filter. The test results show that the proposed filter has robust performance compared with the KF.

1. Introduction

Indoor pedestrian navigation (PN), which provides the position and the heading information of the target person in indoor environment, has received a topic [1, 2].

In order to provide the accurate position information of the person in indoor environment, many approaches have been proposed. For example, in [3, 4], the radio frequency identification (RFID)-based technologies have been proposed provide object self-localization. Yang and Shao and Ma et al. proposed in [5, 6] an autonomous positioning system operating on WiFi, which is also able to achieve the indoor localization in indoor environment. However, it should be pointed out that although the RFID- and WiFi-based methods mentioned above are able to provide the position information in indoor environment, the accuracy of such approaches is on meter-level. In order to improve the positioning accuracy, ultrasonic-based approaches are proposed in [7]. However, it should be pointed that although the ultrasonic-based approaches are on centimeter-level, it is easy to be outage. Meanwhile, the ultra wideband (UWB) technology [8] is employed in some approaches. For example, a location detection and tracking of moving targets by a 2D IR-UWB radar system is presented in [9]. It should be emphasized that the methods mentioned above have to employ extra infrastructures; moreover, the sampling time of these methods is larger [1, 10].

In order to overcome the disadvantages of the methods mentioned above, the inertial navigation systems (INS) have been employed for providing the human position in global positioning system- (GPS-) denied areas. One of the famous examples is the navigation shoe proposed in [11], which employs the foot-mounted inertial measurement unit (IMU) to correct the error drift of the INS solution. Based on this model, there are many improving approaches. For example, in [12–14], the magnetic sensor is used to correct the positioning error of the foot-mounted IMU. On the other hand, there are also many approaches for the signal processing in INS. For example, the analysis for microelectromechanical system (MEMS) gyroscope within wide-temperature range is shown in [15].

Based on the INS, the Kalman filter (KF) and its improving methods are widely used to correct the INS solution error [16, 17]. However, despite great progress in the development of the KF approach, the recursive KF-based algorithms demonstrate good performance mostly when the noise statistics are known exactly, and the model perfectly matches the process; otherwise, the KF-based estimators often demonstrate poor performance [18–20]. In order to provide the robust estimation, the unbiased finite impulse response (UFIR) filter has been proposed in [21–23]. Then, there are some improving UFIR filters proposed to improve the performance [24–27]. Compared with the KF-based filters, UFIR filters just need the state vector size and the filtering horizon size, while ignoring the noise statistics [25, 28, 29].

In this paper, we propose a self-contained pedestrian navigation by fusing the IMU and compass measurements via UFIR filtering. To the self-contained pedestrian navigation scheme, the compass measurement is used to provide the accurate yaw to improve the accuracy of the attitude transformation matrix for the foot-mounted IMU solution. Moreover, we investigate the UFIR filter based on the self-contained pedestrian navigation scheme, which just needs the state vector size M^U and the filtering horizon size N^U, while ignoring the noise statistics. Finally, a real test has been done to verify the performance of the proposed self-contained pedestrian navigation using the IMU and compass measurements via UFIR filter. The test results show that the proposed filter has robust performance compared with the KF. The remaining part of this paper is organized as follows. Section 2 designs the scheme of the self-contained pedestrian navigation using the IMU and compass measurements. Section 3 presents the UFIR algorithms for the self-contained pedestrian navigation. Testing and results are discussed in Section 4. Finally, Section 5 gives the conclusions.

2. The Self-Contained Pedestrian Navigation Using the IMU and Compass Measurements

In this section, the scheme of self-contained pedestrian navigation using the IMU and compass measurements will be designed. Then, the state and measurement equations based on the scheme which we designed will be investigated.

2.1. The Scheme of Self-Contained Pedestrian Navigation Using the IMU and Compass Measurements. In this subsection, we will introduce the scheme of self-contained pedestrian navigation using the IMU and compass measurements. The architecture of the self-contained pedestrian navigation employing the recent IMU and compass measurements is shown in Figure 1. In this work, we employ inertial measurement unit (IMU) and compass measurements for the indoor self-contained pedestrian navigation. The IMU is fixed on the shoe, and the compass is fixed on the shoulder. In this paper, the IMU is used to provide the acceleration a, angular velocity w, pitch, and the roll, which are used to compute the position and posture of the target person. Compared with the IMU fixed on the shoe, thigh, shank, and waist, the IMU fixed on

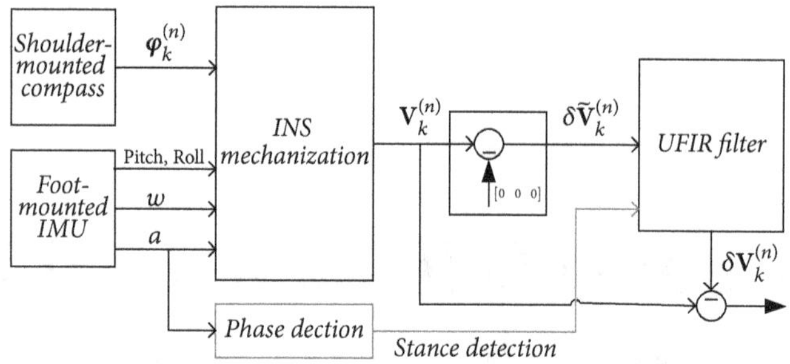

FIGURE 1: The architecture of the self-contained pedestrian navigation using the IMU and compass measurements.

the shoulder has the best performance for yaw measurement. Thus, we employ shoulder-mounted IMU to provide the yaw measurement for the solution, which can improve the accuracy of the attitude transformation matrix for the foot-mounted IMU solution. In theory, the true velocity of the pedestrian shoes should be zero when the shoe is on the floor (so called stance phase). Thus, the velocity measured from the foot-mounted IMU will be the error measurement of the velocity in the stance phase. And when the person is in a stance phase, the unbiased finite impulse response (UFIR) filter works; it employs the measurement of the velocity error derived from the foot-mounted IMU to estimate the INS position error; then, the INS solution is corrected by the error estimation.

Meanwhile, the body frame (so called b-frame) and the navigation frame (so called n-frame (East-North-Up, ENU)) used in this paper are also shown in this figure. Compared with the outdoor navigation, the area of the indoor self-contained pedestrian navigation is very small; thus, the earth's rotation is not considered in this paper.

2.2. The State and Measurement Equations for the UFIR Filter. In this subsection, the state and measurement equations for the UFIR filter will be designed. Based on the self-contained pedestrian navigation scheme proposed in Section 2.1, a UFIR filter with a 15-element vector will be introduced in this paper. The state equation of the UFIR filter used in this paper at time step tt is listed as

$$\underbrace{\begin{bmatrix} \phi_{tt}^{(n)} \\ \delta\mathbf{V}_{tt}^{(n)} \\ \nabla_{tt}^{(n)} \\ \varepsilon_{tt}^{(n)} \end{bmatrix}}_{\mathbf{X}_{tt}} = \underbrace{\begin{bmatrix} \mathbf{I}_{3\times3} & \mathbf{0}_{3\times3} & \mathbf{0}_{3\times3} & \mathbf{I}_{3\times3}T \\ S\left(\mathbf{f}_{tt}^{(n)}\right)T & \mathbf{I}_{3\times3} & \mathbf{I}_{3\times3}T & \mathbf{0}_{3\times3} \\ \mathbf{0}_{3\times3} & \mathbf{0}_{3\times3} & \mathbf{I}_{3\times3} & \mathbf{0}_{3\times3} \\ \mathbf{0}_{3\times3} & \mathbf{0}_{3\times3} & \mathbf{0}_{3\times3} & \mathbf{I}_{3\times3} \end{bmatrix}}_{\mathbf{A}_{tt}} \underbrace{\begin{bmatrix} \phi_{tt-1}^{(n)} \\ \delta\mathbf{V}_{tt-1}^{(n)} \\ \nabla_{tt-1}^{(n)} \\ \varepsilon_{tt-1}^{(n)} \end{bmatrix}}_{\mathbf{X}_{tt-1}} + \omega_{tt-1}, \tag{1}$$

where $[\phi_{tt}^{(n)}, \delta\mathbf{V}_{tt}^{(n)}]^T$ denotes attitude and velocity error vectors in n-frame at time step tt and $[\nabla_{tt}^{(b)}, \varepsilon_{tt}^{(b)}]^T$ denotes the biases for accelerometers and gyroscopes in b-frame at time

Data: $\mathbf{Z}_{tt}, \mathbf{X}_0, \mathbf{P}_0, \mathbf{Q}, \mathbf{R}_{tt}$
Result: $\widehat{\mathbf{X}}_{tt}$
begin
  for $tt = 1 : \infty$ **do**
    $\widehat{\mathbf{X}}_{tt|tt-1} = \mathbf{A}_{tt}\widehat{\mathbf{X}}_{tt-1}$
    $\mathbf{P}_{tt|tt-1} = \mathbf{A}_{tt-1}\mathbf{P}_{tt-1}\mathbf{A}_{tt-1}^{T} + \mathbf{Q}$
    $\mathbf{K}_{tt} = \mathbf{P}_{tt|tt-1}\mathbf{H}_{tt}^{T}(\mathbf{H}_{k}\mathbf{P}_{tt|tt-1}\mathbf{H}_{tt}^{T} + \mathbf{R}_{tt})^{-1}$
    $\widehat{\mathbf{X}}_{tt} = \widehat{\mathbf{X}}_{tt|tt-1} + \mathbf{K}_{tt}(\mathbf{Z}_{tt} - \mathbf{H}_{k}\widehat{\mathbf{x}}_{tt|tt-1})$
    $\mathbf{P}_{tt} = (\mathbf{I} - \mathbf{K}_{tt}\mathbf{H}_{tt})\mathbf{P}_{tt|tt-1}$
  end for
end

ALGORITHM 1: Kalman filter algorithm for the self-contained scheme.

step tt, respectively. All these 5 components mentioned above have 3 elements each:

$$S\left(\mathbf{f}_{tt}^{(n)}\right) = \begin{bmatrix} 0 & a_{Dtt}^{(n)} & -a_{Ntt}^{(n)} \\ -a_{Dtt}^{(n)} & 0 & a_{Ett}^{(n)} \\ a_{Ntt}^{(n)} & -a_{Ett}^{(n)} & 0 \end{bmatrix}; \quad (2)$$

here, $\left[a_{Ett}^{(n)}\ a_{Ntt}^{(n)}\ a_{Dtt}^{(n)}\right]$ is the acceleration in n-frame (East-North-Up, ENU) at time step tt; T denotes the sampling time; ω_{tt} is a system noise at time step tt with the covariance $\mathbf{Q}_{tt}$.

The measurement equation is listed in

$$\underbrace{\left[\delta\widetilde{\mathbf{V}}_{tt}^{(n)}\right]}_{\mathbf{Z}_{tt}} = \underbrace{\left[\mathbf{0}_{3\times3}\ \mathbf{I}_{3\times3}\ \mathbf{0}_{3\times3}\ \mathbf{0}_{3\times3}\right]}_{\mathbf{H}_{tt}} \underbrace{\begin{bmatrix} \phi_{tt}^{(n)} \\ \delta\mathbf{V}_{tt}^{(n)} \\ \nabla_{tt}^{(n)} \\ \varepsilon_{tt}^{(n)} \end{bmatrix}}_{\mathbf{X}_{tt}} + \eta_{tt}, \quad (3)$$

where $\delta\widetilde{\mathbf{V}}_{tt}^{(n)}$ is the observations for the velocity error in n-frame at time step tt; η_{tt} is a measurement noise with covariance is $\mathbf{R}_{tt}$.

Data: $\mathbf{Z}_{tt}, M^{U}, N^{U}$
Result: $\widehat{\mathbf{X}}_{tt}$
begin
  for $tt = N^{U} : \infty$ **do**
    $m = tt - N^{U} + 1, \quad s = m + M^{U} - 1$
    $\mathbf{H}_{s,m} = \begin{bmatrix} \mathbf{H}_{m+3}\mathbf{A}_{m+3}\mathbf{A}_{m+2}\mathbf{A}_{m+1} \\ \mathbf{H}_{m+2}\mathbf{A}_{m+2}\mathbf{A}_{m+1} \\ \mathbf{H}_{m+1}\mathbf{A}_{m+1} \\ \mathbf{H}_{m} \end{bmatrix}$
    $\mathbf{A}_{s,0}^{t} = \mathbf{A}_{s}\mathbf{A}_{s-1}\cdots\mathbf{A}_{m+1}$
    $\mathbf{G}_{s} = \mathbf{A}_{s,0}^{t}(\mathbf{H}_{s,m}^{T}\mathbf{H}_{s,m})^{-1}(\mathbf{A}_{s,0}^{t})^{T}$
    if $tt = N$ **then**
      $\mathbf{Z}_{s,m} = \left[\mathbf{Z}_{m+3}^{T}\ \mathbf{Z}_{m+2}^{T}\ \mathbf{Z}_{m+1}^{T}\ \mathbf{Z}_{m}^{T}\right]^{T}$
      $\widetilde{\mathbf{X}}_{s} = \mathbf{A}_{s,0}(\mathbf{H}_{s,m}^{T}\mathbf{H}_{s,m})^{-1}\mathbf{H}_{s,m}^{T}\mathbf{Z}_{s,m}^{T}$
    end if
    if $tt > N$ **then**
      $\widetilde{\mathbf{X}}_{s} = \widehat{\mathbf{X}}_{tt}$
    end if
    for $kk = s + 1 : tt$ **do**
      $\widetilde{\mathbf{X}}_{kk|kk-1} = \mathbf{A}_{kk}\widetilde{\mathbf{X}}_{kk-1} + \omega_{kk}$
      $\mathbf{G}_{kk} = [\mathbf{H}_{kk}^{T}\mathbf{H}_{kk} + (\mathbf{A}_{kk}\mathbf{G}_{kk-1}\mathbf{A}_{kk}^{T})^{-1}]^{-1}$
      $\mathbf{K}_{kk} = \mathbf{G}_{kk}\mathbf{H}_{kk}^{T}$
      $\widetilde{\mathbf{X}}_{kk} = \widetilde{\mathbf{X}}_{kk|kk-1} + \mathbf{K}_{kk}(\mathbf{Z}_{kk} - \mathbf{H}_{kk}\widetilde{\mathbf{X}}_{kk|kk-1})$
    end for
    $\widehat{\mathbf{X}}_{tt} = \widetilde{\mathbf{X}}_{tt}$
  end for
end
† M^{U} is the state vector size
† N^{U} is the filtering horizon size

ALGORITHM 2: UFIR filter algorithm for the self-contained scheme.

3. UFIR Algorithms for the Self-Contained Pedestrian Navigation

In this section, the KF and the UFIR filter for the indoor self-contained pedestrian navigation based on the scheme proposed in Section 2.1 will be discussed.

3.1. KF Algorithm. As one of the most used data fusion algorithm, the Kalman filter (KF) and its improving methods have been widely used in many fields [30–34]. And its pseudo code is listed as Algorithm 1. It should be pointed out that the Kalman filter and its improving methods need the accurate model description and noise description, especially the accurate $\mathbf{Q}$ and $\mathbf{R}_{tt}$, to maintain the performance. However, it is not easy to get the information mentioned above in someplace.

3.2. UFIR Filtering Algorithm. Compared with KF, the UFIR filter does not need the accurate $\mathbf{Q}$ and $\mathbf{R}_k$ to keep its accuracy [26, 35]. Based on the self-contained scheme proposed in Section 2.1, the UFIR filtering algorithm is listed in Algorithm 2. Thus, we can say that the UFIR filter is more robust than the KF filter.

4. Test and Discussion

In this paper, we employ a real indoor test to verify the performance of the proposed self-contained pedestrian navigation using the IMU and compass measurements via UFIR filter. The real test was done in the Machine Building of the University of Jinan, Jinan, China. In this section, the real test will be designed and we will investigate the corresponding results. Firstly, the real test will be designed. Then, the performances of the KF and UFIR filter will be compared.

4.1. Setting. The test platform used in the real test consists of one 9 degree of freedom (DOF) IMU, one compass, one computer, and one encoder. The 9-DOF IMU is fixed on the shoe; it employs ADXL203, ADXRS620, and HMC5983 as accelerometer, gyroscope, and magnetometer, respectively,

Table 1: The parameters for the tests.

	$\mathbf{X}_0$	$\mathbf{P}_0$	$\mathbf{Q}$	$\mathbf{R}$	N^U
Group 1	$\mathbf{0}_{12\times1}$	$\mathbf{I}_{12\times12}$	$\mathbf{I}_{12\times12}$	$\mathbf{I}_{3\times3}$	13
Group 2	$\mathbf{0}_{12\times1}$	$\mathbf{I}_{12\times12}$	$10^{-2}\mathbf{I}_{12\times12}$	$\mathbf{I}_{3\times3}$	13

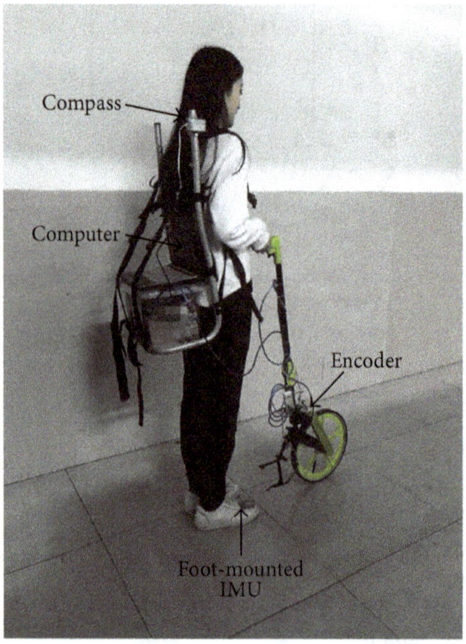

Figure 2: The prototype of the test platform.

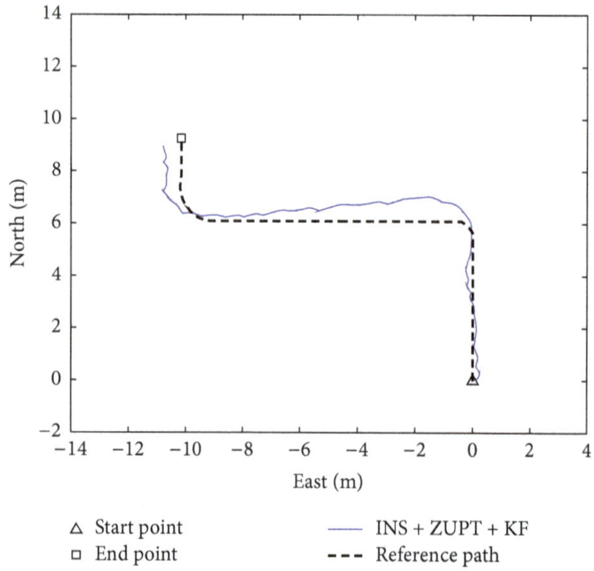

Figure 3: Trajectory measured by INS + ZUPT + KF for the first group of possible parameters (Group 1).

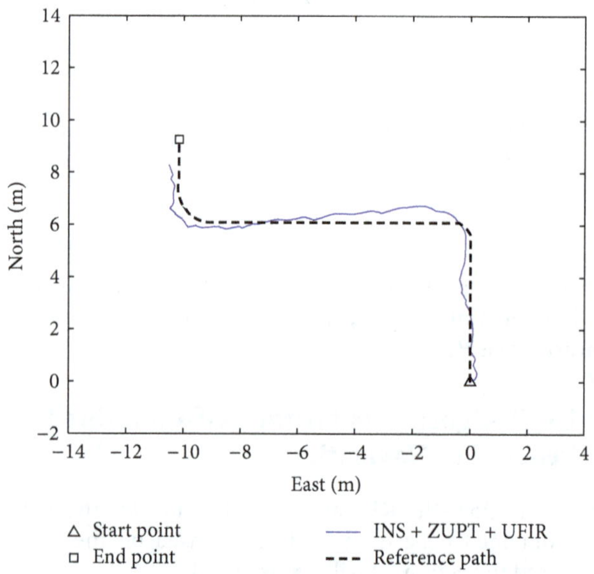

Figure 4: Trajectory measured by INS + ZUPT + UFIR for the first group of possible parameters (Group 1).

which is able to provide the human position. Then, one HMC5983-based compass which is fixed on the shoulder is used to provide more accurate yaw measurement, which helps improve the accuracy of the yaw for the foot-mounted IMU. The computer is used to collect the sensor data. The encoder used in the real test is able to provide the reference velocity of the person. The sampling time T used in (1) is 0.03 s. From Algorithm 2, we can see that the performance of the UFIR filter is just need M^U and N^U. From (1), we can get that $M^U = 12$ and we employ $N^U = 13$ in this paper. The prototype of the test platform is shown in Figure 2.

4.2. The Performance of the UFIR Filter. The position error of the UFIR filter for self-contained pedestrian navigation using the IMU and compass measurements will be discussed in this section. In this subsection, we employ two groups of parameters which listed in Table 1.

(1) INS Position Errors-Group 1. Trajectories measured by INS + ZUPT + KF and INS + ZUPT + UFIR for the first group of possible parameters (Group 1) are shown in Figures 3 and 4. From the figures, firstly, we can see that the proposed self-contained pedestrian navigation scheme is able to provide the person position without any auxiliary equipment. Secondly, we can see that the trajectories estimated by the KF and the UFIR filter are similar.

The absolute average position errors by the KF and UFIR filter in Test 1 (Group 1) are listed in Table 2. From the table, we can see that the position errors of the KF and UFIR filter are similar. Thus, we can see that the performances of KF and UFIR filter are similar in Test 1 (Group 1).

(2) INS Position Errors-Group 2. We now repeat the experiment for the second group of possible parameters (Group

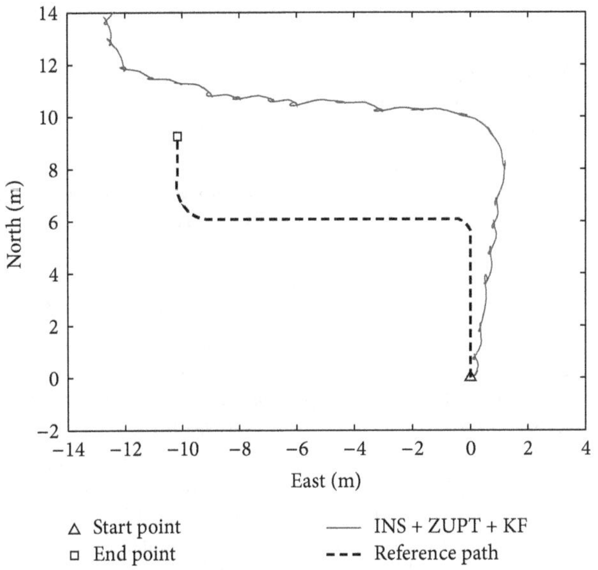

Figure 5: Trajectory measured by INS + ZUPT + KF for the second group of possible parameters (Group 2).

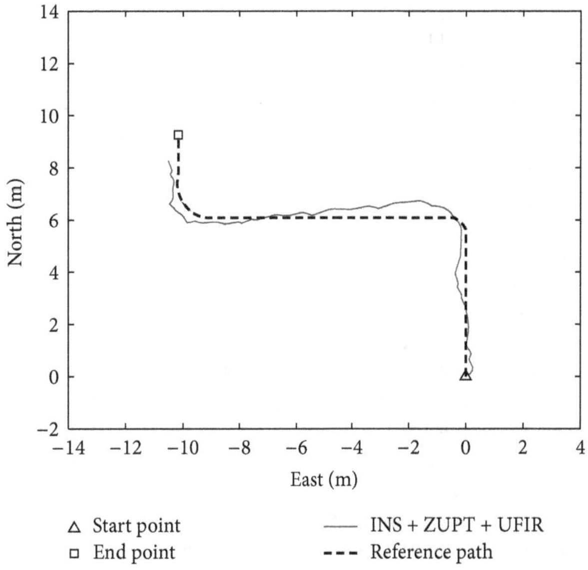

Figure 6: Trajectory measured by INS + ZUPT + UFIR for the second group of possible parameters (Group 2).

Table 2: Absolute average position errors by different filters in Test 1 (Group 1).

Model	Absolute average position errors [m]	
	East direction	North direction
KF	0.25	0.29
UFIR	0.22	0.31

Table 3: Absolute average position errors by different filters in Test 2 (Group 2).

Model	Absolute average position errors [m]	
	East direction	North direction
KF	0.88	2.79
UFIR	0.22	0.31

2). The trajectories measured from UWB only model using least square algorithm and the INS/UWB loosely coupled integrated model using the KF and UFIR filter for the second group of possible parameters (Group 2) are shown in Figures 5 and 6. And the absolute average position errors by the KF and UFIR filter in Test 2 (Group 2) are listed in Table 3. From the figures, we can see easily that the KF has been diverged, while the performance of the UFIR filter is still good.

5. Conclusion

In this paper, a self-contained pedestrian navigation by fusing the IMU and compass measurements via UFIR filtering and both the implement and test are proposed. To the self-contained pedestrian navigation scheme, the compass measurement is used to provide the accurate yaw to improve the accuracy of the attitude transformation matrix for the foot-mounted IMU solution. Moreover, we investigate the UFIR filter based on the self-contained pedestrian navigation scheme, which just needs the state vector size M^U and the filtering horizon size N^U, while ignoring the noise statistics. Finally, a real test has been done to verify the performance of the proposed self-contained pedestrian navigation using the IMU and compass measurements via UFIR filter. The test results show that the proposed filter has robust performance compared with the KF.

Conflicts of Interest

The authors declare no conflicts of interest.

Acknowledgments

This work was supported in part by the China Postdoctoral Science Foundation (no. 2017M622204) and the Doctoral Foundation of the University of Jinan (no. XBS1503).

References

[1] Y. Xu, X. Chen, J. Cheng, Q. Zhao, and Y. Wang, "Improving tightly-coupled model for indoor pedestrian navigation using foot-mounted IMU and UWB measurements," in *Proceedings of the 2016 IEEE International Instrumentation and Measurement Technology Conference, I2MTC 2016*, Taiwan, May 2016.

[2] Y. Zhuang, H. Lan, Y. Li, and N. El-Sheimy, "PDR/INS/WiFi integration based on handheld devices for indoor pedestrian navigation," *Micromachines*, vol. 6, no. 6, pp. 793–812, 2015.

[3] A. Aguilar-Garcia, S. Fortes, E. Colin, and R. Barco, "Enhancing RFID indoor localization with cellular technologies," *EURASIP Journal on Wireless Communications and Networking*, vol. 2015, no. 1, article no. 219, 2015.

[4] J. J. Pomarico-Franquiz and Y. S. Shmaliy, "Accurate self-localization in rfid tag information grids using fir filtering,"

IEEE Transactions on Industrial Informatics, vol. 10, no. 2, pp. 1317–1326, 2014.

[5] C. Yang and H.-R. Shao, "WiFi-based indoor positioning," *IEEE Communications Magazine*, vol. 53, no. 3, pp. 150–157, 2015.

[6] R. Ma, Q. Guo, C. Hu, and J. Xue, "An improved WiFi indoor positioning algorithm by weighted fusion," *Sensors*, vol. 15, no. 9, pp. 21824–21843, 2015.

[7] X. Chen, Y. Xu, Q. Li, J. Tang, and C. Shen, "Improving ultrasonic-based seamless navigation for indoor mobile robots utilizing EKF and LS-SVM," *Measurement*, vol. 92, pp. 243–251, 2016.

[8] Y. Xu, Y. S. Shmaliy, Y. Li, and X. Chen, "UWB-Based Indoor Human Localization with Time-Delayed Data Using EFIR Filtering," *IEEE Access*, vol. 5, pp. 16676–16683, 2017.

[9] V.-H. Nguyen and J.-Y. Pyun, "Location detection and tracking of moving targets by a 2D IR-UWB radar system," *Sensors*, vol. 15, no. 3, pp. 6740–6762, 2015.

[10] Y. Xu, X. Chen, and Y. Wang, "Two-mode navigation method for low-cost inertial measurement unit-based indoor pedestrian navigation," *Simulation*, vol. 92, no. 9, pp. 839–848, 2016.

[11] E. Foxlin, "Pedestrian tracking with shoe-mounted inertial sensors," *IEEE Computer Graphics and Applications*, vol. 25, no. 6, pp. 38–46, 2005.

[12] A. R. Jimenez, F. Seco, J. C. Prieto, and J. Guevara, "Indoor Pedestrian navigation using an INS/EKF framework for yaw drift reduction and a foot-mounted IMU," in *Proceedings of the 2010 7th Workshop on Positioning, Navigation and Communication, WPNC'10*, pp. 135–143, Germany, March 2010.

[13] Z.-Q. Zhang and X. Meng, "Use of an inertial/magnetic sensor module for pedestrian tracking during normal walking," *IEEE Transactions on Instrumentation and Measurement*, vol. 64, no. 3, pp. 766–783, 2015.

[14] Y. Xu, X. Chen, Q. Li, and J. Tang, "Indoor pedestrian navigation using two low-cost IMU-based framework," *Journal of Chinese Inertial Technology*, vol. 23, no. 6, pp. 714–717, 2015.

[15] H. Cao, H. Li, X. Shao et al., "Sensing mode coupling analysis for dual-mass MEMS gyroscope and bandwidth expansion within wide-temperature range," *Mechanical Systems and Signal Processing*, vol. 98, pp. 448–464, 2018.

[16] Y. Xu and X. Chen, "Online cubature Kalman filter Rauch–Tung–Striebel smoothing for indoor inertial navigation system/ultrawideband integrated pedestrian navigation," *Proceedings of the Institution of Mechanical Engineers, Part I: Journal of Systems and Control Engineering*, vol. 232, no. 4, pp. 390–398, 2017.

[17] S. Zhao, Y. S. Shmaliy, P. Shi, and C. K. Ahn, "Fusion Kalman/UFIR Filter for State Estimation With Uncertain Parameters and Noise Statistics," *IEEE Transactions on Industrial Electronics*, vol. 64, no. 4, pp. 3075–3083, 2017.

[18] X. Qing, H. R. Karimi, Y. Niu, and X. Wang, "Decentralized unscented Kalman filter based on a consensus algorithm for multi-area dynamic state estimation in power systems," *International Journal of Electrical Power & Energy Systems*, vol. 65, pp. 26–33, 2015.

[19] H. R. Karimi, N. A. Duffie, and S. Dashkovskiy, "Local capacity h∞ control for production networks of autonomous work systems with time-varying delays," *IEEE Transactions on Automation Science and Engineering*, vol. 7, no. 4, pp. 849–857, 2010.

[20] Y. Wei, J. Qiu, H. R. Karimi, and M. Wang, "Filtering design for two-dimensional Markovian jump systems with state-delays and deficient mode information," *Information Sciences*, vol. 269, pp. 316–331, 2014.

[21] Y. Xu, C. K. Ahn, Y. S. Shmaliy, X. Chen, and Y. Li, "Adaptive robust INS/UWB-integrated human tracking using UFIR filter bank," *Measurement*, vol. 123, pp. 1–7, 2018.

[22] Y. S. Shmaliy, S. Zhao, and C. K. Ahn, "Unbiased finite impulse response filtering: an iterative alternative to Kalman filtering ignoring noise and initial conditions," *IEEE Control Systems Magazine*, vol. 37, no. 5, pp. 70–89, 2017.

[23] Y. S. Shmaliy, S. Khan, and S. Zhao, "Ultimate iterative UFIR filtering algorithm," *Measurement*, vol. 92, pp. 236–242, 2016.

[24] C. K. Ahn and Y. S. Shmaliy, "New receding horizon fir estimator for blind smart sensing of velocity via position measurements," *IEEE Transactions on Circuits and Systems II: Express Briefs*, vol. 65, no. 1, pp. 135–139, 2018.

[25] Y. Xu, H. R. Karimi, Y. Li, F. Zhou, and L. Bu, "Real-time accurate pedestrian tracking using extended finite impulse response filter bank for tightly coupling recent inertial navigation system and ultra-wideband measurements," *Proceedings of the Institution of Mechanical Engineers, Part I: Journal of Systems and Control Engineering*, vol. 232, no. 4, pp. 464–472, 2017.

[26] C. K. Ahn, Y. S. Shmaliy, and S. Zhao, "A New Unbiased FIR Filter with Improved Robustness Based on Frobenius Norm with Exponential Weight," *IEEE Transactions on Circuits and Systems II: Express Briefs*, 2017.

[27] S. Zhao, Y. S. Shmaliy, C. K. Ahn, and P. Shi, "Real-Time Optimal State Estimation of Multi-DOF Industrial Systems Using FIR Filtering," *IEEE Transactions on Industrial Informatics*, vol. 13, no. 3, pp. 967–975, 2017.

[28] H. R. Karimi and H. Gao, "Mixed H2 / H∞ output-feedback control of second-order neutral systems with time-varying state and input delays," *ISA Transactions®*, vol. 47, no. 3, pp. 311–324, 2008.

[29] H. R. Karimi, P. J. Maralani, B. Lohmann, and B. Moshiri, "h∞ control of parameter-dependent state-delayed systems using polynomial parameter-dependent quadratic functions," *International Journal of Control*, vol. 78, no. 4, pp. 254–263, 2005.

[30] S. Eftekhar Azam, E. Chatzi, and C. Papadimitriou, "A dual Kalman filter approach for state estimation via output-only acceleration measurements," *Mechanical Systems and Signal Processing*, vol. 60, pp. 866–886, 2015.

[31] Z. Chen, H. Zou, H. Jiang, Q. Zhu, Y. C. Soh, and L. Xie, "Fusion of WiFi, smartphone sensors and landmarks using the kalman filter for indoor localization," *Sensors*, vol. 15, no. 1, pp. 715–732, 2015.

[32] B. Allotta, L. Chisci, R. Costanzi et al., "A comparison between EKF-based and UKF-based navigation algorithms for AUVs localization," in *Proceedings of the MTS/IEEE OCEANS 2015 - Genova*, Italy, May 2015.

[33] Y. Li, H. R. Karimi, Q. Zhang, D. Zhao, and Y. Li, "Fault Detection for Linear Discrete Time-Varying Systems Subject to Random Sensor Delay: A Riccati Equation Approach," *IEEE Transactions on Circuits and Systems I: Regular Papers*, 2017.

[34] Y. Li, H. R. Karimi, C. K. Ahn, Y. Xu, and D. Zhao, "Optimal residual generation for fault detection in linear discrete time-varying systems with uncertain observations," *Journal of The Franklin Institute*, vol. 355, no. 7, pp. 3330–3353, 2018.

[35] S. Zhao, Y. S. Shmaliy, B. Huang, and F. Liu, "Minimum variance unbiased FIR filter for discrete time-variant systems," *Automatica*, vol. 53, pp. 355–361, 2015.

13

An Application-Level QoS Control Method based on Local Bandwidth Scheduling

Yong Wang, Fu Xu, Zhibo Chen, Yu Sun, and Haiyan Zhang

School of Information Science and Technology, Beijing Forestry University, Beijing 100083, China

Correspondence should be addressed to Fu Xu; xufu@bjfu.edu.cn

Academic Editor: Stefano Vitturi

Quality of service (QoS) is an important performance indicator for Web applications and bandwidth is a key factor affecting QoS. Current methods use network protocols or ports to schedule bandwidth, which require tedious manual configurations or modifications of the underlying network. Some applications use dynamic ports and the traditional port-based bandwidth control methods cannot deal with them. A new QoS control method based on local bandwidth scheduling is proposed, which can schedule bandwidth at application level in a user-transparent way and it does not require tedious manual configurations. Experimental results indicate that the new method can effectively improve the QoS for applications, and it can be easily integrated into current Web applications without the need to modify the underlying network.

1. Introduction

Improving bandwidth efficiency is a focus in academic research and industrial innovation. Internet adopts a best-effort service model, which lacks the capability of QoS guarantee inherently [1]. Although traditional methods such as Resource Reservation Protocol (RSVP) [2] and Differentiated Service (DiffServ) [3] can improve the QoS for Web applications effectively, they require modifying the underlying network [4]. For example, Neto proposed a multicast-aware RSVP for class-based networks [5] and FU Qi proposed a service-aware multipath QoS strategy, which achieved the fair use of different types of traffic channels and wireless links [6]. All these methods require modifying the underlying network. Allowing for the diversity of network types, these modifications are not always feasible [7, 8]. Some QoS methods such as Traffic Control (TC) [9] do not require modifying the underlying network, which schedule bandwidth based on network protocols or ports. Manual configurations are inevitable when using these methods. This process is very cumbersome and requires highly specialized expertise. Even network experts cannot handle this task easily. Furthermore, some applications such as uTorrent [10] use dynamic ports and traditional port-based bandwidth control methods cannot deal with them.

With the rapid emergence of Web applications, the demand for application-level QoS control on local computers is growing. The disorderly competition in bandwidth usage may cause bandwidth-sensitive applications working abnormally. Although the traditional fixed bandwidth allocation strategy can avoid this problem and guarantee the QoS, it can lead to bandwidth idling.

A new QoS control method is proposed in this research, which uses an improved token bucket algorithm and can schedule bandwidth at application level in a user-transparent manner. This method dynamically allocates bandwidth based on the bandwidth requirements of applications, which achieves a good balance between bandwidth usage efficiency and QoS guarantee. It can maximize the bandwidth usage while guaranteeing the QoS for applications running in the same computer. In addition, the new method does not require modifying the underlying network and can avoid the problem of tedious configurations.

2. Related Work

2.1. DiffServ. DiffServ is often used to ensure the QoS of backbone networks, which classifies data flows into different levels according to their QoS requirements. High-level data flows are preferentially transmitted than those at low levels

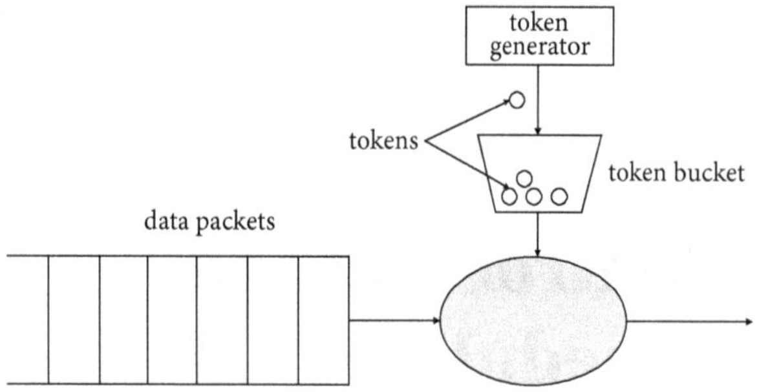

FIGURE 1: Flowchart of token bucket algorithm.

when encountering congestions. More details on DiffServ can be found in RFC2474 [11].

DiffServ has become the current mainstream QoS architecture because of its good scalability, simplicity, and operability. However, DiffServ is not an end-to-end method and it cannot be used without modifying the underlying network.

2.2. Token Bucket Algorithm. As a classical bandwidth control method, token bucket algorithm [12] has good capability in congestion processing, which consists of three components: token, token bucket, and token generator (Figure 1). Token controls the transmission of data packets and token generator generates tokens to fill into the token bucket. When there are enough tokens in the bucket, data packets can be transmitted. Otherwise, they will be buffered or discarded.

2.3. Review of Existing Methods for Bandwidth Control. There are lots of researches in QoS guarantee. Most of them schedule bandwidth at router or network layer, which are not application-level based scheduling methods, so that they cannot avoid the confliction of bandwidth usage in the local computer.

Lin N. proposed a QoS control method based on DiffServ and token bucket algorithm [13], which designed three services of QoS: Guaranteed Service (GS), Control Load Service (CLS), and Best Effort Service (BES). Protocols and ports in the packet headers were used to distinguish these services. This method can guarantee the QoS for applications with GS type. It cannot deal with the applications using dynamic ports and requires modifying the underlying network and making port-based manual configurations.

Kidambi proposed a token bucket based method, in which bandwidth was equally divided among data flows [14]. This method can guarantee the QoS for applications but has the problem of idle bandwidth due to the fixed bandwidth allocation strategy.

Cucinotta proposed a QoS control method for real-time applications, which can guarantee the QoS for real-time applications [15]. This method uses fixed bandwidth allocation strategy, which has the similar problem with Kidambi's method.

Hierarchical token bucket [16] method can avoid idle bandwidth, but it requires tedious manual configurations based on network protocols or ports. It is not an application-level method and cannot schedule the bandwidth for applications using dynamic ports.

3. The New Application-Level QoS Control Method

Bandwidth is a key performance indicator for many Web applications. For example, a VOIP application using G.711 codec requires at least 64 KB/s bandwidth to guarantee good call quality [17]. In the new method, priorities are assigned to applications according to their bandwidth requirements. An improved token bucket algorithm is designed, which can dynamically schedule bandwidth based on the priorities of applications. The following two steps describe the new method in detail.

3.1. Application-Level Bandwidth Scheduling. Popular operating systems (OS) such as Linux, Windows, and MacOS usually support some priority mechanisms. These mechanisms are more like a gentleman's agreement and many applications do not follow them strictly, so that these mechanisms cannot solve the problem in bandwidth control. Moreover, priorities assigned by OS are based on CPU timeslicing instead of bandwidth [18], so that they cannot guarantee the QoS for Web applications.

To achieve application-level bandwidth scheduling, the network packets should be associated with their corresponding processes. The hooking mechanism [19] provided by OS can be used to do this. For example, WFP (Windows Filtering Platform) [20] hooking mechanism can be used to filter network packets on Windows. Other operating systems such as Linux or MacOS also have similar mechanisms.

The new application-level bandwidth scheduling method consists of four steps. Let us take Windows as an example.

(1) Build a priority database to store the fingerprints of applications. The fingerprint can be the MD5 hash of an application, the application name, or other tags that can uniquely identify the application. In this research, MD5 hashes are used to generate the fingerprints. Except the fingerprints, the bandwidth requirements and application priorities are also stored in the priority database.

(2) Hook the packet sending and receiving functions in network protocol libraries and use WFP interfaces to obtain the corresponding process IDs (PIDs) for these packets. Retrieve the application's full paths through PIDs and generate fingerprints using these paths.

(3) Query the priority database using the fingerprints to retrieve the application's priorities. If an application is not configured in the priority database, it will be assigned the lowest priority.

(4) Schedule bandwidth at application level using the method described in Section 3.2.

3.2. Bandwidth Dynamic Scheduling Method. There are two typical bandwidth scheduling methods. One method is to let applications freely compete for bandwidth, which can maximize bandwidth usage and has no problem of idle bandwidth. The drawback is that it cannot guarantee the

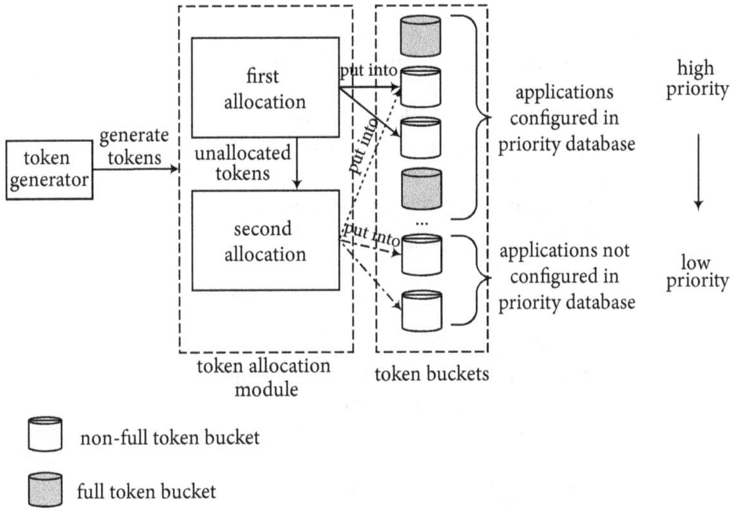

FIGURE 2: Flowchart of bandwidth dynamic scheduling method.

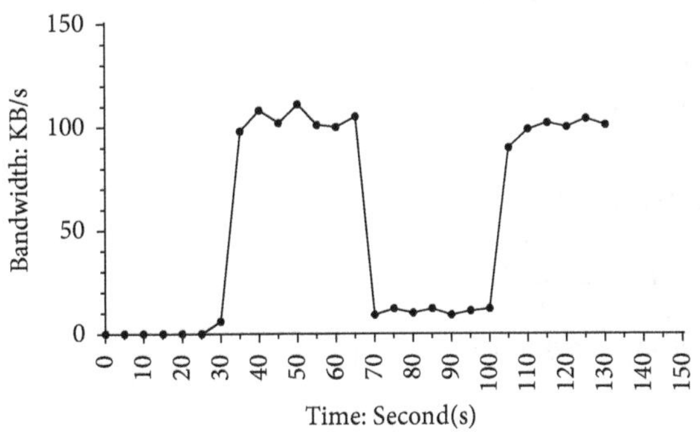

FIGURE 3: Upload bandwidth control.

QoS for applications. The other method is to preallocate bandwidth for applications. Linksys P-WRT1900ACS router [21] uses this method, which preallocates bandwidth for each port or IP to ensure the bandwidth will not be overconsumed by some applications. The second method can guarantee the QoS for applications, but it has the problem of idle bandwidth and cannot maximize the bandwidth usage.

Tokens are privately owned by applications in traditional token bucket algorithm, which are not shared with other applications. This mechanism can guarantee the QoS for applications but has the problem of idle bandwidth. An improved token bucket algorithm is proposed in this research, which designs a borrowing mechanism to make tokens sharable among applications. A new bandwidth dynamic scheduling method based on the improved token bucket algorithm is proposed, which can avoid the problem of idle bandwidth while guaranteeing the QoS for applications. Figure 3 is the flowchart of the new method, which consists of three components: token generator, token allocation module, and token buckets. Token generator generates tokens and puts them into token buckets through token allocation module. Each application has a privately owned token bucket and all applications share the same token generator and token allocation module. The creation and termination of a process can be monitored by hooking process management functions (such as CreateProcess [22] and TerminateProcess [23]). When a process is started or terminated, its corresponding token bucket is built or destroyed simultaneously. A packet can be transmitted only if its corresponding token bucket has enough tokens. Otherwise, the packet will be buffered and suspended to transmit until there are enough tokens.

To achieve bandwidth dynamic scheduling through the above mechanism, the following three questions should be answered: (1) What packet size each token corresponds to? (2) What is the token generation rate? (3) What is the capacity of each token bucket? For question (1), each 1500-byte packet is associated with a token allowing that the Ethernet MTU (Maximum Transmission Unit) is such size [24]. For question (2), the token generation rate should match the available bandwidth of the current computer. The available bandwidth can be calculated by monitoring the peak transmission speed within a period of time. Allowing the non-real-time characteristic of popular operating systems, the token generation rate is set to 1.2 times the actual bandwidth to avoid the possible token generation delays. For question (3), if an application is listed in the priority database, its token bucket capacity is set to 1.2 times the configured bandwidth to match the total token generation rate. If it is not listed in the priority database, the capacity is set to a preconfigured value.

Token allocation module in Figure 2 is the pivot of bandwidth dynamic scheduling method, which consists of two allocations. In the first allocation, the allocation module will check the token bucket of each running process listed in priority database in descending order of priority. If the bucket is not full, put tokens into it according to the configured bandwidth in the priority database. Otherwise, the tokens will be reserved to the second allocation. In the second allocation, a token borrowing mechanism is designed to realize token sharing.

```
Input: The tokens generated by generator (L_token)
Output: The first token allocation policy (L_allocation)
(1) FirstAllocation (L_token) {
(2)  Obtain the running Web process list L_process
(3)  Query the priorities of processes in L_process in the
(4)     priority database
(5)  Sort L_process in descending order of priority
(6)  Initialize the first allocation policy L_allocation
(7)  FOR i=0 to COUNT(L_process)-1 {
(8)    Mark the bucket of i_th process in L_process as B_i
(9)    IF (f_currentSize(B_i) < f_maxSize(B_i)) {
(10)    // Non-full bucket
(11)    Get the bandwidth V_i for the corresponding
(12)       process in the priority database
(13)    IF (V_i != null) { // Process exists in the priority database
(14)    // T_i is the remaining capacity of B_i
(15)     T_i = f_maxSize(B_i) - f_currentSize(B_i)
(16)     Initialize the allocation policy (allocation) for B_i
(17)     // Set the bucket member of allocation (token ID)
(18)     allocation.BUCKET = B_i
(19)     // Set the TOKEN member of allocation (token count)
(20)     // Add the smaller one between 1.2*V_i and T_i to the
(21)     // current token bucket
(22)     allocation.TOKEN = allocation.TOKEN+MIN(T_i, 1.2*V_i)
(23)     Add allocation into the first allocation policy L_allocation
(24)     Delete the allocated tokens from L_token
(25)    }
(26)    ELSE { // All processes in priority database checked
(27)     // A null value of V_i means that the corresponding process
(28)     // of B_i does not exist in the priority database and the
(29)     // remaining processes do not exist in the database
(30)     // either.These processes won't be allocated any tokens
(31)     // in the first allocation.
(32)     RETURN L_allocation
(33)    }
(34)   }
(35) }
(36) RETURN L_allocation // All processes checked
(37) }
```

Algorithm 1: Bandwidth dynamic scheduling method (the first allocation).

The *maxBorrow* parameter is introduced to control how much bandwidth an application can borrow from other applications, which can be set to a larger value when wanting to borrow more bandwidth or set to zero to disable borrowing. *MaxBorrow* will gradually decrease with the transmissions of packets. When the *maxBorrow* for a process decreases to zero, the *maxBorrow* for all the processes with higher or equal priority will be reset to their initial values, while the *maxBorrow* for the other processes will remain unchanged. This means that, when detecting a zero value of *maxBorrow* for some process, the token borrowing privilege will always be granted to all the processes with higher or equal priority simultaneously, and the processes with lower priority will not get such privilege.

The first allocation guarantees that every running process listed in the priority database can obtain the configured bandwidth. The following data structure is introduced to specify the allocation process.

```
struct allocation{

    BUCKET          // ID of token bucket
    TOKEN           // The token count

}
```

Algorithm 1 is the pseudo code for the first allocation. Lines starting with // are comments and all the other lines are valid code. The number at the beginning of each line is the line number. The input of the algorithm is the generated tokens (L_{token}) and the output is the first token allocation policy ($L_{allocation}$).

In Algorithm 1, firstly obtain the running process list ($L_{process}$, *line (2)*), then query the priority database to retrieve the priorities of processes in $L_{process}$ (*lines (3)-(4)*) and sort these processes in descending order of priority (*line (5)*). Initialize the allocation policy ($L_{allocation}$, *line (6)*) and loop through $L_{process}$ (*lines (7)-(35)*). Use functions $f_{\_currentSize}$ and $f_{\_maxSize}$ to get the current and maximal capacity of a token bucket (B_i), respectively, and check whether the bucket is full. If it is not full (*lines (9)-(10)*), query the configured bandwidth (V_i) in the priority database (*lines (11)-(12)*). If V_i is not null, get the remaining capacity of B_i and mark it as T_i (*lines (13)-(15)*). Initialize the allocation policy for B_i (*line (16)*), assign B_i to its BUCKET member and add MIN(T_i, $1.2*V_i$) to its TOKEN member (*lines (17)-(22)*). After that, add it to the first allocation policy ($L_{allocation}$) and delete the allocated tokens from L_{token} (*lines (23)-(24)*). A null value of V_i represents that its corresponding process is not listed in the priority database. Since $L_{process}$ is looped through in descending order of priority, when finding a process not listed in the priority database, all the remaining processes are not listed in this database either. In this case, stop looping and return $L_{allocation}$ directly (*lines (26)-(33)*). If all the processes in $L_{process}$ have been checked, return $L_{allocation}$ (*line (36)*).

Through the first allocation, all the running processes listed in the priority database have attained tokens and token-based packet transmission privileges. Combining the priority mechanism, the QoS for these processes can be guaranteed.

The unallocated tokens (marked as L'_{token}) in the first allocation will be further allocated to other processes in the second allocation. The token sharing mechanism is introduced to maximize the bandwidth usage during the second allocation. Algorithm 2 is the pseudo code, whose input is the unallocated tokens (L'_{token}) and the output is the second token allocation policy ($L'_{allocation}$).

In Algorithm 2, firstly obtain all the running Web processes not listed in the priority database ($L'_{process}$, *lines (2)-(3)*). These processes do not get any tokens in the first allocation, so that tokens will be allocated to them firstly. Loop through $L'_{process}$ to retrieve the token bucket for each process (B_i, *lines (5)-(6)*). When a nonfull token bucket is found (*line (7)*), get its remaining capacity (T_i, *lines (8)-(9)*). Initialize *allocation* for B_i and set its members (*lines (10)-(15)*), then add it into the second allocation policy (*line (16)*) and remove the allocated tokens from L'_{token} (*line (17)*). Since all the processes in $L'_{process}$ have the same and lowest priority, evenly allocate tokens for them (*lines (20)-(33)*). Firstly, calculate the average count of allocated tokens (*lines (20)-(24)*), then update the token count for the second allocation policy (*lines (25)-(33)*). If all the tokens have been allocated (*line (34)*), return $L'_{allocation}$ directly (*line (35)*). Otherwise, allocate tokens to all the running Web processes (L_p, *lines (37)-(65)*). Loop through L_p to retrieve each token bucket (M_j, *lines (42)-(43)*). If the bucket is not full and its *maxBorrow* ($M_{j\_maxBorrow}$) is greater than zero, allocate tokens for it and add the smallest value among the unallocated token count (COUNT(L'_{token})), the remaining capacity of this bucket (T_j) and its token borrowing count ($M_{j\_maxBorrow}$) to its current bucket count (*lines (44)-(54)*). Add it into the second allocation policy (*line (55)*), and decrease the token borrowing count (*lines (56)-(57)*) and the unallocated token count (*line (58)*), respectively. If all the tokens have been allocated (*line (59)*), return $L'_{allocation}$ (*line (60)*). L'_{token} will be discarded and a null allocation policy will be returned if no bucket satisfies all the above conditions (*line (66)*).

It can be seen that from Algorithm 2, processes not listed in the priority database will be firstly allocated tokens using an average allocation policy, which guarantees that they have equal rights to use bandwidth. Since these processes have the lowest priority, their token bucket capacity is set to a preconfigured value. It is possible that there are still unallocated tokens after this average allocation. These tokens will be further allocated among all the running Web processes regardless of their existence in the priority database. This can maximize the bandwidth usage and avoid the problem of idle bandwidth.

It should be noted that other tool functions and facilities are also needed except those listed in Algorithms 1 and 2. For example, a token generating function should be used to generate tokens periodically; a periodic timer should be used to reset the *maxBorrow* value for each process; a queue should be constructed to buffer the packets in low priorities that cannot be transmitted immediately. These functions and facilities are omitted for the sake of brevity.

4. Experimental Results

Three experiments (Experimental PC settings: Intel i7-3770 CPU, 16G RAM, Windows 7 Professional) were designed to verify the effectiveness of the new method. Experiment 1 verified the capability of bandwidth control. Experiment 2 verified the effectiveness of bandwidth scheduling. Experiment 3 verified the improvement of QoS.

4.1. Capability of Bandwidth Control. *Baidu Netdisk* (a cloud storage application) [25] was used to upload a 1G byte file and its bandwidth usage was illustrated in Figure 3. It can be seen that the upload rate was 100 KB/s at the beginning and it decreased to 10 KB/s when a bandwidth limit of 10 KB/s was applied at the tick of 70 second. It recovered to 100 KB/s gradually after the limit was removed.

Similar results were gained in the file download experiment (Figure 4). These results indicate that the new method has a good capability in bandwidth control.

4.2. Effectiveness Of Bandwidth Dynamic Scheduling. Three applications were used in this experiment: a video conference application (*Fsmeeting*) [26], an online music player (*QQ Music*) [27], and a download manager (*Thunder*) [28]. The priority is *Fsmeeting* > *QQ Music* > *Thunder.* The total download bandwidth for them was set to 200 KB/s and the separate download bandwidth for them was set to 130 KB/s, 50 KB/s, and 20 KB/s, respectively (Table 1). All the three applications were allowed to borrow idle bandwidth. Figure 5 illustrates the experimental results. It can be seen that the three processes consumed the preconfigured bandwidth during the period of 0~50 seconds, which indicates that their

Input: The unallocated tokens after the first allocation (L'_{token})
Output: The second token allocation policy ($L'_{allocation}$)

SecondAllocation (L'_{token}) {
Obtain the running Web processes not listed in
the priority database and mark them as $L'_{process}$
Initialize the second allocation policy $L'_{allocation}$
FOR i=0 to COUNT($L'_{process}$)-1 { *// Retrieve each process*
Mark the bucket of i_{th} process in $L'_{process}$ as B_i
IF ($\text{f}_{\text{currentSize}}(B_i) < \text{f}_{\text{maxSize}}(B_i)$) { *// Bucket is not full*
// T_i is the remaining capacity of B_i
$T_i = \text{f}_{\text{maxSize}}(B_i) - \text{f}_{\text{currentSize}}(B_i)$
Initialize the allocation policy (*allocation*) for B_i
// Set the BUCKET member of allocation (token ID)
allocation.BUCKET = B_i
// Set the TOKEN member of allocation (token count).
// This value may be modified in the following steps.
allocation.TOKEN = *allocation*.TOKEN + T_i
Add *allocation* into the second allocation policy $L'_{allocation}$
Delete allocated tokens from L'_{token}
}
}
// Calculate how many tokens each process can get
// using average allocation policy
IF (COUNT($L'_{allocation}$) > 0) {
AVE = COUNT(L'_{token}) / COUNT($L'_{allocation}$)
}
// Loop through $L'_{allocation}$ to update the token count for
// each allocation
FOR k=0 to COUNT($L'_{allocation}$)-1 {
Mark the k_{th} item of $L'_{allocation}$ as A_k
// update the token count using the average value
A_k.TOKEN = A_k.TOKEN +
MIN($\text{f}_{\text{maxSize}}(A_k) - \text{f}_{\text{currentSize}}(A_k)$, AVE)
Delete the allocated tokens from L'_{token}
}
IF (COUNT(L'_{token}) == 0) { *// Token allocation finished*
RETURN $L'_{allocation}$
}
ELSE { *// Still have unallocated tokens*
// Try to allocate the remaining tokens to all the running
// Web processes regardless of their existence in the
// priority database through token borrowing
Get all the running Web process L_p and sort it in descending order of priority
FOR j=0 to COUNT(L_p)-1 {
Mark the bucket of the j_{th} process in L_p as M_j
IF ($\text{f}_{\text{currentSize}}(M_j) < \text{f}_{\text{maxSize}}(M_j)$) { *// Bucket is not full*
Get the *maxBorrow* of the current bucket ($M_{j\_maxBorrow}$)
IF ($M_{j\_maxBorrow}$ > 0) { *// Can borrow more tokens*
// T_j is the remaining capacity of M_j
$T_j = \text{f}_{\text{maxSize}}(M_j) - \text{f}_{\text{currentSize}}(M_j)$
Initialize *allocation* for M_j
// Set the BUCKET member of allocation (token ID)
allocation.BUCKET = M_j
// Set the TOKEN member of allocation (token count)
allocation.TOKEN = *allocation*.TOKEN +
MIN(COUNT(L'_{token}), T_j, $M_{j\_maxBorrow}$)
Add *allocation* into the second allocation policy
Subtract the count of allocated tokens from the token
borrowing parameter ($M_{j\_maxBorrow}$)
Delete the allocated tokens *from* L'_{token}
IF (COUNT(L'_{token}) == 0) { *// Token allocation finished*

ALGORITHM 2: Continued.

```
    RETURN L'_allocation
    }
   }
  }
 }
}
RETURN null
}
```

ALGORITHM 2: Bandwidth dynamic scheduling method (the second allocation).

TABLE 1: Bandwidth and priorities configurations for the three applications.

Application	Bandwidth	Priority
Fsmeeting	130KB/s	High
QQ Music	50KB/s	Medium
Thunder	20KB/s	Low

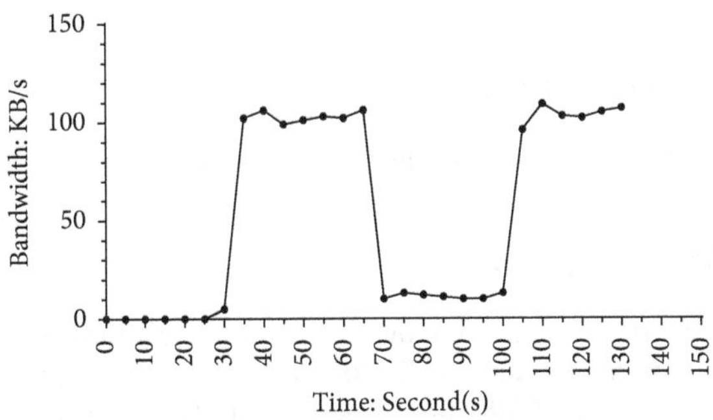

FIGURE 4: Download bandwidth control.

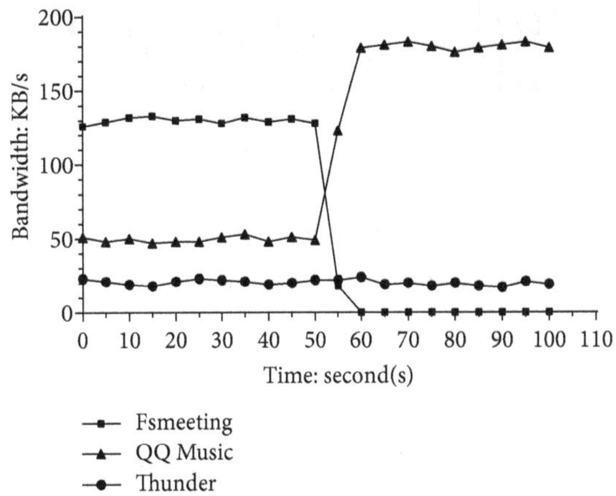

FIGURE 5: Bandwidth usage for three applications.

bandwidth can be scheduled correctly. After the tick of 50 seconds, *Fsmeeting* was killed and its bandwidth usage decreased to 0 KB/s. The bandwidth usage of *QQ Music* gradually increased to 180 KB/s, while the bandwidth usage of *Thunder* is kept unchanged during this period. These experimental results can verify that the new method has good effectiveness in bandwidth sharing and dynamic scheduling, which can maximize the bandwidth usage without the problem of idle bandwidth.

4.3. Verification of QoS Improvement. A VOIP application (*MicroSIP* [29]) was used to verify the improvement of QoS. The reason for choosing a VOIP application is that such applications are very sensitive to bandwidth and their QoS can be measured through a relatively easy method named Mean Opinion Score (MOS) [30]. Strictly speaking, the QoS for VOIP applications can be influenced by many factors, for example, bandwidth, network delay, packet loss, etc. And bandwidth is not the only influencing factor. In this experiment, the other factors were assumed unchanged and bandwidth was assumed to be the only influencing factor.

Currently there is no relevant research on the impact of local bandwidth scheduling for QoS guarantee in the local computer. Therefore, in this research there was no comparison experiment with the existing methods. Instead, two experiments were designed to verify the improvement of QoS when using the new method or not. Three applications *Baidu Netdisk*, *Thunder*, and *MicroSIP* were used in the two experiments. Table 2 lists their configurations in bandwidth and priorities.

The steps of the two experiments are as follows: download a 1G byte file using *Baidu Netdisk* and *Thunder* simultaneously, and during this period use *MicroSIP* to make calls. The first experiment used the new QoS method and the second one did not use it. Figures 6 and 7 illustrate the MOS scores and the bandwidth usage in two cases.

A piece of music instead of human voice was used when making calls to avoid man-made influence. The call quality was scored through MOS which ranged from 1 to 5. The higher the call quality, the higher the MOS score. Thirty participants were invited to scored their MOS. Figure 6 illustrates the MOS in two experiments. The average MOS is 3.91 and 3.18 when using the new QoS method or not, respectively. The former has a 23% higher score than the latter, which concludes that the QoS can be improved significantly when using the bandwidth dynamic scheduling method.

Figure 7 illustrates the bandwidth usage of *MicroSIP* in the two experiments. It can be seen that the bandwidth usage cannot be guaranteed when not using the QoS method. In this case, three applications competed bandwidth freely and *MicroSIP* did not have any bandwidth guarantee. When using the QoS method, the bandwidth for *MicroSIP* was

TABLE 2: Bandwidth and priority configurations in two cases.

Application	Using method		Not using method	
	Bandwidth	Priority	Bandwidth	Priority
MicroSIP	100 KB/s	High	null	null
Baidu Netdisk	80 KB/s	Medium	null	null
Thunder	null	Low	null	null

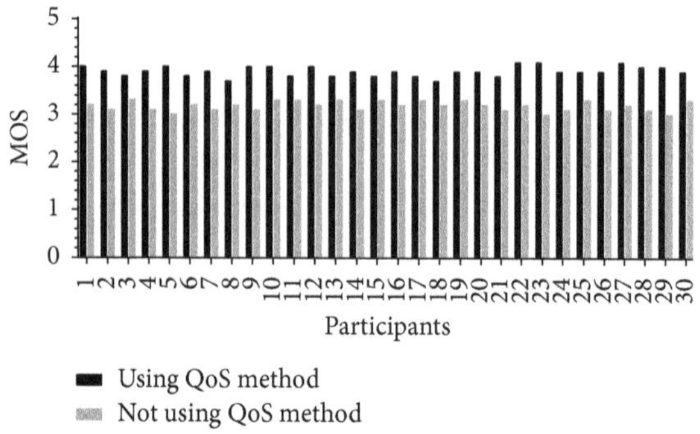

FIGURE 6: MOS scores of the two experiments.

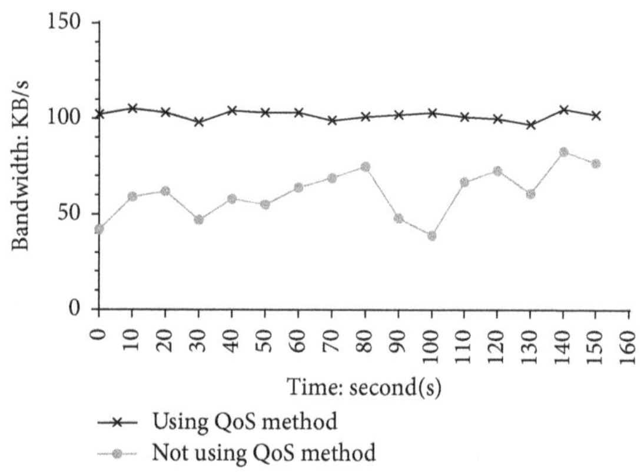

FIGURE 7: Bandwidth usage for MicroSIP in two cases.

generally stable at 100 KB/s, although the other two applications were performing significant bandwidth-consuming operations. This concludes that, from another perspective, the local bandwidth scheduling method can guarantee the bandwidth for Web applications and can improve their QoS effectively.

5. Conclusion

A new application-level QoS control method based on local bandwidth scheduling was proposed and experimental results verified its effectiveness. The new method has three advantages. (1) It can schedule bandwidth without tedious manual configurations. The configurations for commonly used scenarios can be built in advance and most users can reuse them and make their own extensions. This will simplify the configuration and reduce lots of workload. (2) It schedules bandwidth at application level and can control the bandwidth for applications using dynamic ports. (3) It can be easily integrated into current Web applications without modifying the underlying network.

It should be noted that the existing QoS guarantee methods which work at router or network layer are effective and practically verified. Our method is not a competitive or replaceable relationship with these methods. To the opposite, it is an organic supplement to them and can improve QoS further at application level. The new method only solved the problem of bandwidth scheduling in local computer, which cannot avoid the excessive bandwidth consumption caused by other computers in the same local area network. Further research can be made to solve the problem in this scenario.

Conflicts of Interest

The authors declare that they have no conflicts of interest.

Acknowledgments

This work is supported by the National Natural Science Foundation of China (61772078), Key R&D Program of Beijing (D171100001817003), and the Fundamental Research Funds for the Central Universities (YX2014-17).

References

[1] A. Sinaeepourfard and H. Mohamed Hussain, "Comparison of VOIP and PSTN services by statistical analysis," in *Proceedings of the IEEE Student Conference on Research and Development*, pp. 459–461, Piscataway, NJ, USA, December 2012.

[2] "Resource ReSerVation Protocol (RSVP), IETF website [EB/OL]," 2017, https://tools.ietf.org/html/rfc2205.

[3] C. Metz, "Differentiated Services," *IEEE MultiMedia*, vol. 7, no. 3, pp. 84–90, 2000.

[4] P. Papadimitriou and V. Tsaoussidis, "On transport layer mechanisms for real-time QoS," *Journal of Mobile Multimedia Rinton Press*, vol. 1, no. 4, pp. 342–363, 206.

[5] A. Neto, E. Cerqueira, A. Rissato, E. Monteiro, and P. Mendes, "A resource reservation protocol supporting QoS-aware multicast trees for next generation networks," in *Proceedings of the IEEE Symposium on Computers and Communications*, pp. 707–714, Santiago, Portugal, July 2007.

[6] Q. Fu, "A traffic-aware multipath QoS routing strategy," *Jisuanji Xuebao/Chinese Journal of Computers*, vol. 37, no. 10, pp. 2153–2164, 2014.

[7] A. Ayyasamy and K. Venkatachalapathy, "Context aware adaptive fuzzy based QoS routing scheme for streaming services over MANETs," *Wireless Networks*, vol. 21, no. 2, pp. 421–430, 2014.

[8] P. Sondi, D. Gantsou, and S. Lecomte, "Design guidelines for quality of service support in Optimized Link State Routing-based mobile ad hoc networks," *Ad Hoc Networks*, vol. 11, no. 1, pp. 298–323, 2013.

[9] W. Lingrui, *Research about Network Traffic Control on Linux*, University of Science and Technology of China, Beijing, China, 2014.

[10] http://www.utorrent.com, 2017.

[11] https://www.ietf.org/rfc/rfc2474.txt, 2017.

[12] Y. Zhang, S. Liu, R. Zhang et al., "A new multi-service token bucket-shaping scheme based on 802.11e," in *Proceedings of the International Conference on Identification, Information, and Knowledge in the Internet of Things*, pp. 691–696, Zhejiang, China, October 2015.

[13] N. Lin and N. Suo, "Hierarchical and dynamic traffic shaping algorithm based on token buckets," *Modern Computer*, vol. 6, no. 1, pp. 7–9, 2011.

[14] J. Kidambi, D. Ghosal, and B. Mukheqee, "Dynamic token bucket (DTB): A fair bandwidth allocation algorithm for high-speed networks," *Journal of High Speed Networks*, vol. 9, no. 2, pp. 67–87, 2000.

[15] T. Cucinotta, L. Palopoli, L. Abeni, D. Faggioli, and G. Lipari, "On the integration of application level and resource level QoS control for real-time applications," *IEEE Transactions on Industrial Informatics*, vol. 6, no. 4, pp. 479–491, 2010.

[16] D. G. Balan and D. A. Potorac, "Linux HTB queuing discipline implementations," in *Proceedings of the IEEE International Conference on Networked Digital Technologies*, pp. 122–126, Australia, 2009.

[17] N. Aoki, "A technique of lossless steganography for G. 711," *IEICE Transactions on Communications*, vol. E90-B, no. 11, pp. 3271–3273, 2007.

[18] R. Mohanty, H. Behera S, and K. Patwari, "Priority based dynamic round robin algorithm with intelligent time slice for soft real time systems," *International Journal of Advanced Computer Science & Applications*, vol. 16, no. 1, pp. 54–60, 2011.

[19] *Hook*, Wikipedia website, 2017, https://en.wikipedia.org/wiki/Hooking.

[20] *Windows Filtering Platform*, 2017, https://msdn.microsoft.com/en-us/library/windows/desktop/aa366510(v=vs.85).aspx/.

[21] *Linksys P-WRT1900ACS*, 2017, https://www.linksys.com/cn/p/P-WRT1900ACS/.

[22] *CreateProcess Function*, 2017, https://msdn.microsoft.com/en-us/library/windows/desktop/aa366510(v=vs.85).aspx.

[23] *TerminateProcess Function*, 2017, https://msdn.microsoft.com/en-us/library/windows/desktop/ms686714(v=vs.85).aspx.

[24] *Maximum Transmission Unit*, 2017, https://en.wikipedia.org/wiki/Maximum_transmission_unit.

[25] *Baidu Netdisk*, 2017, https://pan.baidu.com.

[26] *Fsmeeting*, 2017, http://www.fsmeeting.com.

[27] *QQ Music*, 2017, https://y.qq.com.

[28] *Thunder*, 2017, http://dl.xunlei.com.

[29] *MicroSIP*, 2017, http://www.microsip.org.

[30] M. Viswanathan and M. Viswanathanb, "Measuring speech quality for text-to-speech systems: development and assessment of a modified mean opinion score (MOS) scale," *Computer Speech and Language*, vol. 19, no. 1, pp. 55–83, 2005.

14

Abnormal Event Detection in Wireless Sensor Networks based on Multiattribute Correlation

Mengdi Wang, Anrong Xue, and Huanhuan Xia

School of Computer Science and Communication Engineering, Jiangsu University, Zhenjiang 212013, China

Correspondence should be addressed to Anrong Xue; xuear@ujs.edu.cn

Academic Editor: Mengxing Huang

Abnormal event detection is one of the vital tasks in wireless sensor networks. However, the faults of nodes and the poor deployment environment have brought great challenges to abnormal event detection. In a typical event detection technique, spatiotemporal correlations are collected to detect an event, which is susceptible to noises and errors. To improve the quality of detection results, we propose a novel approach for abnormal event detection in wireless sensor networks. This approach considers not only spatiotemporal correlations but also the correlations among observed attributes. A dependency model of observed attributes is constructed based on Bayesian network. In this model, the dependency structure of observed attributes is obtained by structure learning, and the conditional probability table of each node is calculated by parameter learning. We propose a new concept named attribute correlation confidence to evaluate the fitting degree between the sensor reading and the abnormal event pattern. On the basis of time correlation detection and space correlation detection, the abnormal events are identified. Experimental results show that the proposed algorithm can reduce the impact of interference factors and the rate of the false alarm effectively; it can also improve the accuracy of event detection.

1. Introduction

Abnormal event detection is one of the main problems in wireless sensor networks [1]. In wireless sensor networks, abnormal events are usually complex, because an event usually involves multiple observed attributes, and it is difficult to describe an abnormal event pattern [2]. Existing anomaly detection algorithms detect an abnormal event by comparing a single attribute threshold [3, 4] or by considering the spatiotemporal correlations of sensor readings [2, 5–8]. However, some important information may be hidden in the correlations among different attributes [9].

In [3], an adaptive distributed event detection method is proposed, which dynamically adjusts the decision threshold based on the trust value of the sensor nodes and uses the moving average filter to tolerate the transient faults of the sensor nodes. Although this method is fault-tolerant, it is still possible to misjudge the event nodes into faulty nodes. Particularly when the event range is large, the accuracy of detection will decrease significantly. Besides, this method computes a trust value for each sensor node, so it can only be applied to univariate applications. Paper [5] models the event region based on Dynamic Markov Random Field. This method can effectively capture the dynamic changes of local area; since the method needs to exchange information of space-time neighbor constantly, the detection efficiency is low. Besides, the detection of the events lacks a global perspective, which may lead to misjudgment of abnormal events. Paper [6] proposed an event detection scheme based on spatiotemporal correlations. In this method, the sensor nodes are divided into multiple working groups; the time correlation of the sensor data is used to eliminate low frequency errors. Different working groups cooperate to determine whether the anomalies represent an event. However, this method only constructs the model based on the single sensing attribute and does not consider the relations between the multisensory attribute and the abnormal event.

The attributes of the sensor readings usually contain time information, sensor topology information, and other attributes directly sensed by the sensor (e.g., temperature, humidity, and light intensity). When abnormal events occur in the network, events often show temporal correlation, spatial correlation, and attributes correlation [9]. In most cases, event detection methods that take the spatiotemporal correlation of the data into account are susceptible to both sensor failures and external environmental noises. For observed attributes, a simple threshold comparison is insufficient to determine whether an abnormal event occurs. For instance, in an indoor fire monitoring application, the increase of the temperature and smoke concentration may be caused by cooking, rather than a fire accident.

In order to improve the accuracy of abnormal event detection in wireless sensor networks with multiple attributes and reduce the influence of environmental noises and sensor failures on the event detection results, this paper proposes a new method called Abnormal Event Detection based on Multiattribute Correlation (MACAED). First, considering that Bayesian network can effectively represent the dependencies among variables, a Bayesian network is used to establish the dependency model of observed attributes. In this model, the dependency structure of abnormal events is obtained by structure learning. Each node learns the parameters to get a conditional probability table. Then, the attribute correlation confidence is introduced to judge whether the attribute correlation mode of the point is an abnormal mode. Based on the sliding window model, the degree of temporal correlation was calculated; the spatial similarity was calculated by using the neighbor node information. Finally, the anomaly events were detected by three kinds of attribute correlation.

2. Attribute Dependency Model

In wireless sensor networks, abnormal events usually show the following three characteristics:

(1) For a single sensor node, the anomaly event will continue for a period of time once the event occurs; the adjacent time of the data shows a certain degree of similarity [7]. In addition, abnormal events will inevitably affect the physical environment of network monitoring, and the sensor data will change accordingly, showing a special mode.

(2) For a number of sensor nodes, sensor nodes in the event region will exhibit spatial similarity when abnormal events occur [10]; in other words, the readings of adjacent nodes exhibit similar patterns.

(3) When the abnormal events occur in the monitoring area, the sensed attributes of the sensor readings show a certain degree of relevance, and this correlation appears as probability relations [9].

According to the three kinds of characteristics of abnormal events in wireless sensor networks and the experience that Bayesian network can effectively represent the probability relationship among attributes, we construct the attribute dependency model. The attribute correlation confidence is proposed to measure the degree of similarity between the measured points and the anomalies in observed attribute probability model.

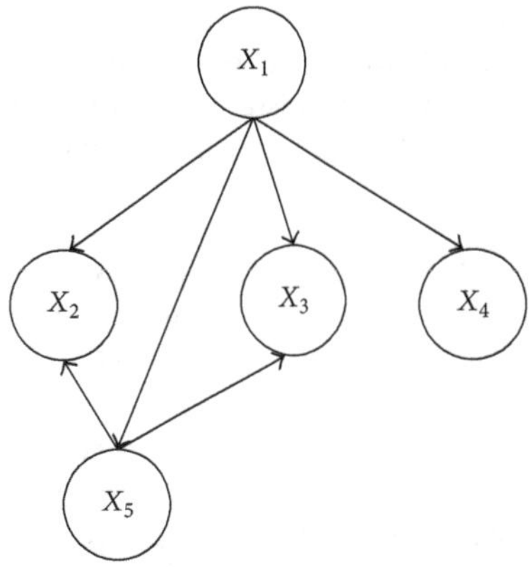

FIGURE 1: An example of attribute dependency model.

2.1. Bayesian Network. Bayesian network is a product of probability theory and graph theory. It is a directed acyclic graph with probabilistic annotations, which can represent the probability dependencies among random variables. It has a solid mathematical foundation [11]. On the one hand, the Bayesian network can reveal the structure of the problem intuitively by using graph theory. On the other hand, the Bayesian network can utilize the structure of the problem according to the principle of probability theory, which reduces the computational complexity of reasoning. In view of this, this paper establishes a dependency model of observed attributes based on the Bayesian network; each attribute is represented by a unique node, and the probabilistic dependencies are represented by arcs between nodes.

2.2. Formal Description. The attribute dependency model is represented by a triplet $B = (D, G, \theta)$, where D is the sample dataset that contains observed attributes, $D = \{d_1, d_2, \ldots, d_m\}$; G denotes a directed acyclic graph, which qualitatively describes the dependencies among attributes, $G = (X, U)$, where X is a set of nodes representing observed attributes, corresponding to the elements in D, and U is the directed edge set representing the dependencies among the attributes; θ is the set of conditional probability distributions for each node, which quantitatively describes the dependencies among attributes, $\theta = \{P(X_i \mid \pi(X_i))\}$, where X_i is the ith node in G and $\pi(X_i)$ is the set of parent nodes of node X_i. Figure 1 is an example of an attribute dependency model.

2.3. Structure Learning. For WSNs with large number of variables and implicit dependencies among variables, it is difficult to obtain a reasonable network structure relying on a priori information and expert knowledge, and the probability is subjective, so we learn the Bayesian network structure from training samples. This paper utilizes a strategy of scoring and searching. Specifically, we use a scoring function to evaluate the matching degree between a specific network structure and the training sample and select the appropriate search strategy to search the network structure with the highest scoring value.

Given a sample dataset $D(d_1, d_2, \ldots, d_m)$, let Bayesian network G take all the variables in the node set $X(X_1, X_2, \ldots, X_m)$ as nodes, and instantiate all the variables of X using the attribute value d_i in the sample dataset. The variable X_i has r_i possible values $(x_{i1}, x_{i2}, \ldots, x_{ir_i})$. Let the parent variable set of X_i be Π_i, w_{ij} denotes the jth instantiation value of the parent node Π_i with respect to D, and N_{ijk} denotes the number of instances in which the value X_{ik} of the variable X_i is taken and is instantiated into w_{ij} by Π_i, $N_{ij} = \sum_{k=1}^{r_i} N_{ijk}$. The Bayesian scoring criterion is used to compute the likelihood ratios of the two Bayesian network structures G_1 and G_2. Since $p(G_1 \mid D)/p(G_2 \mid D) = p(G_1, D)/p(G_2, D)$, we only need to compare the joint probability $p(G_1, D)$ and $p(G_2, D)$. This can be calculated by using the formula [12]

$$p(G, D) = p(G)\, p(D \mid G) = p(G) \cdot \prod_{i=1}^{n} \prod_{j=1}^{q_i} \frac{(r_i - 1)!}{(N_{ij} + r_i - 1)!} \cdot \prod_{k=1}^{r_i} N_{ijk}!, \tag{1}$$

where $p(G)$ is the prior probability and the arrangement order of Π_i is $(1, \ldots, q_i)$. Maximizing the joint probability $p(G, D)$ in (1)

$$\max_G \{p(G, D)\} = p(G) \prod_{i=1}^{n} \max_{\Pi_i} \left[\prod_{j=1}^{q_i} \frac{(r_i - 1)!}{(N_{ij} + r_i - 1)!} \prod_{k=1}^{r_i} N_{ijk}! \right]. \tag{2}$$

It can be seen that, for each variable X_i, it is only necessary to maximize

$$\max_{\Pi_i} \{g(i, \Pi_i)\} = \max_{\Pi_i} \left[\prod_{j=1}^{q_i} \frac{(r_i - 1)!}{(N_{ij} + r_i - 1)!} \prod_{k=1}^{r_i} N_{ijk}! \right]. \tag{3}$$

In the initial stage of constructing the network structure, it is assumed that each node has no parent node. The nodes which meet the posterior probability maximization formula are recursively added to the parent set of nodes. When $p(G, D)$ is no longer increased, stop adding to the parent node set; then the network structure G' is obtained. For the current sample dataset D, G' is the optimal network structure under the Bayesian scoring standard.

2.4. Parameter Learning. According to the trained network structure, the parameter of each node in the network is learned to get the corresponding conditional probability table. The conditional probability table contains the probability relations among the variables. Using the maximum likelihood estimation method, suppose $(x_1, x_2, \ldots, x_n)$ is a set of possible values of random variable set $(X_1, X_2, \ldots, X_n)$, and the probability of $(X_1, X_2, \ldots, X_n)$ falling in the neighborhood of $(x_1, x_2, \ldots, x_n)$ (n-dimensional cubes with side length $dx_1, dx_2, \ldots, dx_n$, resp.) is approximated as $\prod_{i=1}^{n} f(x_i; \theta) dx_i$, where $\prod_{i=1}^{n} f(x_i; \theta)$ is the joint probability density of $(X_1, X_2, \ldots, X_n)$, θ is the structural parameters, and $\theta \in \Theta$. The maximum likelihood estimation value $\hat{\theta}$ of θ is calculated through $\max_{\theta \in \Theta} L(x_1, \ldots, x_n; \theta)$. The conditional probability table for each node is obtained from the sample data and prior knowledge.

2.5. Attribute Correlation Confidence. Attribute correlation confidence is a concept we proposed to measure the fitting degree between the sensor reading and the abnormal event pattern. It is equal to the ratio of the joint probability distribution between the measured point and the abnormal point. Let $(y_1, y_2, \ldots, y_n)$ be the sensor reading at the current time. For an abnormal event E_i, the joint probability of all node variables $P(X_1, X_2, \ldots, X_n \mid E_i)$ is calculated according to the Bayesian network structure and the conditional probability table. Since in Bayesian network, not every node has an arc to the all the rest nodes, the conditional probability only depends on the direct parent node. In other words, given the values of parent variables, the probability of nondescendant node is conditionally independent of the parent node. So the calculation of joint probability $P(X_1, X_2, \ldots, X_n \mid E_i)$ can be simplified by using the chain rule [11],

$$p(x) = \prod_{i=1}^{n} p\left(x_i \mid x_{pa(i)}\right) \tag{4}$$

in which $x_{pa(i)}$ represents the parent node of x_i.

After calculating $P(X_1, X_2, \ldots, X_n \mid E_i)$, we can get the probability pattern of the reading in an event. According to the formula,

$$\alpha = \max_{i \in I} \frac{P(X_1 = y_1, \ldots, X_n = y_n)}{P(X_1 = x_1, \ldots, X_n = x_n \mid E_i)}, \tag{5}$$

the attribute correlation confidence α of the tested point is calculated. The higher the probability, the more the possibility for the anomaly to represent an abnormal event.

3. Abnormal Event Detection Algorithm Based on Multiattribute Correlation

In this paper, we propose a detection algorithm based on multiattribute correlation, which is divided into three phases: attribute correlation pattern decision, temporal similarity detection, and spatial similarity detection.

3.1. Description of Abnormal Event. For an abnormal event, define event information $Info = \{Tm, Loc, Attr, Parm, E_i\}$, where Tm is the time of occurrence of abnormal events, Loc is the location of abnormal events, and $Attr$ is the attribute set that an event involves. $Parm$ is the parameter set, which includes temporal similarity threshold ε, spatial similarity threshold δ, and attribute correlation confidence threshold φ. For different application environments, the values of each item in $Parm$ can be adjusted to achieve the best detection result adaptively. E_i represents the event type, $i = 0$ means no abnormal events occurred, $i > 0$ means that abnormal events occurred, and the higher the value i is, the more severity the abnormal event has.

3.2. Temporal Similarity Detection. The data sampling frequency of most wireless sensor networks is relatively high and data change range at the adjacent time is relatively small, so the sensor data is time-correlated. Combining with sliding window model and the attribute dependency model obtained, candidate anomalies that may represent abnormal events are detected.

Let s be the size of the sliding window, and for each data sequence t_i within the window, calculate the similarity between t_i and the current time series t

$$q(t_i,t) = \frac{1}{\left(1+\sqrt{\sum_{k=1}^{m}\left(x_k^{t_i}-x_k^{t}\right)^2}\right)}. \quad (6)$$

Considering that the data sequence that is closest to the current time is most correlated, the average similarity between the current time data and the data in the window is calculated by the weighted summation method

$$\overline{q(t)} = \frac{\sum_{i=1}^{ww} w_i q(t_i,t)}{s}, \quad (7)$$

where the weight is $w_i = 1/(t-t_i)$. If the average similarity is smaller than the threshold ε and the confidence degree of the attribute correlation is greater than or equal to the threshold φ, it means that not only does the data sequence of the current time significantly deviate from the historical data, but also the relationship among the attributes is in accordance with the probability relation when the abnormal event occurs, which needs a further spatial correlation detection. In other cases, it will be filtered as a noise.

3.3. Spatial Similarity Detection. The similarity between the candidate anomaly and the neighbor node's data sequence is calculated. If the candidate anomaly and the neighbor node's data sequence satisfy certain similarity degree, it indicates that the abnormal event occurs in the region where the candidate anomaly is located and needs to be uploaded to the sink node.

The similarity between the candidate anomaly and the neighbor node sequence is calculated according to the following formula:

$$q(x_t,y_t) = \frac{1}{\left(1+\sqrt{\sum_{k=1}^{m}\left(x_k^{t}-y_k^{t}\right)^2}\right)}. \quad (8)$$

If the spatial similarity $q(x_t, y_t)$ is greater than or equal to the threshold δ, it indicates that both nodes have detected an abnormal event at the same time and mark the candidate anomaly nodes and their neighbor nodes as abnormal event nodes. On the contrary, it indicates that no neighbor nodes detect abnormal information at this time, and the candidate anomaly belongs to noise data, which is also filtered out.

3.4. Description of MACAED Algorithm. Based on the calculation of attribute correlation confidence and the detection of temporal and spatial correlation of sensor data, an abnormal event detection algorithm based on multiattribute correlation is proposed. The pseudocode of the algorithm is shown in Algorithm 1.

In the pseudocode of Algorithm 1, rows (2)~(3) train the Bayesian network through the scoring-searching method and choose the network structure M with the highest score as the observed attribute dependency model, rows (4)~(26) detect abnormal events in real time, where rows (9)~(10) proceed parameter learning for each sensor in order to update the probability distribution in attribute dependency model, rows (10)~(14) compute the attribute correlation confidence of observed attributes, row (15) calculates the average similarity between the current time readings and the readings within the window, row (18) calculates the average similarity between the current node and the adjacent node readings, and rows (17)~(24) determine whether the current reading represents abnormal events readings.

3.5. Time Complexity Analysis. Let n be the number of observed attributes, which corresponds to the number of nodes in Bayesian network; m is the number of instances, that is, the number of readings; r is the number of possible values for each observed attribute; N is the number of nodes in WSN; s is the size of sliding window. For the structure learning part, the time complexity is $O(mn^4r)$ [12]. For abnormal event detection part, it contains two layers: outer layer loops $O(m-s-1)$ times and inner loops $O(N)$ times. The parameter learning consists of a cycle of $O(n)$ times. The time correlation detection consists of a cycle of $O(s)$ times. The spatial correlation detection consists of a cycle of $O(N)$ times. The total time complexity of the algorithm is $O(mn^4r)+O(m-s-1)O(N)O(n+s+N)$. Since, for most wireless sensor networks, the value of n is small (less than 10) and sliding window s and the number of possible values of each attribute r are relatively small (in this experiment, $s = 10$; $r = 9$), the influence of these values on the total time complexity can be ignored, so the total time complexity can be simplified to $O(m) + O(mN^2) = O(mN^2)$.

4. Experimental Results and Analysis

4.1. Datasets. We test the performance of the MACAED algorithm by means of conducting simulation experiments on Matlab 2014a. The experiments are run on a PC with an Intel Core i3-2120 @3.30 GHZ Cpu, 4 GB memory, and Windows 7 operating system. For the instance of detecting fire event, the performance tests are based on the processed data of Intel Lab Data [13] from Intel Berkeley Lab. Except for the real data field, we insert the fire events and interference events data field into the dataset manually.

The experiment dataset contains the records of 54 sensors deployed in the IBRL lab during the time span from February 28th to April 5th in 2004. The MicaDot sensors collect temperature, humidity, light intensity, and voltage value every 31 seconds. Sensor node deployment is shown in Figure 2.

```
Input. WSN data set D
Output. Abnormal event Information Info
(1) standardize D into values between 0 and 1
(2) divide D into K subsets, choose the first set to learn Bayesian network
(3) choose the network with highest score as attribute
    dependency model M
(4) for t = s + 1 to epoch//epoch is incremental tick
(5)     if t%period = 0//period is parameter update period
(6)         flag = true; //flag represents update parameter or not
(7)     end
(8)     for id = 1 to N//id is the id of WSN, N is the number of sensors
(9)         learn parameter for each sensor node
(10)        if dataPointer [id] < group_length
            //prevent the id exceed the length of group
(11)            if groupData_time [id] < t
                //prevent a break caused by data loss
(12)                compute α from M
(13)            end
(14)        end
(15)    compute q(t_i, t)
(16)    if q(t_i, t) < ε && α ≥ φ
(17)        compute q(x_t, y_t);
(18)        if q(x_t, y_t) ≥ δ
(19)            report Info to sink node;
(20)        else
(21)            filter as noise;
(22)        end
(23)    end
(24)    end
(25)    flag = false;
(26) end
```

Algorithm 1: Abnormal event detection algorithm based on multiattribute correlation.

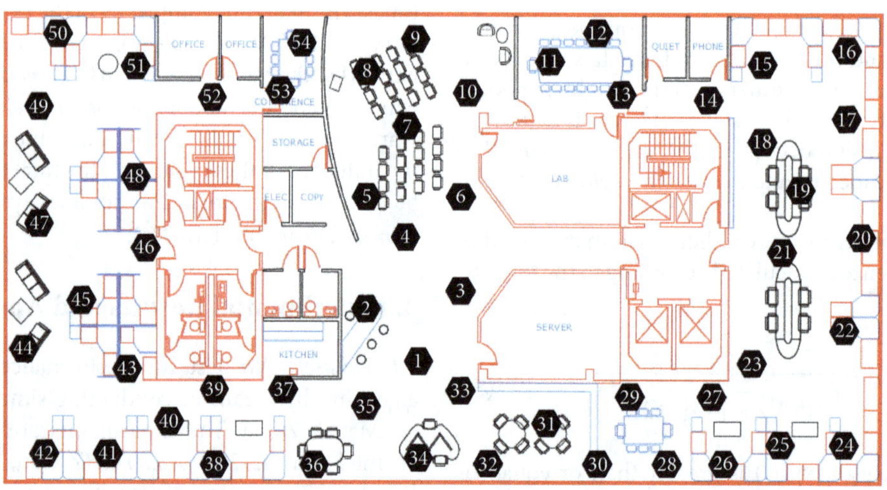

Figure 2: Location of sensor nodes deployed in IBRL lab.

4.2. Data Preprocessing. In our experiment, we choose the records within 24 hours in February 28th as our test data; we preprocess the raw data as follows:

(1) Since the unit of measurement attributes directly sensed by each sensor is different and the changing range of different attributes is wide, so the raw data needs to be standardized and mapped to $[0, 1]$; in this way, the relative distance can be calculated.

(2) Since the change of each attribute value is continuous and periodic, in order to facilitate the calculation, the experimental datasets are discretized, and the values of each attribute are divided into 10 intervals.

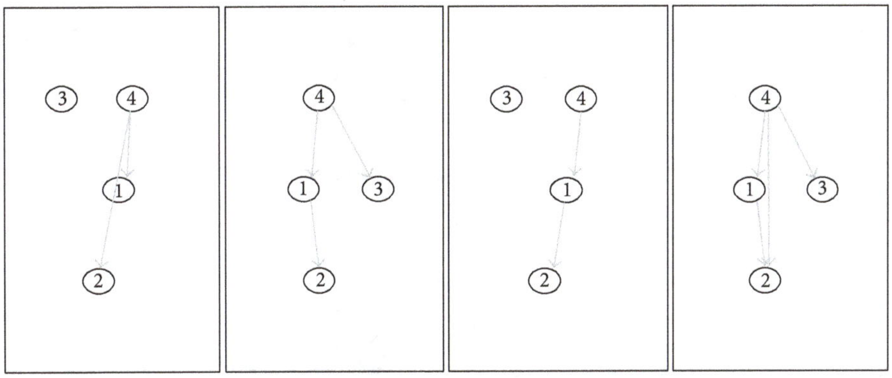

FIGURE 3: Network structure under Bayesian scoring criterion.

(3) For some parts of the raw IBRL datasets have missing values and the failure nodes (both node 5 and node 15 have no records; node 28 only has 3 attribute records), the NaN is used in this experiment to fill the missing values, and these values will be discussed in different situations, not for computation.

(4) In order to verify the performance of our algorithm on detecting abnormal events, abnormal readings that represent abnormal events are added in the dataset. In addition, the readings of the abnormal events with the interference are added (e.g., opening heater in the room will make the temperature rise).

4.3. Experimental Parameters. Temperature T, humidity H, light intensity L, and voltage V are numbered with 1, 2, 3, 4. In order to obtain relatively stable Bayesian network structure, we set the maximum number of parent nodes in structure learning *max_fan_in* = 2, learning step length *step* = 10, and the number of instances *ncases* = 1000. The optimal parameter learning cycle *period* = 600. Bayesian networks with four different scores are showed in Figure 3; the higher the score is, the more stable the network structure is. Thus, we choose the structure whose score = 74 as an attribute dependency model in this experiment.

In this method, the sliding window size has a direct impact on the detection results. The precision, the recall, and the $F1$-measure of anomaly detection under different sliding window sizes are experimented. The experimental results are shown in Figure 4.

From Figure 4, we can find that the recall decreases with the increase of the sliding window width; however, the overall change is not obvious. But the precision declines relatively faster, leading to the quick decrease of $F1$ value. This is because, with the increase of window width, the historical data increases, and the calculated average value declines ceaselessly, which means that the possibility of becoming candidate anomalies is higher. Considering that the sliding window width is small and the amount of uploaded data is small, so we set the sliding window size $s = 10$; in this way, we will make full use of historical data.

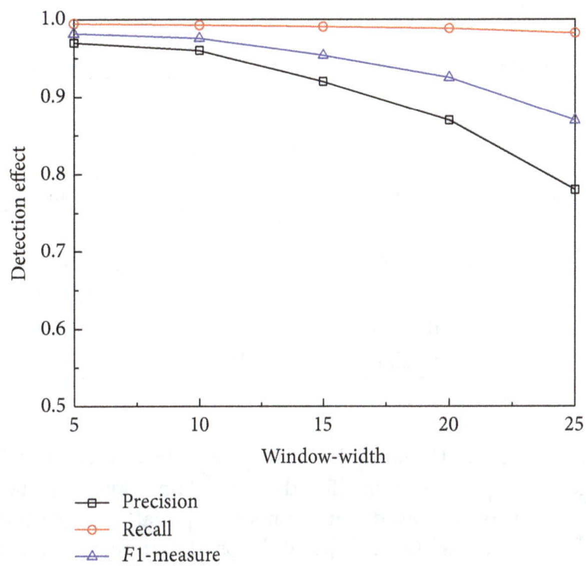

FIGURE 4: Influence of sliding window size on the test results.

There are different requirements for the threshold settings when the environment of wireless sensor networks differs. We change the value of three different thresholds and test the accuracy of the anomalies under the change of single threshold; the results are shown in Figure 5.

From Figure 5 it can be concluded that it gets the highest detection accuracy when temporal similarity threshold $\varepsilon = 0.1$, spatial similarity threshold $\delta = 0.2$, and attribute correlation confidence threshold $\varphi = 0.5$.

4.4. Contrast Experiment. In the contrast experiment, we still use the IBRL dataset, in which the number of sensor nodes is 54, and the deployment of nodes is shown in Figure 2. We use (T, H, L, V) to represent four different attributes: temperature, humidity, light intensity, and voltage. Since there are no interference factors in the dataset, we add some false abnormal events artificially, which are shown in Table 1.

The contrast algorithms include the Adaptive Fault-Tolerant Event Detection (AFTED) algorithm proposed in

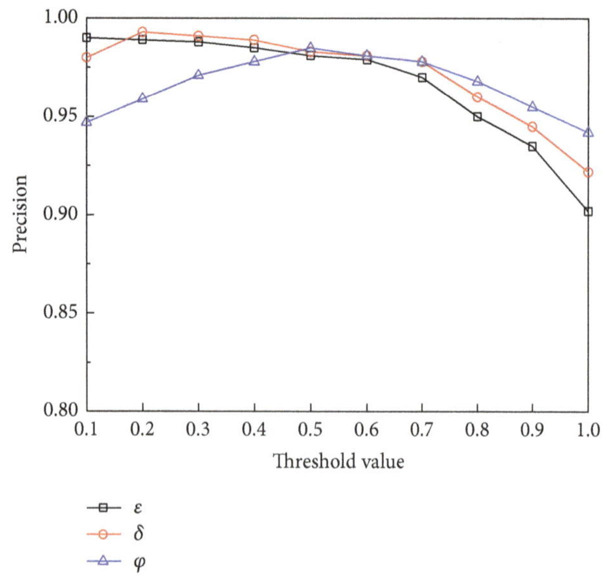

FIGURE 5: Influence of the three thresholds on the test results.

TABLE 1: False abnormal events.

Number	Event name	Attributes	Id of nodes
1	Cooking	T, H, L, V	2, 37
2	Air-condition	T, H, V	5, 17, 24, 36, 44
3	Heater	T, V	11, 12, 13
4	Bath heater	T, L, V	53, 54
5	Humidifier	H	27, 28, 29, 30

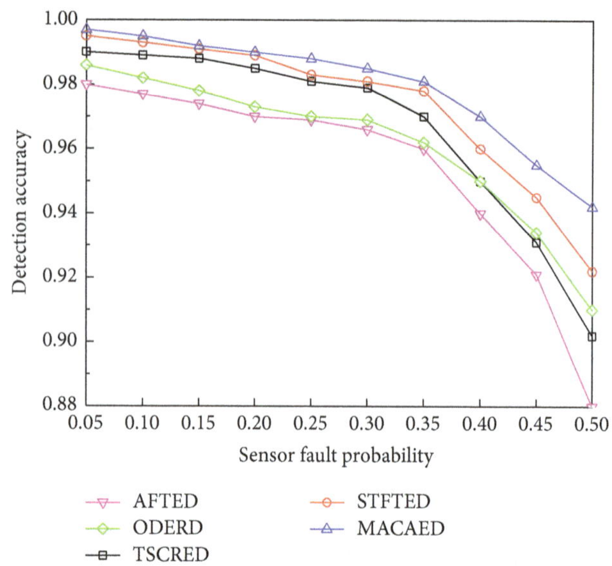

FIGURE 6: Detection accuracy of five algorithms.

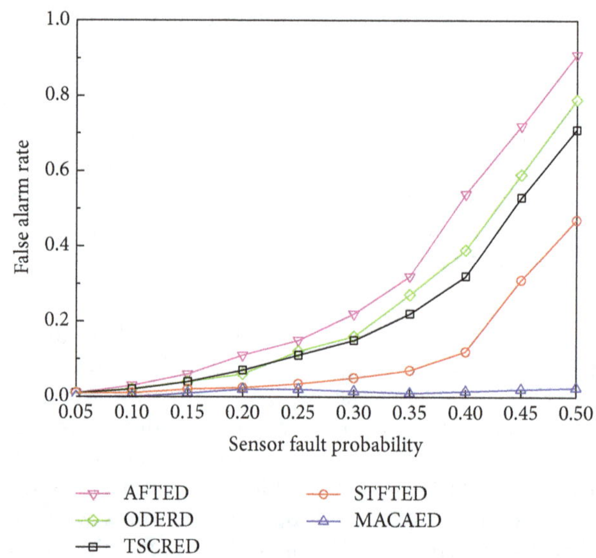

FIGURE 7: False alarm rate of five algorithms.

[3], the Online Dynamic Event Region Detection (ODERD) algorithm proposed in [5], the Real-Time Event Detection Approach based on Temporal-Spatial Correlations (TSCRED) presented in [6], and the Spatiotemporal Correlation based Fault-Tolerant Event Detection (STFTED) scheme proposed in [8]. And we compare the detection accuracy, false alarm rate, and detection time of abnormal events.

In the proposed algorithm, we use the same parameter settings as the previous experiments. In AFTED algorithm, we set the window size for tolerating transient faults $M_{\text{AFTED}} = 4$, and the threshold for filtering transient faults $\delta_{\text{AFTED}} = 0.75$, which have been verified to be the most appropriate in their experiment. In ODERD algorithm, since we only focus on the static abnormal event detection, the parameters controlling the shift and deformation of event regions are set to 0 s. To compare these algorithms in an equivalent level, we set the sliding window size of TSCRED and STFTED to 10, which is the same as the proposed algorithm. Besides, all of the sensor nodes have the same communication range $R = 4$. And each event region is assumed to be a circle with radius $l = 2R$.

The results of the proposed algorithm compared with the other four algorithms in detection accuracy are shown in Figure 6. It can be seen from Figure 6 that when the node failure rate goes from 0.05 to 0.3, the detection accuracies of the five algorithms are similar, reaching 0.96 or more; this is because most of the noise is filtered out in the time correlation detection phase. When the node failure rate is greater than 0.3, the detection accuracies of the five algorithms decrease significantly, but the MACAED algorithm is significantly better than the other four algorithms. The reason is that all the five algorithms have the spatial correlation detection stage. With the increase of the failure rate, the faulty nodes are easily affected by the neighbor nodes which have not detected the abnormal events, and they are converted into the normal state, therefore misjudging that no abnormal events occurred.

As for the false alarm rate, these compared results are shown in Figure 7. It can be seen that MACAED has a significantly lower false alarm rate than the other four algorithms as the node failure rate increases. This is due to the fact that MACAED fully considers the impact of attribute correlations on abnormal event detection. By calculating the attribute

Table 2: Running time of five algorithms.

Algorithms	Time(s)
AFTED	8.381
ODERD	7.647
TSCRED	7.435
STFTED	10.917
MACAED	12.546

correlation confidence, the fitting degree between the data records and the abnormal event attribute dependency model can be determined, so the abnormal event and interference factor can be distinguished effectively.

The running time of the five algorithms is shown in Table 2.

It can be seen from Table 2 that the MACAED algorithm consumes the longest time. The reason is that the MACAED algorithm needs to train the network structure at the beginning. This process takes about 5 s on average. If the trained network structure is saved as the known result, the detection phase needs 12.546 − 5 = 7.546 s, which is very close to TSCRED algorithm and ODRED algorithm.

5. Conclusion

In this paper, we present a new approach to detect abnormal events in wireless sensor networks. We construct a dependency model of observed attributes based on Bayesian network and propose a new method to measure the dependency of the attributes. Combining with the temporal correlation detection based on sliding window and the spatial correlation detection based on neighbor node information, the influence of noise and interference event factors on event detection results is effectively reduced. Experimental results show that the algorithm proposed in this paper can effectively eliminate the influence of interference events. It not only reduces the false alarm rate of abnormal events but also improves the accuracy of event detection compared with the other four algorithms.

Conflicts of Interest

The authors declare that there are no conflicts of interest regarding the publication of this paper.

Acknowledgments

This work is sponsored by Innovative Fund Project of Science and Technology Enterprises of Jiangsu Province in China no. BC2014212.

References

[1] Y. Zhang, N. Meratnia, and P. Havinga, "Outlier detection techniques for wireless sensor networks: a survey," *IEEE Communications Surveys and Tutorials*, vol. 12, no. 2, pp. 159–170, 2010.

[2] A. De Paola, S. Gaglio, G. L. Re, F. Milazzo, and M. Ortolani, "Adaptive distributed outlier detection for WSNs," *IEEE Transactions on Cybernetics*, vol. 45, no. 5, pp. 888–899, 2015.

[3] S.-J. Yim and Y.-H. Choi, "An adaptive fault-tolerant event detection scheme for wireless sensor networks," *Sensors*, vol. 10, no. 3, pp. 2332–2347, 2010.

[4] Y. Wang, D. Wang, F. Chen, and W. Fang, "Efficient event detection using self-learning threshold for wireless sensor networks," *Wireless Networks*, vol. 21, no. 6, pp. 1783–1799, 2014.

[5] T. Wu and Q. Cheng, "Online dynamic event region detection using distributed sensor networks," *IEEE Transactions on Aerospace and Electronic Systems*, vol. 50, no. 1, pp. 393–405, 2014.

[6] F. Li and Z. Feng, "An efficient real-time event detection approach based on temporal-spatial correlations in wireless sensor networks," in *Proceedings of the International Conference on Computer Science and Network Technology (ICCSNT '11)*, pp. 1245–1249, December 2011.

[7] J. Yin, D. H. Hu, and Q. Yang, "Spatio-temporal event detection using dynamic conditional random fields," in *Proceedings of the 21st International Joint Conference on Artificial Intelligence (IJCAI '09)*, pp. 1321–1326, IEEE, Pasadena, Calif, USA, July 2009.

[8] K. Liu, Y. Zhuang, Z. Wang, and J. Ma, "Spatiotemporal correlation based fault-tolerant event detection in wireless sensor networks," *International Journal of Distributed Sensor Networks*, vol. 2015, Article ID 643570, 2015.

[9] C. Luo, J.-G. Lou, Q. Lin et al., "Correlating events with time series for incident diagnosis," in *Proceedings of the 20th ACM SIGKDD International Conference on Knowledge Discovery and Data Mining (KDD '14)*, pp. 1583–1592, August 2014.

[10] K.-Z. Liu, Y. Zhuang, S.-L. Zhou, and S.-J. Liu, "Event detection method based on belief model for wireless sensor networks," *Journal of Beijing University of Posts and Telecommunications*, vol. 38, no. 1, pp. 61–66, 2015.

[11] D. Heckerman, "Bayesian networks for data mining," *Data Mining and Knowledge Discovery*, vol. 1, no. 1, pp. 79–119, 1997.

[12] G. F. Cooper and E. Herskovits, "A Bayesian method for the induction of probabilistic networks from data," *Machine Learning*, vol. 9, no. 4, pp. 309–347, 1992.

[13] IBRL, "Intel Lab Data[EB/OL]," 2004 http://db.lcs.mit.edu/labdata/labdata.html.

15

UFIR Filtering for GPS-Based Tracking over WSNs with Delayed and Missing Data

Karen Uribe-Murcia, Yuriy S. Shmaliy, and Jose A. Andrade-Lucio

Department of Electronics Engineering, Universidad de Guanajuato, 36885 Salamanca, GTO, Mexico

Correspondence should be addressed to Yuriy S. Shmaliy; shmaliy@ugto.mx

Academic Editor: Sos Agaian

In smart cities, vehicles tracking is organized to increase safety by localizing cars using the Global Positioning System (GPS). The GPS-based system provides accurate tracking but is also required to be reliable and robust. As a main estimator, we propose using the unbiased finite impulse response (UFIR) filter, which meets these needs as being more robust than the Kalman filter (KF). The UFIR filter is developed for vehicle tracking in discrete-time state-space over wireless sensor networks (WSNs) with time-stamped data discretely delayed on k-step-lags and missing data. The state-space model is represented in a way such that the UFIR filter, KF, and H_∞ filter can be used universally. Applications are given for measurement data, which are cooperatively transferred from a vehicle to a central station through several nodes with k-step-lags. Better tracking performance of the UFIR filter is shown experimentally.

1. Introduction

Accurate target tracking is one of the key problems in urban areas [1], which especially arises in smart cities design [2]. If a target is equipped with the Global Positioning System (GPS) tracker, then measurement data can be transferred to a central station through one or several nodes of a wireless sensor network (WSN) [3]. The problem which arises here is associated with information latency and missing data [4] due to the following main causes: high maneuverability of the target [1], failures in measurements [5], network congestion [3], non-line-of-sight (NLOS) problems [6, 7], and accidental loss of some collected data [8]. Furthermore, latency naturally occurs due to the limited bandwidth, finite propagation time [9], complexity of very large-scale integration and microelectromechanical systems [10], and time required to complete operations such as signal conditioning and storage [11]. In networks, communication delays go along with data loss called dropout or intermittence [12, 13]. Also the delay between the measurement and its availability to the filter causes the problem of out-of-sequence measurement [6, 14].

Two basic models have been created for delayed data. The delays are assumed to be known when sensors are able to detect the delays or data are time-stamping [15, 16]. In many other applications [17, 18], the delays are considered to be random. The problem becomes more complex in uncertain systems [19]. The best estimate is obtained here by combining delayed and nondelayed data with different probabilities.

The Kalman and H_∞ state estimators are most widely used to deal with latency and associated issues [20]. The linear Kalman filter (KF) is optimal when it matches the system perfectly, noise is white Gaussian and uncorrelated, and the noise statistics are known along with the initial values. When such conditions are not obeyed, the KF may demonstrate poor performance [13, 21]. The robust H_∞ filter bounds the mean square error (MSE) for admissible parameter perturbations and delays [19, 22], which allows for minimizing errors with less information required than for the noise statistics [20, 23].

Another way to achieve better robustness is to process most recent finite data [24] using finite impulse response (FIR) filters [25]. Such filters have been developed during decades by many authors in signal processing [26–33] and control [34–36]. However, only a few authors have proposed FIR solutions for models with delays [37–40]. Let us notice that the available iterative unbiased FIR (UFIR) algorithm

[28, 41–43] is most robust among other FIR solutions owing to an ability to ignore the noise statistics and initial values. This filter is bounded-input bounded-output (BIBO) stable and blind on given horizons of N points, but is still not developed for observations with delayed and missing data.

In this paper, we develop the UFIR filter for GPS-based vehicle tracking over WSNs with time-stamped data discretely delayed and missing data. The rest of the paper is organized as follows. In Section 2, we consider the model and formulate the problem. In Section 3, we develop the UFIR filter for observations with delayed and missing data. Section 4 discusses the estimation errors. Section 5 gives an experimental example of applications to GPS-based tracking and concluding remarks are drawn in Section 6.

2. Tracking Model and Problem Formulation

A typical scenario of GPS-based vehicle tracking in WSNs is sketched in Figure 1. The vehicle current coordinates are measured by the GPS tracker. The time-stamped data are transferred to a central station (CS) via one or several nodes of the WSN. Because each node may discretely delay time-stamped data at least on one-step, the vehicle location is observed in CS with a time varying k_n-step-lag depending on the vehicle location and interaction with the WSN.

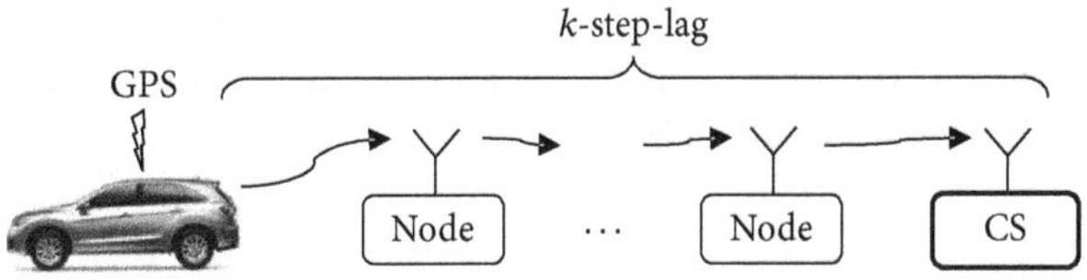

FIGURE 1: Transferring the time-stamped vehicle coordinates measured by a GPS tracker to a central station (CS) via several nodes of a WSN. Latency with a k_n-step-lag is caused by delays in the node.

2.1. Tracking Model. For $k_n \geqslant 0$, the vehicle dynamics and its observation can be represented in discrete-time state-space as

$$x_n = Fx_{n-1} + w_n, \quad (1)$$

$$y_n = Hx_{n-k_n} + v_n, \quad (2)$$

where n is the discrete-time index, $x_n \in \mathbb{R}^K$ is the vehicle state vector, $y_n \in R^M$ is the observation vector, $F \in \mathbb{R}^{K\times K}$ is not singular, and $H \in \mathbb{R}^{M\times K}$. All data are time-stamped, so that k_n is known at each n. Regardless of the delay, the initial state x_{n-1} is supposed to be known. The uncorrelated noise vectors, $w_n \in \mathbb{R}^K$ and $v_n \in \mathbb{R}^M$, are white Gaussian with known covariances, $Q = E\{w_n w_n^T\}$ and $R = E\{v_n v_n^T\}$, and the property $E\{w_n v_q^T\} = 0$ for all n and q.

The UFIR filter can be applied if to transform model (1)-(2) to have no latency. That can be done if to represent x_{n-k_n} using (1) via x_n as

$$x_{n-k_n} = F^{-k_n}\left(x_n - \sum_{i=0}^{k_n-1} F^i w_{n-i}\right) \quad (3)$$

and then substitute (3) into (2) and arrive at

$$y_n = \overline{H}_n x_n + \overline{v}_n, \quad (4)$$

where

$$\overline{H}_n = HF^{-k_n}, \quad (5)$$

$$\overline{v}_n = v_n - H\sum_{i=0}^{k_n-1} F^{-k_n+i} w_{n-i} \quad (6)$$

and the covariance $\overline{R}_n = E\{\overline{v}_n\overline{v}_n^T\}$ of $\overline{v}_n$ is given by

$$\overline{R}_n = R + \overline{H}_n \sum_{i=0}^{k_n-1} F^i Q F^{i^T} \overline{H}_n^T. \quad (7)$$

In compact matrix forms, (6) and (7) can be represented as

$$\overline{v}_n = v_n - H\overline{B}_n W_{p_n,n}, \quad (8)$$

$$\overline{R}_n = R + H\overline{B}_n\overline{Q}_n\overline{B}_n^T H^T, \quad (9)$$

where

$$\overline{B}_n = \left[F^{-1} \; F^{-2} \; \cdots \; F^{-k_n}\right], \quad (10)$$

$$W_{p_n,n} = \left[w_{p_n}^T \; w_{p_n+1}^T \; \cdots \; w_n^T\right]^T, \quad (11)$$

and $\overline{Q}_n = \operatorname{diag}[Q \; Q \; \cdots \; Q]$ has k_n diagonal components. Here, $\overline{B}_n = 0$ and $W_{p_n,n} = 0$ when $k_n = 0$ and $p_n > n$.

Any standard estimation technique can now be applied to models (1) and (4). However, the KF and H_∞ filter were most developed for data with latency. Therefore, below we will introduce in brief only these filters and then compare them to the UFIR filter based on examples of tracking. We will use the following measures: $\hat{x}_n \triangleq \hat{x}_{n|n}$ is the estimate of x_n over data taken from past up to and including time index n, $\hat{x}_n^-$ is the prior estimate, $P_n = E\{(x_n - \hat{x}_n)(x_n - \hat{x}_n)^T\}$ is the error covariance matrix, and $P_n^- = E\{(x_n - \hat{x}_n^-)(x_n - \hat{x}_n^-)^T\}$ is the prior error covariance matrix.

2.1.1. Kalman Filter. For our purposes, we will exploit an alternative form of the KF algorithm given in [44]. This algorithm starts with the prior error covariance matrix

$$P_n^- = FP_{n-1}F^T + Q \quad (12)$$

and then recursively updates the following values:

$$\begin{aligned} P_n &= (P_n^-)^{-1} + \overline{H}_n^T\overline{R}_n^{-1}\overline{H}_n, \\ K_n^{\mathrm{KF}} &= P_n^{-1}\overline{H}_n^T\overline{R}_n^{-1}, \\ \hat{x}_n &= F\hat{x}_{n-1} + K_n^{\mathrm{KF}}\left(y_n - \overline{H}_n F\hat{x}_{n-1}\right), \\ P_{n+1}^- &= FP_n^{-1}F^T + Q, \end{aligned} \quad (13)$$

where $\overline{H}_n$ is given by (5) and $\overline{R}_n$ by (7) for any $k_n \geqslant 0$.

2.1.2. H_∞ Filter. The H_∞ filter has been derived in [44] in the form (13) of the KF using the game theory. For $k_n \geqslant 0$, the H_∞ filtering algorithm becomes

$$P_n = \left(\check{P}_n^-\right)^{-1} - \theta_n S_n + \overline{H}_n^T \check{R}_n^{-1} \overline{H}_n, \tag{14}$$

$$K_n^\infty = P_n^{-1} \overline{H}_n^T \check{R}_n^{-1}, \tag{15}$$

$$\hat{x}_n = F\hat{x}_{n-1} + K_n^\infty \left(y_n - \overline{H}_n F \hat{x}_{n-1}\right), \tag{16}$$

$$\check{P}_{n+1}^- = F P_n^{-1} F^T + \check{Q}, \tag{17}$$

where the user-given symmetric positive definite matrices P_0, $\check{Q}$, and $\check{R}$ have different meanings than in the KF and $\check{P}_n^-$ can be computed via P_0 using (12) with $Q = \check{Q}$. To keep (14) positive definite, the positive definite matrix $S_n \in R^{K\times K}$ is subject to

$$\left(\check{P}_n^-\right)^{-1} - \theta_n S_n + \overline{H}_n^T \check{R}^{-1} \overline{H}_n > 0. \tag{18}$$

If equal weights are required for all errors, matrix S_n must be set identity, $S_n = I$. The performance criterion for this filter is $J_n < 1/\theta_n$, in which a scalar θ_n must be small enough in order for the filter to be efficient. It then follows that the tuning factor θ_n is not allowed to be negative, even though its negative values may reduce errors when the weighting matrices are not maximized. For Gaussian noise, zero θ_n transforms the H_∞ filter to the KF. For any other noise, small $\theta_n > 0$ may result in better robustness.

The problem now formulates as follows. Given (1) and (4) with time varying $k_n \geqslant 0$ and missing data, we would like to develop the UFIR filter and find its fast iterative form for GPS-based tracking of a moving vehicle as shown in Figure 1. We also wish to know how the UFIR filter, KF, and H_∞ filter measure to each other in applications to tracking.

3. UFIR Filter for Tracking with Delayed and Missing Data

To develop the UFIR filter for $k_n > 0$, we extend models (1) and (4) on a horizon $[m, n]$ of N points, from $m = n - N + 1$ to n that referring to [43] yields

$$X_{m,n} = A_N x_m + B_N W_{m,n}, \tag{19}$$

$$Y_{m,n} = C_{m,n} x_m + G_{m,n} W_{m,n} + V_{m,n} \tag{20}$$

with the following extended vectors and matrices:

$$X_{m,n} = \left[x_m^T \; x_{m+1}^T \; \cdots \; x_n^T\right]^T, \tag{21}$$

$$Y_{m,n} = \left[y_m^T \; y_{m+1}^T \; \cdots \; y_n^T\right]^T, \tag{22}$$

$$V_{m,n} = \begin{bmatrix} v_m - C\overline{B}_n W_{p_m,m} \\ v_{m+1} - C\overline{B}_n W_{p_{m+1},m+1} \\ \vdots \\ v_{n-1} - C\overline{B}_n W_{p_{n-1},n-1} \\ v_n - C\overline{B}_n W_{p_n,n} \end{bmatrix}, \tag{23}$$

$$A_N = \left[I \; F^T \; \cdots \; F^{N-1^T}\right]^T, \tag{24}$$

$$B_N = \begin{bmatrix} I & 0 & \cdots & 0 & 0 \\ F & I & \cdots & 0 & 0 \\ \vdots & \vdots & \ddots & \vdots & \vdots \\ F^{N-2} & F^{N-3} & \cdots & I & 0 \\ F^{N-1} & F^{N-2} & \cdots & F & I \end{bmatrix}, \tag{25}$$

$$C_{m,n} \triangleq C_{m,n}(k) = \overline{C}_{m,n}(k) A_N, \tag{26}$$

$$G_{m,n} \triangleq G_{m,n}(k) = \overline{C}_{m,n}(k) B_N, \tag{27}$$

$$\overline{C}_{m,n}(k) = \operatorname{diag}\left[\underbrace{CF^{-k_m} \; CF^{-k_{m+1}} \; \cdots \; CF^{-k_n}}_{N}\right], \tag{28}$$

where k represents a set of $\{k_m, k_{m+1}, \ldots, k_n\}$.

3.1. Batch UFIR Filter Form. The UFIR filtering estimate $\hat{x}_n$ of the vehicle state x_n can be obtained at n in the batch form as [27, 34]

$$\hat{x}_n = \mathscr{H}_{m,n} Y_{m,n}, \tag{29}$$

where $\mathscr{H}_{m,n}$ is the UFIR filter gain and $Y_{m,n}$ is a vector of real data (22), if to satisfy the unbiasedness condition

$$E\{x_n\} = E\{\hat{x}_n\}, \tag{30}$$

in which $E\{z\}$ means averaging of z and x_n can be represented with the last row vector in (19) as

$$x_n = F^{N-1} x_m + B_N^{(N)} W_{m,n}, \tag{31}$$

where $B_N^{(N)}$ is the Nth row vector in (25) given by

$$B_N^{(N)} = \left[F^{N-1} \; F^{N-2} \; \cdots \; F \; I\right]. \tag{32}$$

By combining (29)–(32) and following [27], one arrives at the unbiasedness constraint

$$I = \mathscr{H}_{m,n} \mathscr{C}_{m,n}, \tag{33}$$

in which

$$\mathscr{C}_{m,n} \triangleq \mathscr{C}_{m,n}(k) = \begin{bmatrix} CF^{-N+1-k_m} \\ \vdots \\ CF^{-1-k_{n-1}} \\ CF^{-k_n} \end{bmatrix}. \tag{34}$$

Now multiplying the both sides of (33) with $(\mathscr{C}_{m,n}^T\mathscr{C}_{m,n})^{-1}\mathscr{C}_{m,n}^T\mathscr{C}_{m,n}$ yields the UFIR filter gain

$$\mathscr{H}_{m,n} = \left(\mathscr{C}_{m,n}^T\mathscr{C}_{m,n}\right)^{-1}\mathscr{C}_{m,n}^T \quad (35)$$

and the batch UFIR filtering estimate becomes

$$\hat{x}_n = \left(\mathscr{C}_{m,n}^T\mathscr{C}_{m,n}\right)^{-1}\mathscr{C}_{m,n}^T Y_{m,n} \quad (36a)$$

$$= G_n\mathscr{C}_{m,n}^T Y_{m,n}, \quad (36b)$$

where the *generalized noise power gain* (GNPG) [43] is

$$G_n = \mathscr{H}_{m,n}\mathscr{H}_{m,n}^T = \left(\mathscr{C}_{m,n}^T\mathscr{C}_{m,n}\right)^{-1}. \quad (37)$$

The batch form (36a) may not suite real-time tracking and we go on with its fast iterative algorithm.

3.2. Iterative UFIR Filter Form. Provided (1) and (4), the standard iterative UFIR filtering algorithm [25] can be applied straightforwardly, if to substitute matrix H with $\overline{H}$ given by (5). The UFIR filtering estimate (36a) can then be computed iteratively using recursions

$$G_l = \left[\overline{H}_l^T\overline{H}_l + \left(FG_{l-1}F^T\right)^{-1}\right]^{-1}, \quad (38)$$

$$\hat{x}_l = F\hat{x}_{l-1} + G_l\overline{H}_l^T\left(y_l - \overline{H}_lF\hat{x}_{l-1}\right), \quad (39)$$

beginning with $l = m + K$ and ending when $l = n$. The initial values for (38) and (39) are obtained at $s = m + K - 1$ in the batch forms as

$$G_s = \left(\mathscr{C}_{m,s}^T\mathscr{C}_{m,s}\right)^{-1}, \quad (40)$$

$$\tilde{x}_s = G_s\mathscr{C}_{m,s}^T Y_{m,s}, \quad (41)$$

where $Y_{m,s}$ is a vector (22) of real data. When some data are lost, inaccurate, or unavailable, y_n can be predicted as $y_n = HF\hat{x}_{n-1}$, in which case first data on the horizon $[0, N - 1]$ must be available.

It is known that the linear UFIR filter is BIBO stable and not prone to divergence. However, latency in information delivery may require an ability to predict lost values that inevitably cause extra tracking errors, which we will consider next.

4. Tracking Errors Caused by Latency, $k_n \geqslant 0$

Error produced by the UFIR tracker can be defined as $\epsilon_n = x_n - \hat{x}_n$, where x_n is specified by (31). Provided ϵ_n, the error covariance matrix $P_n = E\{\epsilon_n\epsilon_n^T\}$ can also be represented in two forms.

In the batch form, matrix P_n appears if to substitute x_n with (31) and employ $\hat{x}_n = \mathscr{H}_{m,n}Y_{m,n}$ with $Y_{m,n}$ given by (20). That yields

$$P_n = \left[B_N^{(N)} - \mathscr{H}_{m,n}G_{m,n}\right]\overline{Q}_N\left[B_{m,n}^{(N)} - H_{m,n}G_{m,n}\right]^T + H_{m,n}\overline{R}_NH_{m,n}^T, \quad (42)$$

where $\overline{Q}_N = \text{diag}\left[Q\ Q\ \cdots\ Q\right]$ and $\overline{R}_N = \text{diag}\left[R\ R\ \cdots\ R\right]$ are square matrices with N nonzero diagonal elements. It can be shown that the deterministic case of $\overline{Q}_N = 0$ and $\overline{R}_N = 0$ makes $P_n = 0$ and the UFIR tracker has thus the *deadbeat property*.

4.1. Iterative Computation of P_n. Matrix (42) can also be computed iteratively, if to substitute x_n with (1) and $\hat{x}_n$ with (39). Provided the averaging, the recursion for (42) can be found as

$$P_l = \left(I - G_l\overline{H}_l^T\overline{H}_l\right)P_l^-(\cdots)^T + G_l\overline{H}_l^T\overline{H}_l \times\left(\sum_{i=0}^{k_l-1}F^iQF^{iT}\right)\overline{H}_l^T\overline{H}_lG_l + G_l\overline{H}_l^TR\overline{H}_lG_l, \quad (43)$$

where P_l^- is given by (12), l ranges as in (38) and (39), and P_n is taken when $l = n$. Recursion (43) suggests that the tracking error grows with k_n, because the sum containing Q grows with k_n. However, the same cannot be said about R, which does not accumulate the effect of k_n.

With no latency, the sum in (43) becomes identically zero and one arrives at the error covariance P_l of the standard UFIR filter [43],

$$P_l = \left(I - G_lH^TH\right)P_l^-(\cdots)^T + G_lH^TRHG_l, \quad (44)$$

which also holds for the KF, if to substitute G_nH^T with the Kalman gain K_n.

Note that the minimization of tracking errors will require an optimal number $N_{\text{opt}} - 1$ of iterations for the UFIR filter. At the test stage, the optimal horizon N_{opt} can be found for the known ground truth x_n by minimizing the MSE via the trace of $\mathbf{P}_n$ [30] depicted as tr $\mathbf{P}_n$. Because the ground truth is unavailable in real tracking, N_{opt} can be estimated via the measurement residual as shown in [30].

5. GPS-Based Tracking of a Moving Vehicle

We will now consider the case shown in Figure 1 when the GPS tracker measures the vehicle coordinates of location at each time index n and transfer time-stamped data cooperatively to a CS of a WSN via one or several nodes. We admit that each node may introduce latency and data will thus arrive at the CS with known delay on $k_n > 0$ points. At different time instances, a vehicle may interact with a different number of the nodes that will make the k_n-step-lag time varying. We will base our investigations on data obtained in the Cook county of Illinois and available for free use from the University of Illinois at Chicago. To simplify the problem, in this paper we will consider the case of a constant latency, $k = k_n$.

Concerned with the tracking errors and not with the actual vehicle location, we will conventionally place the start point at zero coordinates as shown in Figure 2.

5.1. State-Space Model. To investigate the trade-off between the estimators, we will suppose that a vehicle is represented with two states in each directions and assign the state vector,

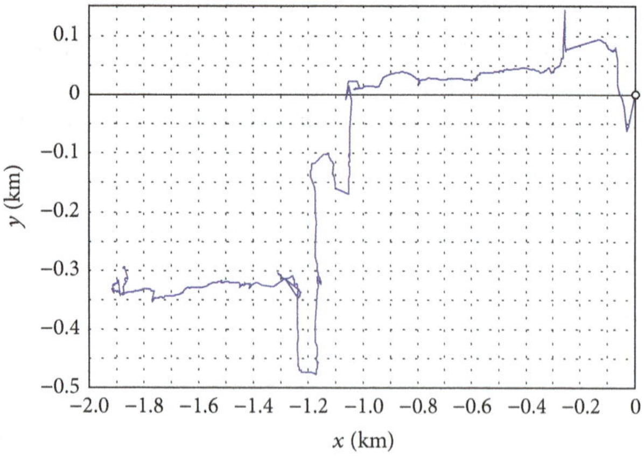

Figure 2: GPS-based vehicle trajectory measured in the north (y) and east (x) coordinates, both in km, with the start point at $\{0, 0\}$. Measurements are provided each second at 858 data points.

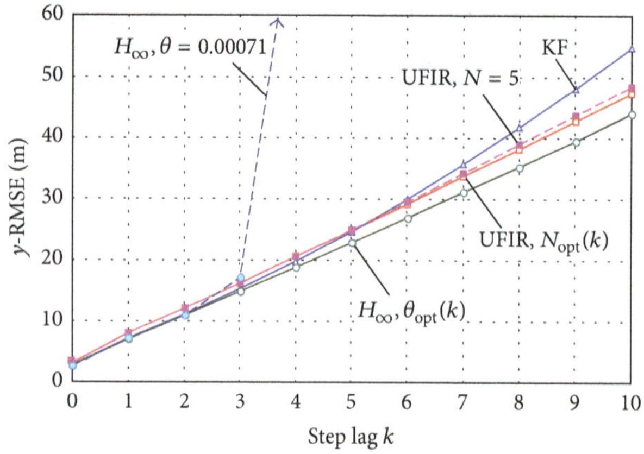

Figure 3: Effect of the k-step-lag on the y-RMSE of the UFIR filter, KF, and H_∞ filter with different tunings.

$K = 4$, as $x_n = [x_{1n}\ x_{2n}\ x_{3n}\ x_{4n}]^T$, where $x_{1n} = x_n$, $x_{2n} = \dot{x}_n$, $x_{3n} = y_n$, and $x_{4n} = \dot{y}_n$. Accordingly, the system matrix attains the form of

$$F = \begin{bmatrix} 1 & \tau & 0 & 0 \\ 0 & 1 & 0 & 0 \\ 0 & 0 & 1 & \tau \\ 0 & 0 & 0 & 1 \end{bmatrix}, \tag{45}$$

where the sampling time is $\tau = 1$ s for the considered database. The GPS tracker provides measurements of the vehicle coordinates, x and y. Therefore, the measurement matrix is

$$H = \begin{bmatrix} 1 & 0 & 0 & 0 \\ 0 & 0 & 1 & 0 \end{bmatrix}. \tag{46}$$

Provided F and H, matrix $\mathscr{C}_{m,s}$ required by the UFIR filter to compute the initial values (40) and (41) for $s = m + 3$ becomes

$$\mathscr{C}_{s,m} = \begin{bmatrix} 1 & -3\tau & 0 & 0 \\ 0 & 0 & 1 & -3\tau \\ 1 & -2\tau & 0 & 0 \\ 0 & 0 & 1 & -2\tau \\ 1 & -\tau & 0 & 0 \\ 0 & 0 & 1 & -\tau \\ 1 & 0 & 0 & 0 \\ 0 & 0 & 1 & 0 \end{bmatrix} F^{-k}. \tag{47}$$

The only tuning factor $N_{opt} = 5$ required by the UFIR filter was found for $k = 0$ by minimizing the derivative of the trace of the mean square value of the residual $y_n - H\hat{x}_n(N)$, as shown in [30]. Because N_{opt} depends on k, we will apply $N_{opt} = 5$ in the worst case for the UFIR filter.

Having no information about the process noise, we observe similar trajectories and estimate the average vehicle speed by about 10 m/s or 36 km/hour. Next, accepting the speed standard deviation of about 20%, we set $\sigma_{w2} = 2$ m/s to the second state along each of the coordinates, ignore the unknown noise in the first state, $\sigma_{w1} = 0$, and describe matrix Q as

$$Q = \sigma_{w2}^2 \begin{bmatrix} \tau^2/2 & \tau/2 & 0 & 0 \\ \tau/2 & 1 & 0 & 0 \\ 0 & 0 & \tau^2/2 & \tau/2 \\ 0 & 0 & \tau/2 & 1 \end{bmatrix}. \tag{48}$$

The GPS standard positioning service provides navigation with an error of less than 15 meters with the probability of 95% in the 2-sigma sense. Referring to this value, we assign the standard deviation of the measurement noise in each direction as $\sigma_v = 15/4 = 3.75$ m and obtain

$$R = \begin{bmatrix} \sigma_v^2 & 0 \\ 0 & \sigma_v^2 \end{bmatrix}. \tag{49}$$

Because the above provided matrices Q and R are overestimated, we set $\check{Q} = Q$ and $\check{R} = R$ for H_∞. It is expected that the tuning factor θ will improve the performance of the H_∞ filter by minimizing the MSE for the maximized errors. However, the ground truth is not available in tracking. Therefore, we will find θ for the measured trajectory and consider it as the best case for H_∞, which is unfeasible.

5.2. Effect of Latency on the Estimation Accuracy. We start with learning the effect of k on the estimation accuracy, which is illustrated in Figure 3 with the root MSEs (RMSEs) in the north direction (y-RMSE). The KF is self-tuned to k. Therefore, we consider its RMSE as a benchmark. A special feature of the KF is that the RMSE grows with k nonlinearly and faster than in the UFIR and H_∞ filters.

The UFIR filter produces a bit more errors than in the KF with small k and lesser with larger k. A special feature is that the UFIR estimate is of low sensitivity to N, in which

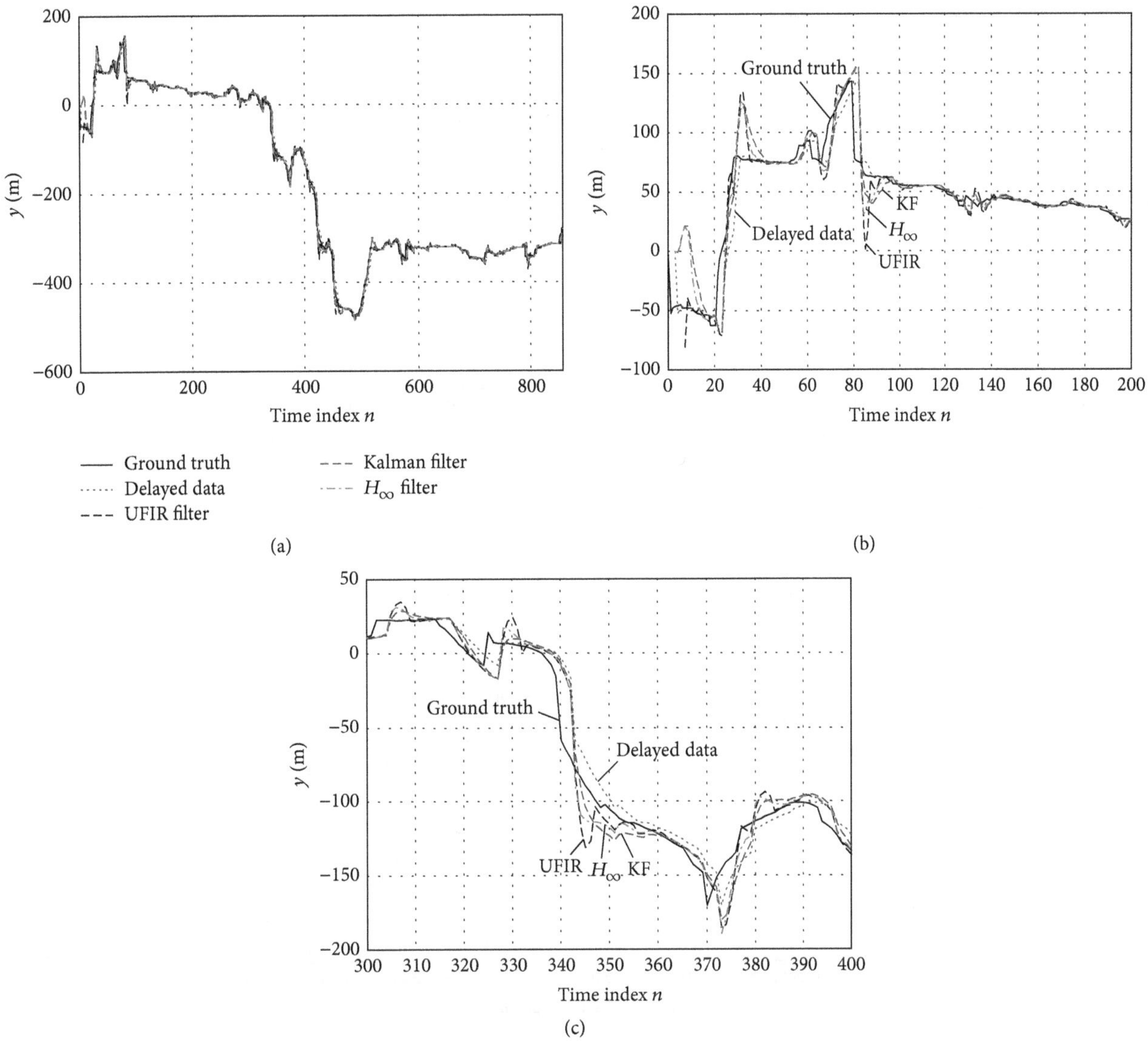

FIGURE 4: Tracking in the north direction y, m for $k = 3$ with UFIR filter ($N_{\text{opt}} = 5$), KF, and H_∞ filter ($\theta = 3.3 \times 10^{-4}$): (a) full time scale, (b) $0 \leqslant n \leqslant 200$, and (c) $300 \leqslant n \leqslant 400$.

optimal value $N_{\text{opt}} = 5$ holds for $0 \leqslant k \leqslant 4$, increases to $N_{\text{opt}} = 6$ for $5 \leqslant k \leqslant 6$, and reaches $N_{\text{opt}} = 7$ when $7 \leqslant k \leqslant 10$. Of practical importance is that setting N optimally for each k does not improve the performance essentially against the worst case when $N_{\text{opt}} = 5$ is set for all k.

The H_∞ filter outperforms both the UFIR filter and KF, provided that θ is set properly for each lag k. However, this filter is highly sensitive to θ, in which optimal value θ_{opt} ranges from 1.8×10^{-2} for $k = 0$ to 1.066×10^{-5} for $k = 10$ in a nonlinear way. Unlike in the UFIR filter, a constant θ is unacceptable for all k. An example is given in Figure 3, where $\theta_{\text{opt}} = 7.1 \times 10^{-4}$ found for $k = 2$ is applied in a wide range of k. As can be seen, it is only when $k = 2$ that the H_∞ filter improves the KF performance. For $k < 2$, there is no improvement and, when $k > 2$, the H_∞ filter rapidly diverges.

5.3. Tracking over Data Delayed on $k = 3$. We now suppose that data are transferred from a vehicle to a CS with $k = 3$ and investigate tracking errors in the north and east directions. Filters will be tuned as follows: UFIR in the worst case of $N_{\text{opt}} = 5$ being valid for $k = 0$; KF as near optimal; and H_∞ being in the best (unfeasible) case of known ground truth.

5.3.1. Tracking in the North Direction. The vehicle trajectory (measured and delayed) and estimates provided by the filters in the north direction are sketched in Figure 4. As can be seen in Figure 4(a), the trajectory is nonlinear and the vehicle sometimes maneuvers rapidly. All filters produce consistent estimates with poorly distinguishable differences (Figure 4(a)). We therefore additionally show two parts of the trajectory on short-time spans of $0 \leqslant n \leqslant 200$ in Figure 4(b) and $300 \leqslant n \leqslant 400$ in

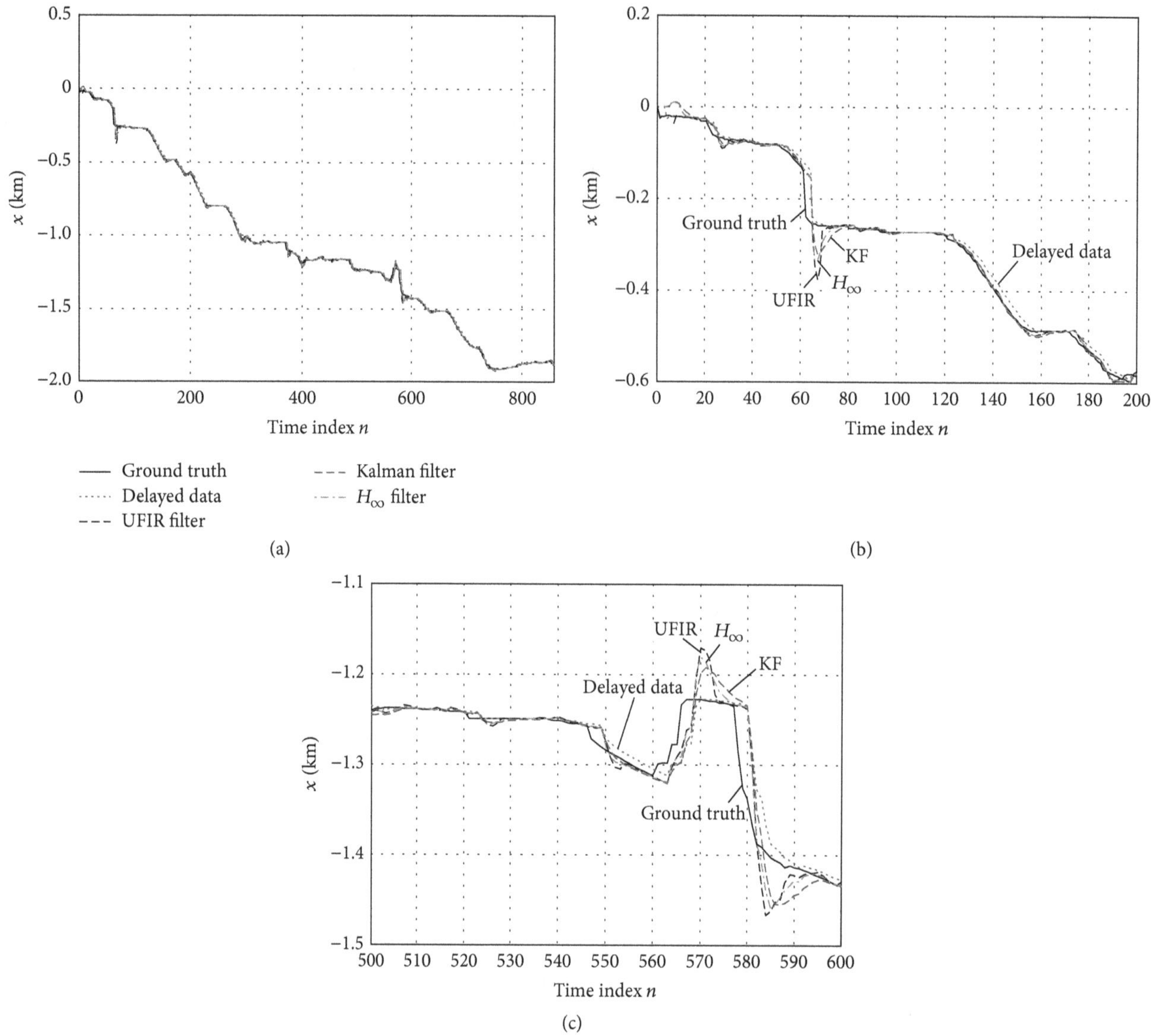

FIGURE 5: Tracking in the east direction x, m for $k = 3$ with UFIR filter ($N_{\text{opt}} = 5$), KF, and H_∞ filter ($\theta = 3.3 \times 10^{-4}$): (a) full time scale, (b) $0 \leqslant n \leqslant 200$, and (c) $500 \leqslant n \leqslant 600$.

Figure 4(c). Several observations can be made from Figure 4:

(i) All filters track well the trajectory when a vehicle travels with a near constant velocity in one direction, as on $140 \leqslant n \leqslant 180$ in Figure 4(b).

(ii) Filters temporarily lose an ability of tracking and go along the delayed data when a vehicle quickly changes the direction, as on $340 \leqslant n \leqslant 343$ in Figure 4(c).

(iii) Responding to fast maneuvers, all filters produce dynamic errors such that the UFIR filter comes up with larger excursions but shorter transients, KF with shorter excursions but longer transients, and H_∞ filter demonstrates inbetween properties; see on $80 \leqslant n \leqslant 100$ in Figure 4(b) and $344 \leqslant n \leqslant 360$ in Figure 4(c).

Because errors are unacceptably large in all filters when they temporarily lose an ability of tracking, a key question arises of how fast each of them returns back to the normal mode. In this regard the UFIR filter looks better with its shortest transient.

5.3.2. Tracking in the East Direction. Tracking in the east direction (Figure 5) does not reveal any essential features. The filters still track well the trajectory when a vehicle travels with a near constant velocity as, for example, in a span of $120 \leqslant n \leqslant 160$ in Figure 5(b). Here, all filters also temporarily lose an ability of tracking and go along the delayed data when the trajectory quickly changes as, for example, in a span of $560 \leqslant n \leqslant 570$ in Figure 5(c). Finally, the UFIR filter still demonstrates larger excursions but shorter transients, KF shorter excursions but longer transients, and H_∞ filter is inbetween as, for example, in

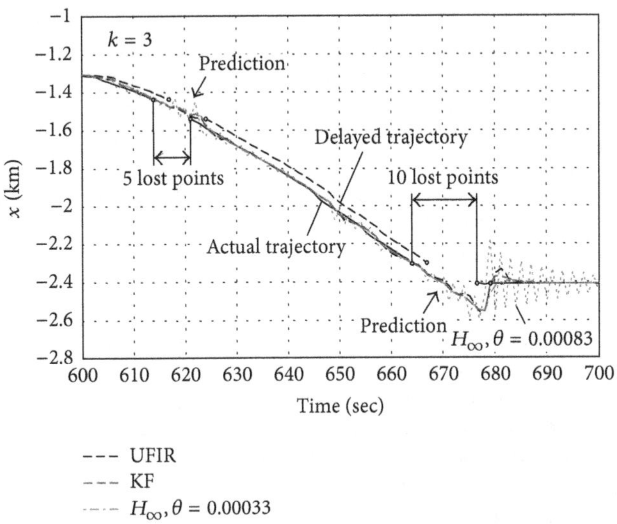

FIGURE 6: Vehicle tracking with temporary missing data in the east direction x by the UFIR filter, KF, and H_∞ filter, all augmented with the prediction option. Tuned improperly, the H_∞ filter becomes unstable and prone to divergence, as in Figure 3.

spans of $60 \leqslant n \leqslant 80$ in Figure 5(b) and $570 \leqslant n \leqslant 580$ in Figure 5(c).

5.4. Tracking with Temporary Lost Data. We finally admit that some data points can be lost during the transmission and remove 5 data points at $n = (615 \cdots 620)$ s and 10 at $n = (665 \cdots 675)$ s as shown in Figure 6 for $k = 3$. To predict lost data, we augment each algorithm with the prediction block as mentioned below (41) and run the filters. As can be seen, the filters act consistently with, however, some specifics. The estimates do not get away essentially from each other and the actual trajectory. However, when a vehicle maneuvers during the prediction, all filters diverge and return back to the actual trajectory with similar transients as in Figures 4 and 5. The latter again speaks in favor of the UFIR filter, which has shorter transients.

6. Conclusions

The UFIR filter developed in this paper for GPS-based vehicle tracking over WSNs with time-stamped discretely delayed and missing data has demonstrated better performance than the KF and H_∞ filter. The main benefits of using the UFIR filter are that it (1) does not require any information about noise and initial conditions, (2) becomes blind on given horizons, and (3) has shorter transients. The latter can be considered as an important practical advantage in all situations when the trajectory changes rapidly and estimators temporarily loses an ability of tracking. Applications to GPS-based vehicle tracking with known discretely delayed and missed data have proved a better performance of the UFIR filter.

Conflicts of Interest

The authors declare that there are no conflicts of interest regarding the publication of this paper.

References

[1] Y. Bar-Shalom, X. R. Li, and T. Kirubarajan, *Estimation with Applications to Tracking and Navigation*, John Wiley & Sons, Hoboken, NJ, USA, 2001.

[2] H. Song, R. Srinivasan, T. Sookoor, and S. Jeschke, *Smart Cities: Foundations, Principles, and Applications*, Wiley, Hoboken, NJ, USA, 2017.

[3] F. Zhao and L. J. Guibas, *Wireless Sensor Networks: An Information Processing Approach*, Elsevier, Amsterdam, The Netherlands, 2004.

[4] M. Malek-Zavarei and M. Jamshidi, *Time-Delay Systems: Analysis, Optimization and Application*, North-Holland Publishing, Amsterdam, The Netherlands, 1987.

[5] H. Wang, T. Kirubarajan, and Y. Bar-Shalom, "Precision large scale air traffic surveillance using IMM/assignment estimators," *IEEE Transactions on Aerospace and Electronic Systems*, vol. 35, no. 1, pp. 255–266, 1999.

[6] Y. Bar-Shalom, "Update with out-of-sequence measurements in tracking: exact solution," *IEEE Transactions on Aerospace and Electronic Systems*, vol. 38, no. 3, pp. 769–778, 2002.

[7] S. Zhang and Y. Bar-Shalom, "Optimal update with multiple out-of-sequence measurements with arbitrary arriving order," *IEEE Transactions on Aerospace and Electronic Systems*, vol. 48, no. 4, pp. 3116–3132, 2012.

[8] Z. Wang, F. Yang, D. W. C. Ho, and X. Liu, "Robust H_∞ filtering for stochastic time-delay systems with missing measurements," *IEEE Transactions on Signal Processing*, vol. 54, no. 7, pp. 2579–2587, 2006.

[9] A. Ray, "Introduction to networking for integrated control systems," *IEEE Control Systems Magazine*, vol. 9, no. 1, pp. 76–79, 1989.

[10] M. Sharma, R. Gautam, and M. A. Khan, *Design and Modeling of Low Power VLSI Systems*, IGI Global, Herchey, Pa, USA, 2016.

[11] D. C. Swanson, *Signal Processing for Intelligent Sensor Systems with MATLAB*, Taylor & Francis Group, Boca Raton, Fla, USA, 2nd edition, 2011.

[12] M. Sahebsara, T. Chen, and S. L. Shah, "Optimal H_∞ filtering in networked control systems with multiple packet dropout," *IEEE Transactions on Automatic Control*, vol. 52, no. 8, pp. 1508–1513, 2007.

[13] B. Sinopoli, L. Schenato, M. Franceschetti, K. Poolla, M. I. Jordan, and S. S. Sastry, "Kalman filtering with intermittent observations," *Institute of Electrical and Electronics Engineers Transactions on Automatic Control*, vol. 49, no. 9, pp. 1453–1464, 2004.

[14] K. Zhang, X. R. Li, and Y. Zhu, "Optimal update with out-of-sequence measurements," *IEEE Transactions on Signal Processing*, vol. 53, no. 6, pp. 1992–2004, 2005.

[15] T. Kailath, A. H. Sayed, and B. Hassibi, *Linear Estimation*, Prentice-Hall, Englewood Cliffs, NJ, USA, 1999.

[16] H. Zhang, G. Feng, G. Duan, and X. Lu, "H_∞ filtering for multiple-time-delay measurements," *IEEE Transactions on Signal Processing*, vol. 54, no. 5, pp. 1681–1688, 2006.

[17] J. D. Gibson, *Mobile Communications Handbook*, Taylor & Francis Group, Boca Raton, Fla, USA, 3rd edition, 2013.

[18] M. Chitre, S. Shahabudeen, and M. Stojanovic, "Underwater acoustic communications and networking: recent advances and future challenges," *Marine Technology Society Journal*, vol. 42, no. 1, pp. 103–116, 2008.

[19] Z. Wang, D. W. C. Ho, and X. Liu, "Robust filtering under randomly varying sensor delay with variance constraints," *IEEE Transactions on Circuits and Systems II: Express Briefs*, vol. 51, no. 6, pp. 320–326, 2004.

[20] M. Liu and H. Chen, "H_∞ state estimation for discrete-time delayed systems of the neural network type with multiple missing measurements," *IEEE Transactions on Neural Networks and Learning Systems*, vol. 26, no. 12, pp. 2987–2998, 2015.

[21] A. K. Singh, P. Date, and S. Bhaumik, "A modified Bayesian filter for randomly delayed measurements," *IEEE Transactions on Automatic Control*, vol. 62, no. 1, pp. 419–424, 2017.

[22] R. M. Palhares, C. E. de Souza, and P. L. Dias Peres, "Robust H_∞ filtering for uncertain discrete-time state-delayed systems," *IEEE Transactions on Signal Processing*, vol. 49, no. 8, pp. 1696–1703, 2001.

[23] K. M. Nagpal and P. P. Khargonekar, "Filtering and smoothing in an H_∞ setting," *IEEE Transactions on Automatic Control*, vol. 36, no. 2, pp. 152–166, 1991.

[24] A. H. Jazwinski, *Stochastic Processes and Filtering Theory*, Academic Press, New York, NY, USA, 1970.

[25] Y. S. Shmaliy, S. Zhao, and C. K. Ahn, "Unbiased finite impulse response filtering: an iterative alternative to Kalman filtering ignoring noise and initial conditions," *IEEE Control Systems Magazine*, vol. 37, no. 5, pp. 70–89, 2017.

[26] Y. S. Shmaliy, "Unbiased FIR filtering of discrete-time polynomial state-space models," *IEEE Transactions on Signal Processing*, vol. 57, no. 4, pp. 1241–1249, 2009.

[27] Y. S. Shmaliy, "Linear optimal FIR estimation of discrete time-invariant state-space models," *IEEE Transactions on Signal Processing*, vol. 58, no. 6, pp. 3086–3096, 2010.

[28] Y. S. Shmaliy, "An iterative Kalman-like algorithm ignoring noise and initial conditions," *IEEE Transactions on Signal Processing*, vol. 59, no. 6, pp. 2465–2473, 2011.

[29] Y. S. Shmaliy, "Suboptimal FIR filtering of nonlinear models in additive white Gaussian noise," *IEEE Transactions on Signal Processing*, vol. 60, no. 10, pp. 5519–5527, 2012.

[30] F. Ramirez-Echeverria, A. Sarr, and Y. Shmaliy, "Optimal memory for discrete-time FIR filters in state-space," *IEEE Transactions on Signal Processing*, vol. 62, no. 3, pp. 557–561, 2014.

[31] S. Zhao, Y. S. Shmaliy, F. Liu, and S. H. Khan, "Unbiased, optimal, and in-betweens: The trade-off in discrete finite impulse response filtering," *IET Signal Processing*, vol. 10, no. 4, pp. 325–334, 2016.

[32] S. Zhao, Y. S. Shmaliy, and F. Liu, "Fast Kalman-like optimal unbiased FIR filtering with applications," *IEEE Transactions on Signal Processing*, vol. 64, no. 9, pp. 2284–2297, 2016.

[33] S. Zhao, Y. S. Shmaliy, and F. Liu, "Fast computation of discrete optimal FIR estimates in white Gaussian noise," *IEEE Signal Processing Letters*, vol. 22, no. 6, pp. 718–722, 2015.

[34] W. H. Kwon and S. Han, *Receding Horizon Control: Model Predictive Control for State Models*, Springer, London, UK, 2005.

[35] Z. Quan, S. Han, and W. H. Kwon, "A robust FIR filter for linear discrete-time state-space signal models with uncertainties," *IEEE Signal Processing Letters*, vol. 14, no. 8, pp. 553–556, 2007.

[36] C. K. Ahn, "Strictly passive FIR filtering for state-space models with external disturbance," *International Journal of Electronics and Communications*, vol. 66, no. 11, pp. 944–948, 2012.

[37] C. K. Ahn, P. Shi, and L. Wu, "Receding horizon stabilization and disturbance attenuation for neural networks with time-varying delay," *IEEE Transactions on Cybernetics*, vol. 45, no. 12, pp. 2680–2692, 2015.

[38] I. Y. Song, D. Y. Kim, V. Shin, and M. Jeon, "Receding horizon filtering for discrete-time linear systems with state and observation delays," *IET Radar, Sonar & Navigation*, vol. 6, no. 4, pp. 263–271, 2012.

[39] I. Y. Song, V. Shin, and M. Jeon, "Distributed fusion receding horizon filtering for uncertain linear stochastic systems with time-delay sensors," *Journal of The Franklin Institute*, vol. 349, no. 3, pp. 928–946, 2012.

[40] H. D. Choi, C. K. Ahn, M. T. Lim, and M. K. Song, "Dynamic output-feedback H_∞ control for active half-vehicle suspension systems with time-varying input delay," *International Journal of Control, Automation, and Systems*, vol. 14, no. 1, pp. 59–68, 2016.

[41] Y. S. Shmaliy, S. H. Khan, S. Zhao, and O. Ibarra-Manzano, "General unbiased FIR filter with applications to GPS-based steering of oscillator frequency," *IEEE Transactions on Control Systems Technology*, vol. 25, no. 3, pp. 1141–1148, 2017.

[42] Y. S. Shmaliy, S. Khan, and S. Zhao, "Ultimate iterative UFIR filtering algorithm," *Measurement*, vol. 92, pp. 236–242, 2016.

[43] Y. S. Shmaliy and D. Simon, "Iterative unbiased FIR state estimation: a review of algorithms," *EURASIP Journal on Advances in Signal Processing*, vol. 2013, article 113, 16 pages, 2013.

[44] D. Simon, *Optimal State Estimation: Kalman, H_∞, and Nonlinear Approaches*, J. Wiley & Sons, Hoboken, NJ, USA, 2006.

16

A 1.25–12.5 Gbps Adaptive CTLE with Asynchronous Statistic Eye-Opening Monitor

Chen Cai, [1,2] **Jian-zhong Zhao**, [2] **and Yu-mei Zhou** [1,2]

[1] *Institute of Microelectronics of Chinese Academy of Sciences, Beitucheng West Road, Chaoyang District, Beijing 100029, China*
[2] *University of Chinese Academy of Sciences, Yuquan Road, Shijingshan District, Beijing 100049, China*

Correspondence should be addressed to Yu-mei Zhou; ymzhou@ime.ac.cn

Academic Editor: George S. Tombras

The equalization of a large attenuation signal and multirate communication in high-speed serial interface is hard to balance. To overcome this difficulty, an adaptive equalization system with optimized eye-opening monitor is proposed. The designed eye-opening monitor is based on the asynchronous statistic eye diagram tracking algorithm, and the eye diagram is obtained by undersampling with the low-speed asynchronous clock. With the eye-opening monitor into the adaptive loop, an adaptive equalization system combined with continuous-time linear equalization (CTLE) is completed. And the inductor peaking technology is used to improve the capacity of compensation. With SMIC 28 nm CMOS process to achieve the overall design, the power consumption and core chip area are 12 mW @ 12.5 Gbps and 0.12 mm^2, respectively. And postsimulation results show that it can offer compensation from 6 to 21 dB for 1.25–12.5 Gbps range of receiving data, which achieves a large range of data rate and channel loss, and its power efficiency is 0.046 pJ/bit/dB for the worst case, which is better than most previous works.

1. Introduction

High-speed serial interface has become the inevitable choice for high-speed data transmission. However, the nonideal factors of the wireline transmission channel will lead to channel noise and frequency attenuation and significantly reduce the quality of the received signal. Various equalization techniques had been used to compensate the loss, among the most popular are decision feedback equalizer (DFE) and CTLE. The DFE can eliminate intersymbol interference (ISI) effectively, but it will increase system complexity, and for high-speed data, the time constraint is apparent. On the contrary, CTLE can realize the compensation of the full-frequency band and has simple structure, and it does not need any clock signal. For that reason, we choose the CTLE to realize the signal equalization.

In addition, a fixed preset equalizer may not work well across a large range of data rate, it is imperative to have the robust adaptation algorithm to achieve a well-behaved adaptive equalizer. The adaptive equalization algorithm based on energy extraction proposed by Won et al. [1] had good compensation effect on high-speed data, but the circuit complexity and power consumption were greatly increased through its algorithm. Choi et al. [2] used the frequency filter to change low-frequency gain adaptively, and the adaptive linear device had a simple circuit structure, but the frequency information shaped by the rectifier was inaccurate, and the stability still needed to be optimized. The data pattern information also had been used to change the equalization coefficient in [3], which had better area and power efficient, but the compensation capacity was limited.

As the optimal compensation for the high-frequency signal and low-frequency signal is difficult to balance, and it is even more difficult to cover a large range of channel loss. In order to overcome these obstacles, we proposed an optimized adaptive equalization system which is based on eye diagram tracking. We use the asynchronous undersampling method to obtain an on chip eye-opening monitor. And the proposed tolerance judgement in the adaptation algorithm could track the change of the data rate, and it can avoid the over equalization results. Besides, the undersampling method alleviates the circuit timing constraints and simplifies the circuit design, and we use the active-inductor technology to improve the performance of the CTLE module. The optimized CTLE combined with the digital eye-opening monitor realizes the automatic equalization for channel changing, and it also can achieve the optimal compensation results for different data rates. The equalization system can compensate the data rate range from 1.25 Gbps to 12.5 Gbps and provide a wide equalization range from 6 dB to 21 dB. And the power efficiency of the core circuit is only 0.046 pJ/bit/dB. It achieves both design flexibility and stability, also with lower power consumption.

2. Nonideal Factors of Channel

The nonideal characteristics of transmission channels mainly include skin effect, dielectric loss, return loss, crosstalk, and all kinds of noises. Among which, the skin effect and dielectric loss are main reasons for high-frequency channel losses [4, 5]. Their impacts can be expressed by

$$C_S = \exp\left[(-h_s * l)(1 + j)\sqrt{f}\right], \quad (1)$$

$$C_d = \exp\left[(-h_d * l)f\right], \quad (2)$$

where h_s and h_d are the coefficients of skin effect and dielectric loss, respectively, f is the frequency, and l is the transmission distance. It indicates that skin effect loss and dielectric loss are both proportional to the channel length. And when the signal frequency is relatively low, the channel attenuation is mainly resulted from the skin effect. With signal frequency increasing, the attenuation caused by the dielectric loss becomes more and more obvious. Taking these nonideal factors into account, we can get the frequency response trend of the actual channel, shown in Figure 1. The actual channel shows low-pass characteristics and suppresses the high-frequency signal significantly.

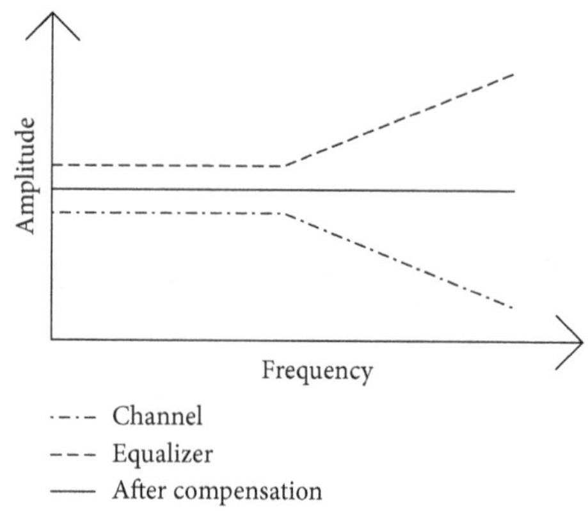

Figure 1: Channel frequency characteristic.

3. Adaptive Equalization Algorithm

3.1. Asynchronous Statistic Eye Diagram Tracking Algorithm. Figure 2 shows two eye diagrams and their statistic results. According to the actual eye diagram, it can be concluded that height of the eye-opening indicates the amplitude of the received signal. In addition, the thickness of the eyelid proves that if the loss signal gets an effective and uniform compensation after passing through the equalization system. With that observation, the eye diagram can be obtained by two steps: sampling and counting. The asynchronous undersampling technique is adopted to achieve eye diagram sampling [6], in which the sampling clock is not synchronous with the periodic input signal and the sample clock is much slower than the data clock. As shown in Figure 3, the asynchronous clock is used to sample the high-speed input signal, and the low-frequency clock is sweeping the sample time across the period of the input waveform. After a certain number of sampling periods, the time domain waveform of input data can be obtained, and the statistical result of sampling data represents the information of input data. In that way, we would realize an on-chip eye-opening monitor. And certain eye characteristics can be extracted as long as a sufficient number of samples are taken.

However, the adaptive algorithm in [6] has a defect that the analysis of the statistic information is not comprehensive enough. Using comparison of Figures 2(a) and 2(b) to explain that, it seems that biggest sample number of Figure 2(b) is much bigger than Figure 2(a). But when we focus on the compensation performance, in fact, Figure 2(b) provides an over equalization, and Figure 2(a) gets the optimal equalization result, which proves that simply searching for the histogram that has the largest peak value to choose the equalization coefficient is not accurate. In others words, not only the height of eye opening shows the amplitude of the compensated signal, the reference voltage value corresponding to the maximum value of statistics results provides useful information too. Therefore, we optimize the algorithm with a tolerance judgement, using the reference voltage to do the second-order decision. The proposed optimized algorithm could further improve the performance of the adaptation system and gets the optimal equalization effects.

3.2. Architecture of Adaptive Equalizer. Figure 4 presents the structure of the proposed adaptive equalizer, which mainly consists of a CTLE, a full differential dynamic comparator, and a DAC to provide reference voltage, and Figure 5 demonstrates the adaptation control flow chart. Based on the asynchronous undersampling technique, we divide the amplitude of an eye diagram into 16 reference levels, named REF_j ($j = 0, 1, 2, \ldots, 15$), and the equalizer coefficients are called EQ_i ($i = 0, 1, 2, \ldots, 15$), which aims to provide a large range of equalization capacities.

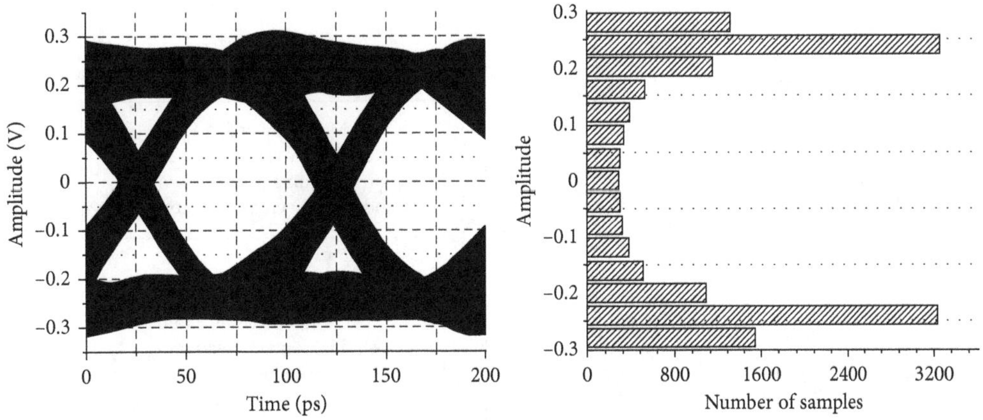

(a)

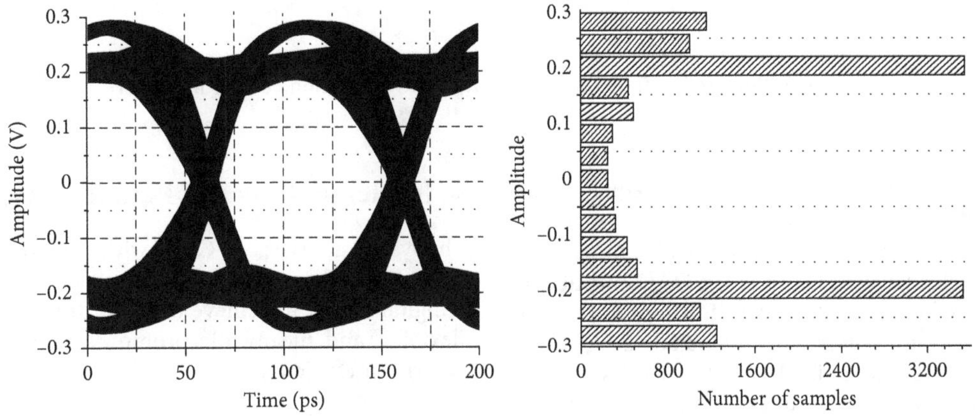

(b)

Figure 2: Eye diagrams and statistic bars. (a) Optimal equalization. (b) Over equalization.

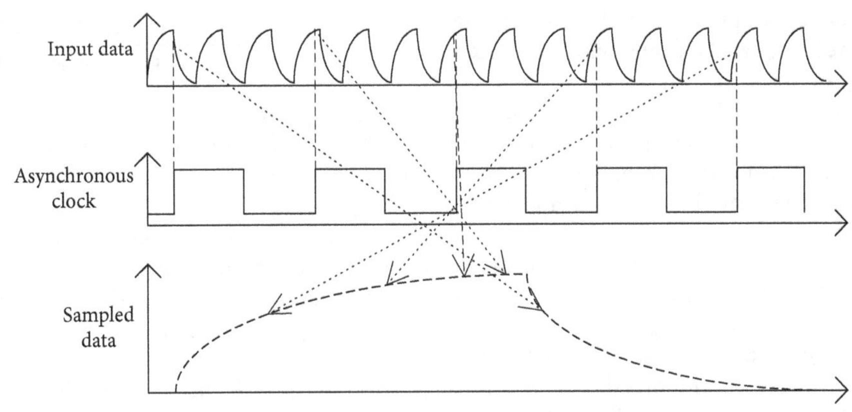

Figure 3: Asynchronous undersampling technique.

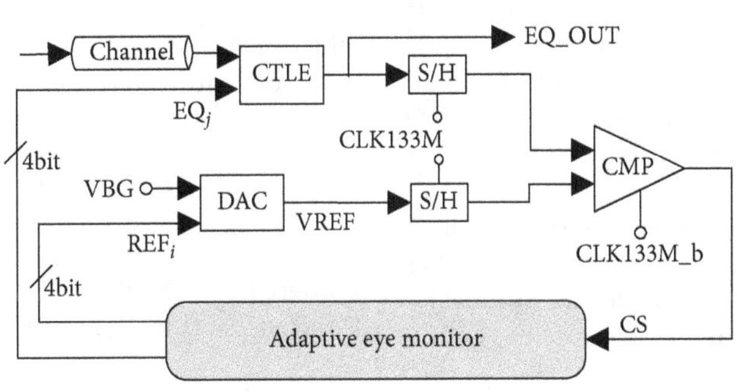

Figure 4: Proposed adaptive equalizer structure.

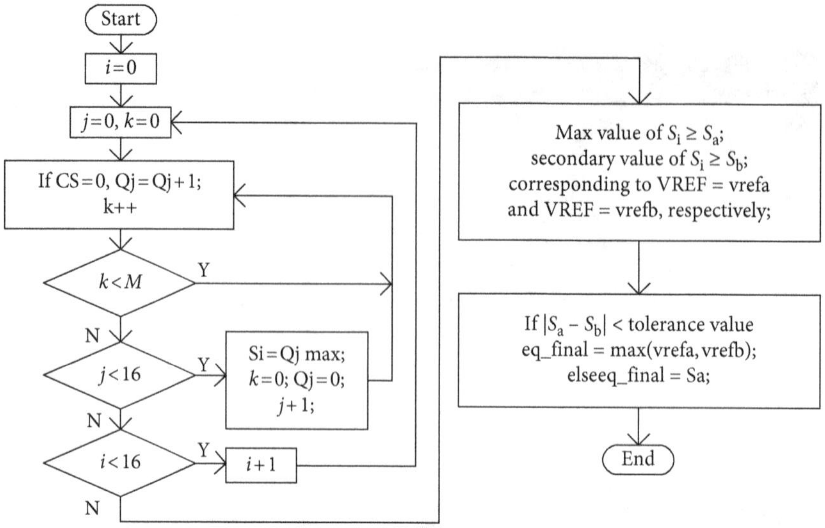

FIGURE 5: Proposed adaptive equalizer structure's adaptation control flow chart.

The operation process of the adaptive equalization system is as follows: Step 1, the adaptive eye-opening monitor sets the original equalization coefficient as EQ_0, the reference level is set to REF_0, and then DAC outputs the corresponding reference voltage VREF. Step 2, the full differential comparator compares equalized data EQ_OUT with VREF in M (for example, $M = 8192$) CLK sampling periods (k is the count number of sample clocks). And the comparison result CS is transmitted to the eye-opening monitor. Step 3, the eye-opening monitor moves the reference value to next level ($j = j + 1$) and repeats sampling operation in Step 2, until $j = 15$, and the whole eye diagram corresponded to EQ_0 is sampled. Step 4, change the equalization coefficient to EQ_1 and repeat operation in Step 2 and Step 3, and those processes are repeated for each equalization coefficient ($i = 0, 1, 2, \ldots, 15$). Then, the eye-opening monitor obtains the counting results in order to complete the eye diagram. Step 5, the eye-opening monitor selects the optimal coefficient of the equalizer and completes the adaptive process.

The algorithm provides a full consideration of asynchronous statistic eye diagram tracking. In Step 5, the eye-opening monitor firstly selects the biggest and secondary peaking value of eye diagram statistical results, which are S_a and S_b, respectively. And they both correspond to a reference voltage (called $vref_a$ and $vref_b$), which indicate the amplitude of eye diagram. If the difference value of S_a and S_b is less than the tolerance value, then it compares the reference voltages $vref_a$ and $vref_b$ and chooses the equalization coefficient corresponding to the bigger reference voltage. The tolerance value is depends on comparator and sampler errors, and we did the simulation of those circuits to set the tolerance value.

The settle time of the proposed adaptation process is calculated by M (sample number) × 16 (reference voltage level number) × 16 (equalization coefficient number) × 7.5 ns (asynchronous clock period)~15 ms.

4. Implementation of Equalizer System

The system consists of a CTLE with the active inductor, a full differential dynamic comparator, and a DAC. We use a digital circuit to achieve the algorithm to improve design flexibility and, meanwhile, promote system performance and algorithm efficiency.

4.1. CTLE. The common structure of conventional CTLE is capacitive degenerated differential pair [7], as shown in Figure 6(a). The transfer function is given by

$$H(s) = \frac{(g_m R_d)(1 + sR_s C_s)}{(1 + sR_s C_s + (g_m R_s/2))/(1 + sR_d C_d)}, \quad (3)$$

where g_m is the transconductance of input difference pairs. And the expressions for the locations of zero and poles can be deduced as $w_z = 1/(R_s C_s)$, $w_{p1} = (1 + g_m R_s/2)/R_s C_s$, $w_{p2} = 1/(R_d C_d)$.

This topology, however, suffers from limited bandwidth and consequently insufficient compensation at high frequencies. It is because w_{p1} exceeds w_z by a factor of $(1 + g_m R_s/2)$, and the DC gain drops by the same amount of factor. In other words, $g_m R_s$ must stay low so as to avoid large DC loss (otherwise the interposed buffers suffer). This issue limits the maximum achievable boost in magnitude and phase. To expand the bandwidth of CTLE, the inductor peaking technology is introduced, shown in Figure 6(b). The transfer function of CTLE with the inductor load is given by

$$H(s) = \frac{g_m R_d}{1 + (g_m R_s/2)} \cdot \frac{1 + (s/w_{z1})}{1 + (s/w_{p1})} \cdot \frac{1 + (s/w_{z2})}{1 + (2\zeta/w_n)s + (s^2/w_n^2)}, \quad (4)$$

where $w_{z2} = 2\zeta w_n$, $\zeta = (R_d/2)(C_d/L_p)^{1/2}$, $w_n = 1/(C_d \cdot L_p)^{1/2}$ and w_{z1} and w_{p1} are unchanged.

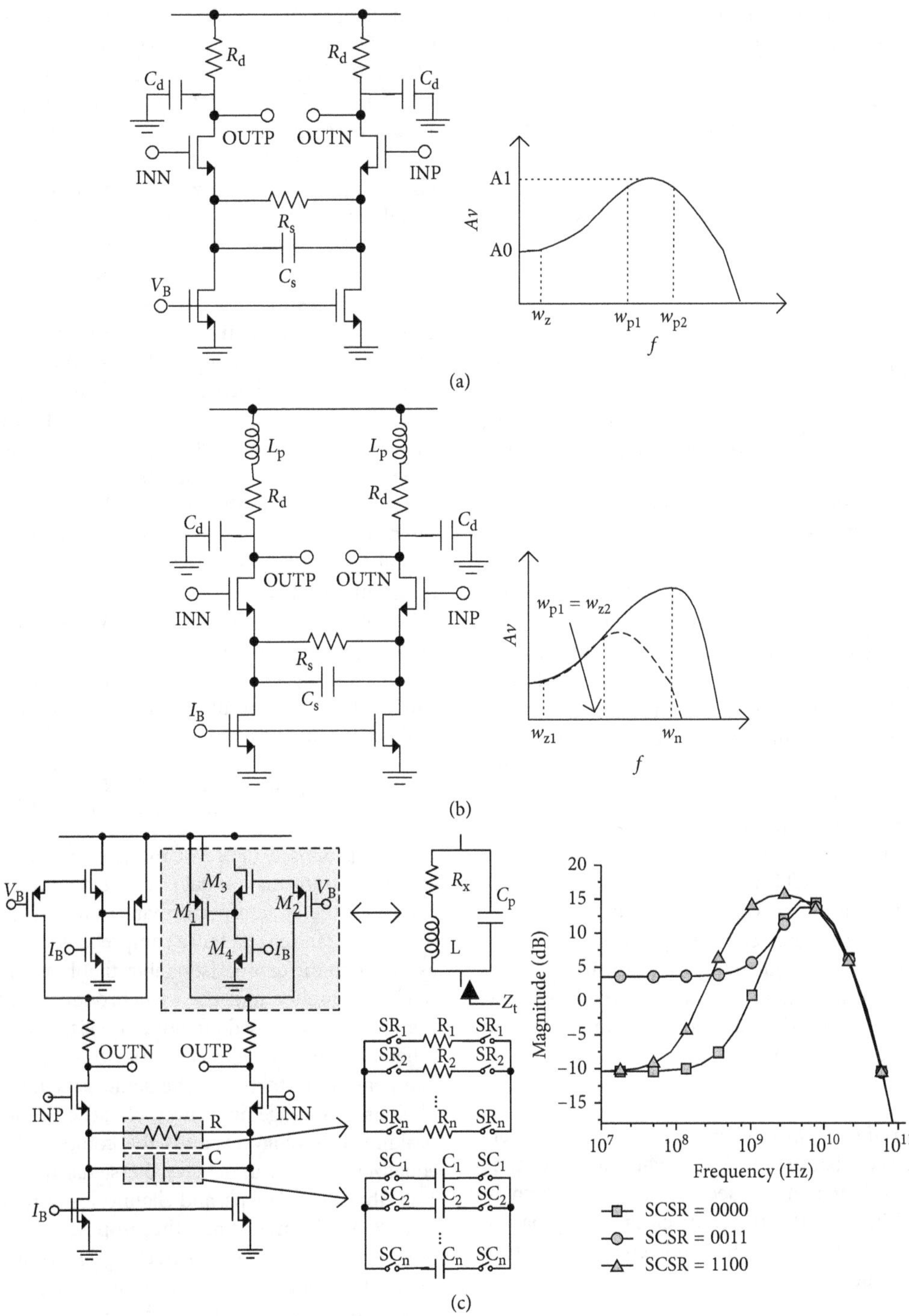

FIGURE 6: CTLE circuit. (a) Conventional CTLE circuit and frequency-response curve. (b) CTLE with inductor peaking. (c) Designed CTLE circuit and frequency-response curve.

Since a passive inductor may take large area, we introduce an optimized structure of CTLE called the active inductor to realize the inductive termination [8, 9], which aims to extend the output bandwidth in the presence of large capacitive loads and save chip area. A PMOS-based active-inductor circuit is used as the load of CTLE in Figure 6(c), which enhances the compensation ability for high-speed data. It uses a MOS resistor (M2, which operates in deep-triode region) through which the output node is coupled to the gate of the PMOS transistor M1. A level shifter, consisting of a source follower M3 and a current source M4, is inserted between M1 and M2 to allow a lower gate bias voltage for M1. The termination impedance of presented CTLE is given by the following equation:

$$Z_t = \frac{1 + j\omega C_{gs} R_2}{(g_{ds} + g_m) - \omega^2 R_2 C_{ds} C_{gs} + j\omega\left(C_{gs} + C_{ds} + R_2 C_{gs} g_{ds}\right)}, \quad (5)$$

$$L = \frac{\left(R_2 C_{gs1}\right)}{\left(g_{ds1} + g_{m1}\right)}, \quad C_p = C_{ds1}, \quad R_x = \frac{1}{\left(g_{ds1} + g_{m1}\right)}, \quad (6)$$

where C_{ds}, C_{gs}, g_{ds}, and g_m are the parameters of M1 and R2 is the equivalent resistor. And the termination impedance can be represented as (6).

The inductive peaking facilitates the equalizing filter design, and the 20 dB compensation for data rate above 10 Gbps becomes feasible. Besides, for a single transmission data rate, as the equalizer is designed for a fixed frequency point, the adjusting of the equalization coefficient only depends on the channel change. While the data rate is changing, it could not achieve an optimal compensation by a coefficient fixed equalizer. Therefore, the RC values both need to be changed through the switches SR and SC to provide different DC gains and peaking frequency. There are 4 resistance values (SR is 2 bit) and 4 capacitance values (SC is 2 bit), and they can combine out 16 sets of different equalization coefficients. The frequency response of designed active-inductor CTLE is shown in Figure 6(c). It can realize a compensation range of 6–21 dB in the effective frequency band, the DC gain can change from −10 dB to 5 dB, and the peaking frequency can scan from 1.25–12.5 GHz (at Nyquist frequency), which can provide an area-efficient alternative for passive inductive terminations and well satisfy the equalization ability.

4.2. Full Differential Comparator and DAC. The full differential comparator consists of the SA in the first stage, and the slave set-reset (SR) latch in the second stage is shown in Figure 7(a), and Figure 7(b) is the DAC circuit using partial pressure resistance structure and hot code control switch to reduce switch number.

5. Postsimulation Results

The equalizer layout in SMIC 28 nm CMOS technology is demonstrated in Figure 8, core circuit power consumption and area are 12 mW @ 12.5 Gbps and 0.12 mm^2, respectively, and the digital implementation is included.

The postsimulation results are shown in Figure 9. The S-parameter curve of a 563 mm PCB channel with 18.1 dB loss at 5 GHz is shown in Figure 9(a), and the eye diagram of the 10 Gbps receiving signal is shown in Figure 9(b); it can be seen that the eye diagram is closed totally. The acquired eye diagram after the adaptive equalizer of 456, 593, 336, and 1355 mm PCB channels are shown in Figures 9(c)–9(f), and according channel loss are −21.4 dB @ 6.25 GHz, −22.3 dB @ 5 GHz, −12 dB @ 5 GHz, and −8 dB @ 620 MHz, respectively. According data rates are 12.5 Gbps, 10 Gbps, 10 Gbps, and 1.25 Gbps, respectively. By comparing Figure 9(d) with Figure 9(e), it shows that, for the fixed data rate, while the channel length is changing, the equalization system can realize well equalization. And by comparing Figures 9(c) and 9(d) with Figure 9(f), it demonstrates that although the data rates are different, the loss data can achieve the optimal compensation results after the equalization system, and the changing of signal frequency can be well tracked by the proposed adaptation algorithm. And the final equalization coefficients for Figures 9(c)–9(f) are SRSC = 0000, 0001, 0110, 1111, respectively. On the one hand, it proves that the equalizer can provide large compensation range; on the other hand, it shows that the adaptation algorithm can avoid the over equalization results with the proposed eye-opening monitor. As seen from the results, receiving loss signal can be well compensated after equalization, and eye diagrams open well.

Compared with other recent work given in Table 1, the advantages of this design can be displayed intuitively. Comparing with the eye-opening monitor in [15], our design can compensate the bigger loss, and the adaptation part is realized on the chip. A proposed FOM is calculated by

$$\text{FOM} = \frac{\text{total power}}{\text{data rate} * \text{channel loss}}. \quad (7)$$

The proposed figure of merit normalizes the power consumption, transmission data frequency, and channel loss, which compares the system performance under the same evaluation index, and can make a more comprehensive measurement. And as shown in the comparison table, thanks to digital eye-opening tracking monitor and the asynchronous sampling technique, the FOM of proposed adaptive equalizer is clearly superior to other designs, which means that this design achieves better compensation ability for the same data rate, and it also has obvious advantage in power efficiency. The active-inductor peaking technology expands the compensation ability of the equalizer, which can achieve the signal frequency range from 1.25 Gbps to 12.5 Gbps, and channel losses range from −6 dB to −21 dB. In the meantime, the proposed optimized asynchronous statistic eye diagram tracking algorithm also ensures to obtain the appropriate compensation effects and avoids over equalization for the different data rate.

6. Conclusion

An adaptive equalization system based on the asynchronous statistic eye diagram tracking algorithm was realized on SMIC 28 nm CMOS technology. As the height of eye diagram and the concentration of statistical data are taking into account, the optimized adaptive algorithm provides a tolerance judgement to track equalized eye opening so that the equalization system can avoid under equalization or over equalization of different signal frequencies. The active-inductor peaking technology also enhances CTLE capacity to provide a wide equalization range. The adaptive equalization system can offer

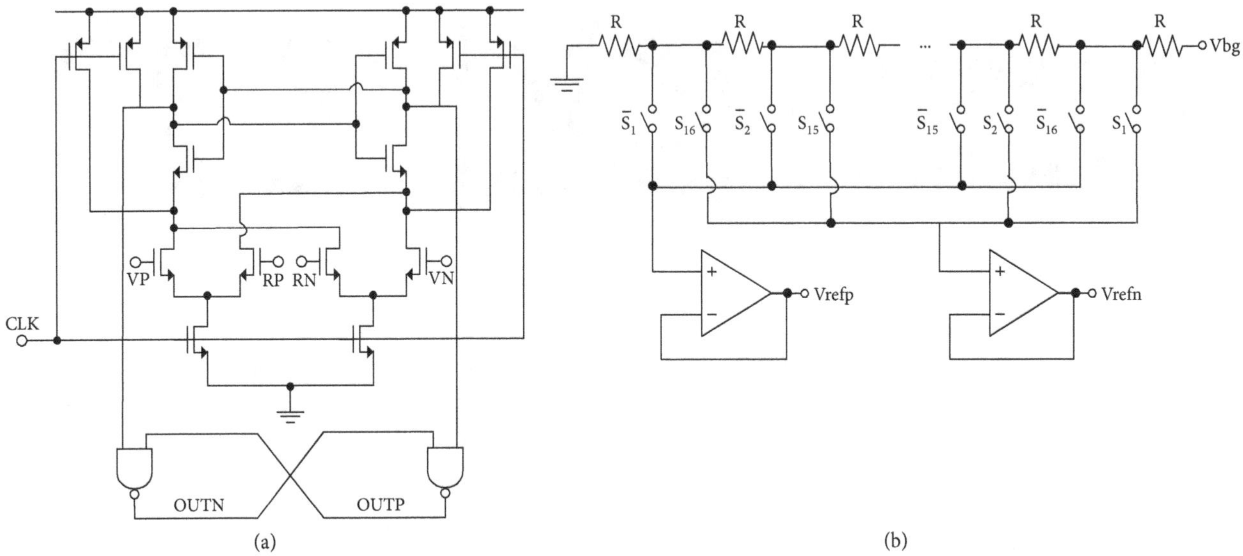

FIGURE 7: Circuit of full differential comparator and DAC.

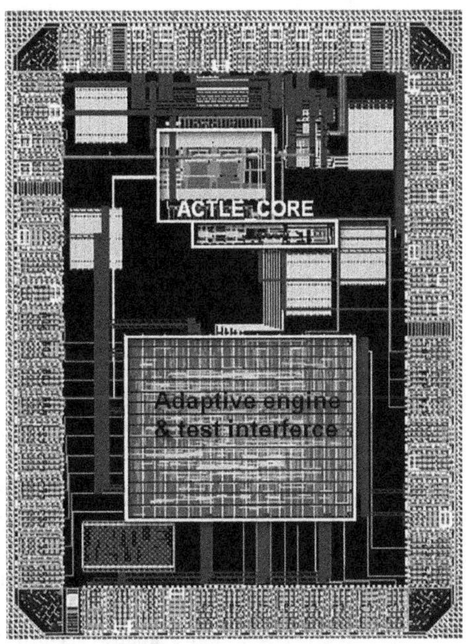

FIGURE 8: Layout of the chip.

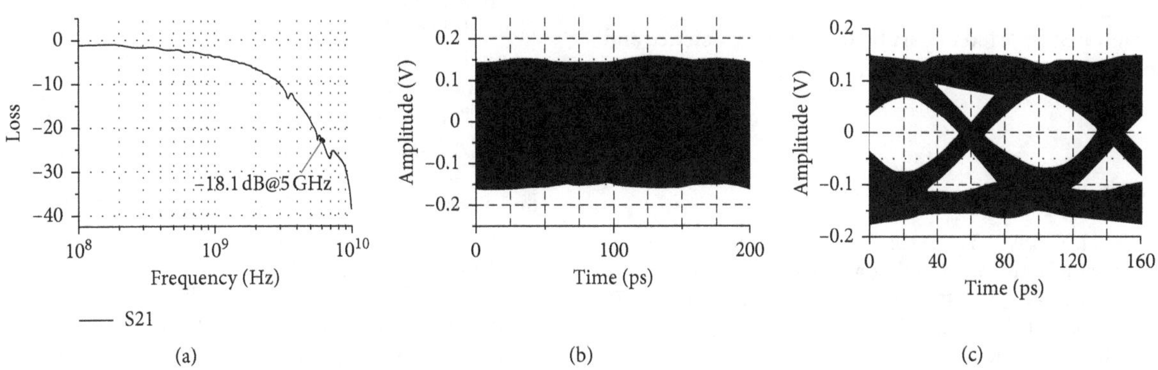

FIGURE 9: Continued.

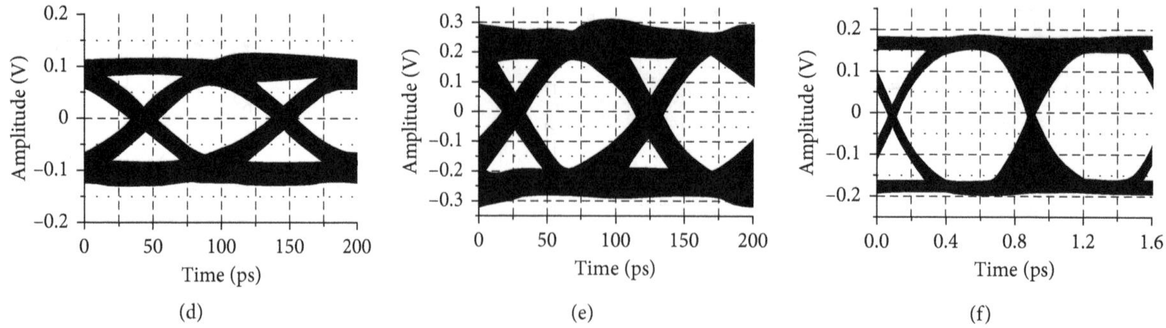

Figure 9: Postsimulation results of adaptive equalization for different data rates and channel losses.

Table 1: Performance comparison.

Reference	[10]	[11]	[12]	[13]	[14]*	[15]*	[6]*	This work
Technology	65 nm	65 nm	180 nm	130 nm	110 nm	90 nm	130 nm	28 nm
Equalization	DFE	CTLE + DFE	CTLE	DFE	CTLE + DFE	DFE	CTLE	CTLE
Adaptation	YES	YES	YES	NO	YES	YES	YES	YES
Data rate (Gb/s)	10	10	5	6.25	11.5	10	5.4	1.25–12.5
Channel loss (dB)	16.2	25	6.8	18	21.7	8.8	6–16	6–21
Supply (V)	1.2	—	1.8	1.2	1.3	1.2	—	0.9/1.8
Core area (mm^2)	0.01	—	—	—	0.014	0.01	0.18	0.12
Power (mW)	24	66	18	1.875	25.35	11	34.99	12
FOM (pJ/bit/dB)	0.148	0.264	0.53	0.167	0.102	0.125	0.405	0.046

*Measurement data and others are based on simulation.

a compensation from 6 dB to 21 dB for 1.25–12.5 Gbps of the receiving signal, and its power efficiency is 0.046 pJ/bit/dB for the worst case. It has low power consumption and strong adaptive capacity so as to greatly optimize the high-speed interface analog front-end design. As the adaptation judgement is realized by the digital controller, it achieves a reusable design of the adaptive equalization system as well.

Conflicts of Interest

The authors declare that there are no conflicts of interest regarding the publication of this paper.

Acknowledgments

This work was supported in part by the National Science and Technology Major Project of China (2014ZX02302002).

References

[1] H. Won, J. Y. Lee, T. Yoon et al., "A 28-Gb/s receiver with self-contained adaptive equalization and sampling point control using stochastic sigma-tracking eye-opening monitor," *IEEE Transactions on Circuits and Systems I: Regular Papers*, vol. 64, no. 3, pp. 664–674, 2017.

[2] J. S. Choi, M. S. Hwang, and D. K. Jeong, "A 0.18-μm CMOS 3.5-gb/s continuous-time adaptive cable equalizer using enhanced low-frequency gain control method," *IEEE Journal of Solid-State Circuits*, vol. 39, no. 3, pp. 419–425, 2004.

[3] S. Shahramian, C. Ting, A. Sheikholeslami et al., "A pattern-guided adaptive equalizer in 65 nm CMOS," in *Proceedings of IEEE International Solid-State Circuits Conference*, pp. 354–356, San Francisco, CA, USA, February 2011.

[4] J. Liu and X. F. Lin, "Equalization in high-speed communication systems," *IEEE Circuits and Systems Magazine*, vol. 4, no. 2, pp. 4–17, 2004.

[5] L. Tang, W. Gai, L. Shi et al., "A 32Gb/s 133mW PAM-4 transceiver with DFE based on adaptive clock phase and threshold voltage in 65nm CMOS," in *Proceedings of IEEE International Solid-State Circuits Conference-(ISSCC)*, pp. 114–116, San Francisco, CA, USA, September 2018.

[6] W. S. Kim, C. K. Seong, and W. Y. Choi, "A 5.4-Gbit/s adaptive continuous-time linear equalizer using asynchronous undersampling histograms," *IEEE Transactions on Circuits and Systems II: Express Briefs*, vol. 59, no. 9, pp. 553–557, 2012.

[7] C. H. Lee, M. T. Mustaffa, and K. H. Chan, "Comparison of receiver equalization using first-order and second-order continuous-time linear equalizer in 45 nm process technology," in *Proceedings of International Conference on Intelligent and Advanced Systems*, pp. 795–800, Kuala Lumpur, Malaysia, July 2012.

[8] J. Song, S. Hwang, H. W. Lee et al., "A 1-V 10-Gb/s/pin single-ended transceiver with controllable active-inductor-based driver and adaptively calibrated cascaded-equalizer for post-LPDDR4 interfaces," *IEEE Transactions on Circuits and Systems I: Regular Papers*, vol. 65, no. 1, pp. 331–342, 2018.

[9] Y. S. M. Lee, S. Sheikhaei, and S. Mirabbasi, "A 10Gb/s active-inductor structure with peaking control in 90nm CMOS," in *Proceedings of IEEE Asian Solid-State Circuits*

Conference, pp. 229–232, Fukuoka, Japan, November 2008.

[10] S. Yuan, Z. Q. Wang, X. Q. Zheng et al., "A 10Gb/s speculative decision feedback equalizer with a novel implementation of adaption in 65nm CMOS technology," in *Proceedings of IEEE International Conference on Electron Devices and Solid-State Circuits*, pp. 1-2, Chengdu, China, June 2014.

[11] M. Dolan and F. Yuan, "An adaptive edge decision feedback equalizer with 4PAM signalling," in *Proceedings of IEEE 60th International Midwest Symposium on Circuits and Systems (MWSCAS)*, pp. 535–538, Boston, MA, USA, August 2017.

[12] S. Agarwal and V. S. R. Pasupureddi, "A 5-Gb/s adaptive CTLE with eye-monitoring for multi-drop bus applications," in *Proceedings of IEEE 57th International Midwest Symposium on Circuits and Systems (MWSCAS)*, pp. 410–413, College Station, TX, USA, August 2014.

[13] X. L. Zhang, Y. Wang, J. C. Cao et al., "A 6.25-Gbps 4-tap low-power decision feedback equalizer in 0.13μm CMOS technology," in *Proceedings of IEEE International Conference of Electron Devices and Solid-state Circuits*, pp. 1-2, Hong Kong, China, June 2013.

[14] Y. Kim, Y. Kim, T. Lee, and L. Kim, "An 11.5 Gb/s 1/4th baud-rate CTLE and two-tap DFE with boosted high frequency gain in 110-nm CMOS," *IEEE Transactions on Very Large Scale Integration Systems*, vol. 23, no. 3, pp. 588–592, 2015.

[15] C. Seong, J. Rhim, and W. Choi, "A 10-Gb/s adaptive look-ahead decision feedback equalizer with an eye-opening monitor," *IEEE Transactions on Circuits and Systems II: Express Briefs*, vol. 59, no. 4, pp. 209–213, 2012.

17

Online Identification of Distribution Line Parameters by PMUs under Accuracy, Positive Sequence, and Noise Considerations

Mustafa M. Al Khabbaz[1] **and Mohamed A. Abido**[2]

[1]*Facilities Planning Department, Saudi Arabian Oil Company (Saudi Aramco), Dhahran 31311, Saudi Arabia*
[2]*Electrical Engineering Department, King Fahd University of Petroleum and Minerals, Dhahran 31261, Saudi Arabia*

Correspondence should be addressed to Mustafa M. Al Khabbaz; am.1@hotmail.com

Academic Editor: Salvatore Favuzza

This paper proposes an inaccuracy mitigation measure to reduce the error associated with distribution line parameters identification. Additionally, it introduces the concept of positive sequence quantities for determining the line resistance, reactive inductance, and shunt admittance. The positive sequence-based analysis is required for asymmetrical related studies such as unbalanced fault analysis. The paper, also, includes the consideration of noisy distribution networks. It compares the performance of three line parameters identification techniques by using different statistical measures. A total of 12,960 different case studies are simulated and analyzed under six main loading scenarios and four categories with changing line parameters. The line parameters are calculated online using voltage and current signals obtained from phasor measurement units (PMUs) placed at the line two terminals. Finally, the study outcomes and the associated recommendations have been summarized for future works considerations.

1. Introduction

Distribution line (DL) parameters identification forms the basis for distribution power system studies, including dynamic and transient stabilities, state estimate, protection setting, etc. The common practice in the industry, till today, is to determine the parameters using values from design datasheets, manufacture specification sheets, and engineer estimation. The latter could base the calculation on conductor dimensions, sag, temperature, tower geometries, and other elements. These elements are used to identify the DL data through different mechanisms such as calculating the geometric mean radius and the geometric mean distance, denoted by GMR and GMD, respectively. Additionally, the official electrical transient analysis program (known as ETAP) model could be utilized to find the DL data, which is an off-line tool. Assumptions and approximations are included in the calculation process steps which reduce the accuracy of results. Basing DL parameter estimation on off-line techniques or preidentified information significantly impacts the accuracy level of the power system studies that depend on these values due to the following:

(1) Conductor resistance and reactance vary with ambient conditions, conductor situation, and power flow.

(2) A number of installed circuits are spliced with other conductors that are different in types and specifications. This represents an inhomogeneity of the line sections.

(3) The overhead conductor arrangement changes due to using different tower configurations and applying the concept of transposition.

(4) Cable installation conditions such as grouping, underground, overhead, cable trays, conduits, and submarine, etc., play a major role in line parameter estimation.

(5) Cable aging could impact the line parameters due to several factors such as degradation, tension, and life cycle.

The above five factors are sources of conductor impedance and admittance identification errors. With the emergence of PMU technology, it is possible to obtain more accurate data about the system conditions with high-frequency samples along with the corresponding time stamp. Accordingly, it is possible to develop more accurate DL impedance parameters estimation by online measurement techniques using the synchronized PMUs. This online analysis can be used to improve power system operations reliability as detailed below:

(a) Power system restoration and reclosing: phasor data is used to bring equipment back into service avoiding the risk of instability or unsuccessful reclosing trials
(b) Automated management of voltage and frequency response: the data is used for better system management to frequency and voltage changes
(c) Wide-area protection: real-time phasor data allows for improved grid events identification and execution of appropriate system protection measures
(d) Planned power system islanding: this is to improve islanding of power system during instability situations
(e) Power plant monitoring and integration: real-time data is used for better integration of different power plants that includes intermittent renewables or distributed power sources

The majority of research works to estimate the power system line parameters are focused on transmission systems. Numerous techniques have been introduced to calculate the transmission parameters using the synchronized measurement devices. A two-port ABCD parameter identification based technique was introduced in [1]. This method utilized two sets of three samples of sending and receiving terminals' voltage and current signals. This was to find three estimates of ABCD parameters. The ABCD method is referred to in this research work as a "two-port circuit measurement technique". In Reference [2], four methods were discussed to identify short transmission line parameters by synchronized measurements. Reference [3] proposed a novel method to identify transmission line parameters for different cases, including short and long, transposed and untransposed lines with balanced and unbalanced load conditions. The positive sequence line parameters considering the effects of the line shunt capacitance were estimated in [4], employing a two-terminal transmission line model. Likewise, Reference [5] aimed to achieve the same objectives where a new estimation method was presented using synchronized phasor measurements at both line ends. The approach in [6] proposes the use of recursive parameter estimation to find the network branch parameters online and off-line. The least-square technique was leveraged in [7] with the objective of obtaining the line parameters iteratively.

Unlike the abundance of publications on transmission line parameters estimation, the work in distribution is limited. The probability theory, which builds on voltage drop linear equivalent model, was used in [8]. The approach objective was to estimate the DL impedance and get precise parameters. Numerous works discussed the uncertainties of network parameters and inaccuracy of measurements. In particular, the DL parameters and measurement uncertainties were analyzed in [9]. A novel power system uncertainty analysis technique was proposed in [10], where a two-step approach based on static weighted least-squares analysis was used. Reference [11] presents a method to estimate distribution line parameters using only conventional SCADA measurements (voltage magnitude and power measurements). It resulted in a negligible deviation between simulation, experiment, and the actual manufacturer specifications. The key outcomes of the DL parameters estimation studies were that the accuracy of line parameters is crucial for a number of applications including the grid control, stability analysis, and fault location studies.

To the best of the authors' knowledge, the applicability of different methods to identify the sequence DL parameters has not been considered before. This paper proposes the use of PMU to identify the DL parameters under the consideration of accuracy, positive sequence, and noise. The concept of symmetrical components is leveraged to extract the positive sequence of the synchronized phasor voltage and current measurement signals. The online synchronized signals obtained from the PMUs will be used in calculating both the phase and positive sequence DL parameters. In Section 2, three techniques have been developed to measure DL resistance, reactive inductance, and shunt admittance. Section 3 describes the used accuracy statistical measures to evaluate and compare the performance of the three techniques. The developed case studies along with their results and discussion are presented in Sections 4 and 5, respectively. Finally, the study recommendations and outcomes are stipulated in Section 6. The main data used to support the findings of this study are included within this article. If additional data is required, it could be requested from the corresponding author with proper justification.

2. Techniques of Distribution Line Parameters Estimation

Three different techniques are discussed in this section with the objectives of identifying the DL parameters. The techniques leverage the PMU voltage and current signals obtained at the two terminals of the line. In order to perform DL parameters estimation, the line is represented in a π-model equivalent circuit as illustrated in Figure 1(a).

The study considers the positive sequence of the voltage and current phasors in addition to the phase values. This aims to explore accuracy enhancement opportunities and compare the results. Additionally, the sequence quantities are required for developing any asymmetrical analysis. The positive sequence equivalent π-model is shown in Figure 1(b).

2.1. Ohm's Formula Technique. The proposed ohm's formula technique (OFT) depends on the ohm's law [12]. Under this method, both phase and positive sequence voltage and current phasors are used. This method requires only single set of voltage and current samples of the phasor voltage and current signals produced by PMUs.

The developed OFT equations to calculate the DL parameters are described below:

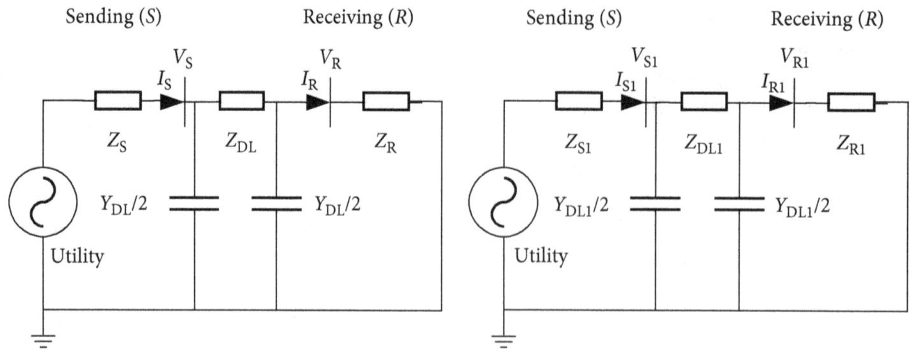

(a) (b)

FIGURE 1: Distribution line equivalent model (π-Type). (a) One-line diagram using the phasor quantities (b) One-line diagram using the positive sequence quantities. The parameters of the above circuits are described as follows: Z_S, equivalent impedance at the source side; Z_{S1}, positive sequence equivalent impedance at the source side; Z_R, equivalent impedance at the receiving end; Z_{R1}, ppositive sequence equivalent impedance at the receiving end; Z_{DL}, distribution line impedance; Z_{DL1}, positive sequence distribution line impedance; Y_{DL}, distribution line admittance; Y_{DL1}, positive sequence distribution line admittance; V_S, phase voltages at sending end; V_{S1}, positive sequence phase voltages at sending end; V_R, phase voltages at receiving end; V_{R1}, positive sequence phase voltages at receiving end; I_S, Phase current at sending end; I_{S1}, positive sequence phase current at sending end; I_R, phase current at receiving end; I_{R1}, positive sequence phase current at receiving end.

$$Z_{DL1} = \frac{2(V_{S1} - V_{R1})}{I_{R1} + I_{S1}}, \qquad Y_{DL1} = \frac{I_{S1} - I_{R1}}{V_{S1}}. \tag{1}$$

2.2. Single Measurement Technique. The proposed single measurement technique (SMT) aims to find DL resistance, reactive inductance and shunt admittance [12]. It uses both the phase and positive sequence of the voltage and current signals that are obtained from PMUs at the steady state. The SMT equations are formulated as follows:

$$Z_{DL1} = \frac{V_{S1}^2 - V_{R1}^2}{V_{R1} I_{S1} + V_{S1} I_{R1}}, \qquad Y_{DL1} = \frac{2(V_{S1} - V_{R1})}{I_{R1} + I_{S1}}. \tag{2}$$

2.3. Two-Port Circuit Measurement Technique. The two-port circuit measurement technique (TPCMT) requires two sets of synchronized measurement samples at different loading conditions [12]. The samples are taken from the DL terminals to calculate the two-port circuit parameter known as A, B, C, and D. The DL impedance and admittance are identified from the ABCD matrix.

The TPCMT is conventionally used to represent transmission lines. Additionally, it provides adequate accuracy for DLs at some cases. Representation of positive sequence TPCMT for DL is shown in Figure 2, where V_{S1}, V_{R1}, I_{R1}, and I_{S1} are the positive sequence of the sending and receiving ends voltage and current signals, respectively.

The following equations form the relation between the sending end and the receiving end quantities:

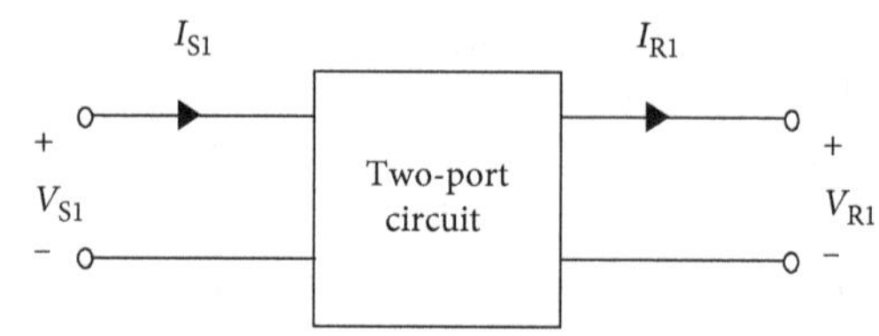

FIGURE 2: Representation of positive sequence two-port circuit for distribution line.

$$V_{S1} = A\,V_{R1} + B\,I_{R1}, \qquad I_{S1} = C\,V_{R1} + D\,I_{R1}, \tag{3}$$

where the parameters A, B, C and D are influenced by the DL resistance, inductance, capacitance, and conductance. The ABDC parameters are complex numbers in which A and D are unit less, B is measured in ohms, and C has a unit of Siemens.

The ABCD parameters of the DL equivalent π-model shown in Figure 1 are obtained by the following equations:

$$\begin{aligned} V_{S1} &= V_{R1} + Z_{DL1}\left(I_{R1} + \frac{V_{R1} Y_{DL1}}{2}\right) \\ &= \left(1 + \frac{Z_{DL1} Y_{DL1}}{2}\right) V_{R1} + Z_{DL1} I_{R1}. \end{aligned} \tag{4}$$

By applying the Kirchhoff current law (known as KCL) at the sending end, the following equation is obtained:

$$I_{S1} = I_{R1} + \frac{Y_{DL1}(V_{R1} + V_{S1})}{2}. \tag{5}$$

Combining the previous two equations yields

$$\begin{aligned} I_{S1} &= I_{R1} + \frac{Y_{DL1} V_{R1}}{2} + \left[\left(1 + \frac{Z_{DL1} Y_{DL1}}{2}\right) V_{R1} + Z_{DL1} I_{R1}\right] \frac{Y_{DL1}}{2} \\ &= Y_{DL1}\left(1 + \frac{Z_{DL1} Y_{DL1}}{4}\right) V_{R1} + \left(1 + \frac{Z_{DL1} Y_{DL1}}{2}\right) I_{R1}. \end{aligned} \tag{6}$$

Comparing the last above formula with the ABCD equations yields

$$A = D = \left(1 + \frac{Z_{DL1}Y_{DL1}}{2}\right) \text{ per unit,} \qquad (7)$$

$$C = Y_{DL1}\left(1 + \frac{Z_{DL1}Y_{DL1}}{4}\right) \text{ ohm.}$$

From the simple DL (only series impedance representation) analysis and derivation B is obtained to be

$$B = Z_{DL1} \text{ ohm.} \qquad (8)$$

The above A, B, C and D equations are solved to find Z_{DL1} and Y_{DL1} which will be as follows:

$$Z_{DL1} = B \text{ ohm,}$$
$$Y_{DL1} = \frac{2(A-1)}{B} \text{ Siemens.} \qquad (9)$$

This method could be extended to accommodate two sets of PMU measurements. The two sets could be obtained from two different redundant PMUs or from two readings recorded at different timing or loading conditions. The ABCD equations for the two sets are as follows:

$$V'_{S1} = A\,V'_{R1} + B\,I'_{R1},$$
$$I'_{S1} = C\,V'_{R1} + D\,I'_{R1},$$
$$V''_{S1} = A\,V''_{R1} + B\,I''_{R1},$$
$$I''_{S1} = C\,V''_{R1} + D\,I''_{R1}. \qquad (10)$$

The samples of the voltages and currents for the receiving and sending ends are as the following:

(i) V'_{S1}, V'_{R1}, I'_{S1}, and I'_{R1} are for the first set

(ii) V''_{S1}, V''_{R1}, I''_{S1}, and I''_{R1} are for the second set

The ABCD parameters are calculated to account for the two sets to be as follows:

$$A = \frac{I'_{R1}V''_{S1} - I''_{R1}V'_{S1}}{I'_{R1}V''_{R1} - I''_{R1}V'_{R1}},$$

$$B = \frac{V''_{R1}V'_{S1} - V'_{R1}V''_{S1}}{I'_{R1}V''_{R1} - I''_{R1}V'_{R1}},$$

$$C = \frac{I'_{R1}I''_{S1} - I''_{R1}I'_{S1}}{I'_{R1}V''_{R1} - I''_{R1}V'_{R1}},$$

$$D = \frac{I'_{S1}V''_{R1} - I''_{S1}V'_{R1}}{I'_{R1}V''_{R1} - I''_{R1}V'_{R1}}. \qquad (11)$$

3. Accuracy Statistical Measures

The accuracy of the proposed methods is evaluated using different statistical measures. This is to ensure that the measures will converge for all case studies analyzed in this paper. That is, in case one statistical measure fails to perform in one of the cases, the evaluation will be achieved by the other measures.

3.1. Percentage Error. The first step toward accepting or rejecting the proposed methods is assessing its accuracy using the percentage error given by the following equation:

$$\text{error}(\%) = \frac{|\text{actual values} - \text{calculated value}|}{\text{actual value}} \times 100. \qquad (12)$$

3.2. Coefficient of Determination. The coefficient of determination (*CoD*), denoted by R^2, is used to indicate the difference of the obtained values by a proposed formula compared to the actual ones. It measures the strength of the proposed formula and benchmarks it with the ideal situation which will result in a coefficient of determination of 100%. It is, also, called the squared error which is the error between the curve obtained by the proposed formula and the actual curve. The range of coefficient of determination varies between 0 and 1. The higher the number means the proposed formula is more descriptive and reflective to the actual values. Figure 3 is an explanatory sketch for calculating the CoD.

The coefficient of determination equation is formulated as follows:

$$\text{CoD} = 1 - \frac{\sum_1^k (y_{\text{calc}} - y_{\text{act}})^2}{\sum_1^k (y_{\text{calc}} - \overline{y_{\text{act}}})^2} = 1 - \frac{\text{SE}}{\text{TV}},$$

$$\text{SE} = \sum_1^k (y_{\text{calc}} - y_{\text{act}})^2,$$

$$\text{TV} = \sum_1^k (y_{\text{calc}} - \overline{y_{\text{act}}})^2. \qquad (13)$$

The parameters are described as follows: CoD, coefficient of determination; SE, total square error between the calculated points and the actual values; TV, total variation between the calculated points and the actual values; $\overline{y_{\text{act}}}$, mean of the actual values; y_{calc}, calculated value; and y_{act}, actual value.

3.3. Other Accuracy Statistical Measures. Other accuracy statistical measures are required to be integrated with the percentage error and CoD. This is due to the fact that the percentage error does not represent the correlation and the CoD has certain shortfalls, especially for small scientific numbers.

The following additional statistical measures are used to evaluate the proposals presented in this paper:

(1) Mean absolute deviation (MAD), which is the summation of the absolute deviation between the actual and calculated values over the number of records (or the length of the range)

(2) Mean square error (MSE), which is considered as the most common error metric. It is mainly the summation of the squared errors over the number of records

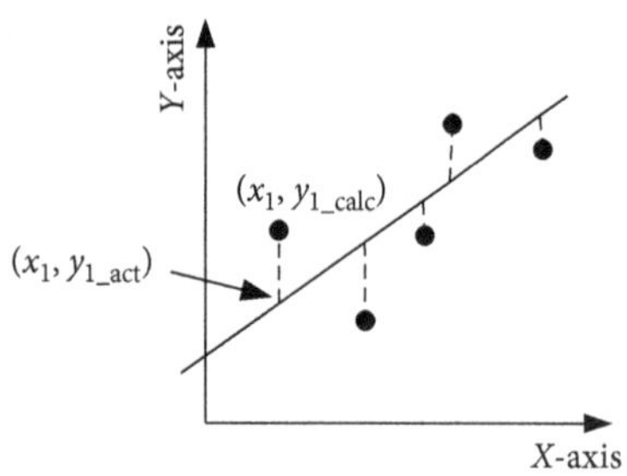

FIGURE 3: Coefficient of determination explanatory sketch.

(3) Root mean square error (RMSE) is obtained by applying the square root to the MSE

(4) Mean absolute percentage error (MAPE) is the average of absolute errors over the actual records

4. Case Studies

A 25-kV distribution system (refer to Figure 4) is modeled in MATLAB/Simulink to verify the effectiveness of the three line parameters identification techniques. A total of 12,960 different case studies have been performed under six main loading scenarios (stated in Table 1) and four categories (presented in Table 2) with changing of the line parameters. The line parameters have been varied in 60 steps. The loading scenarios and categories considered under this study are tabulated in below tables.

The large number of case studies has been developed to test the robustness and accuracy of this paper proposals. The 12,960 simulations differ in the loading conditions, line lengths, noise, and inaccuracy mitigations.

The selected DL is modeled as three-phase DL with a π-type. The model consists of one set of resistance and inductance elements in series connected between sending and receiving terminals. Two sets of shunt capacitances lumped are, also, included at both ends as illustrated in Figure 1. The initial DL parameters are stated in Table 3.

The total series resistance, reactive inductance, and shunt admittance are given by the following formulas, respectively:

$$\begin{aligned} R &= r\ell, \\ X_{\mathrm{DL1}} &= \omega L\ell, \\ Y_{\mathrm{DL}} &= \omega C\ell, \end{aligned} \tag{14}$$

where R, L, and C are the total DL resistance, inductance, and capacitance, and ℓ is the total length of the line.

In MATLAB, two sets of simulated PMUs are placed at both terminals of the selected DL to measure the voltages and currents waveforms simultaneously. The recorded waveforms are in the shape of sinusoidal signals and then converted into phasor equivalents.

5. Results and Discussions

The simulation results of the 12,960 cases are summarized in this section and organized into four categories. Under each category, the resistance, reactive inductance, and shunt admittance are calculated using the three methods for different loading conditions and parameter values. The calculation is based on the voltage and current signals obtained from PMUs that are installed at both ends of the line. Figure 5 shows the voltage and current signals obtained from PMU devices considering noise-free system.

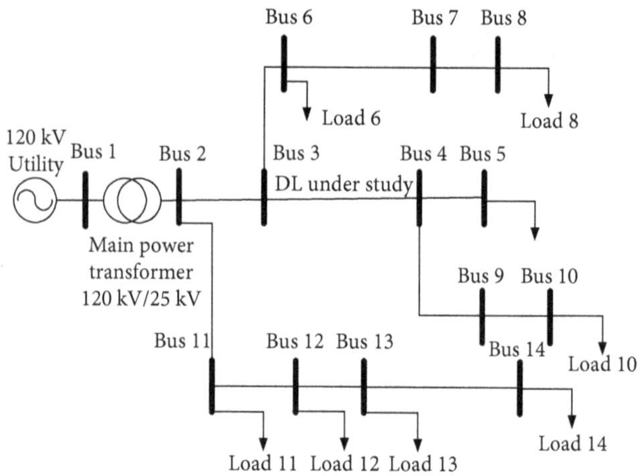

FIGURE 4: The 25 kV 14-bus test distribution network under consideration.

TABLE 1: The six loading scenarios for simulations.

Scenario	Load	
	Active (MW)	Reactive (MVar)
1	1	0.25
2	2	0.5
3	3	0.75
4	4	1
5	5	1.25
6	6	1.5

TABLE 2: The four categories for this study.

Category	Description
1	Phase quantities
2	Positive sequence quantities
3	Phase quantities with noise
4	Phase quantities with the proposed inaccuracy mitigation for noisy systems in category 3

TABLE 3: Initial parameters of the distribution line test circuits.

Parameter	Actual value	Dimension
r	0.1153	(Ohms/km)
l	1.05e-3	(H/km)
c	11.33e-009	(F/km)

where r, l, and c are the resistance, inductance, and capacitance per unit length, respectively.

5.1. Phase Quantities. In this category, the phase quantities of voltage and current are used to perform the analysis. This type of analysis is required for asymmetrical related studies such as unbalanced fault analysis.

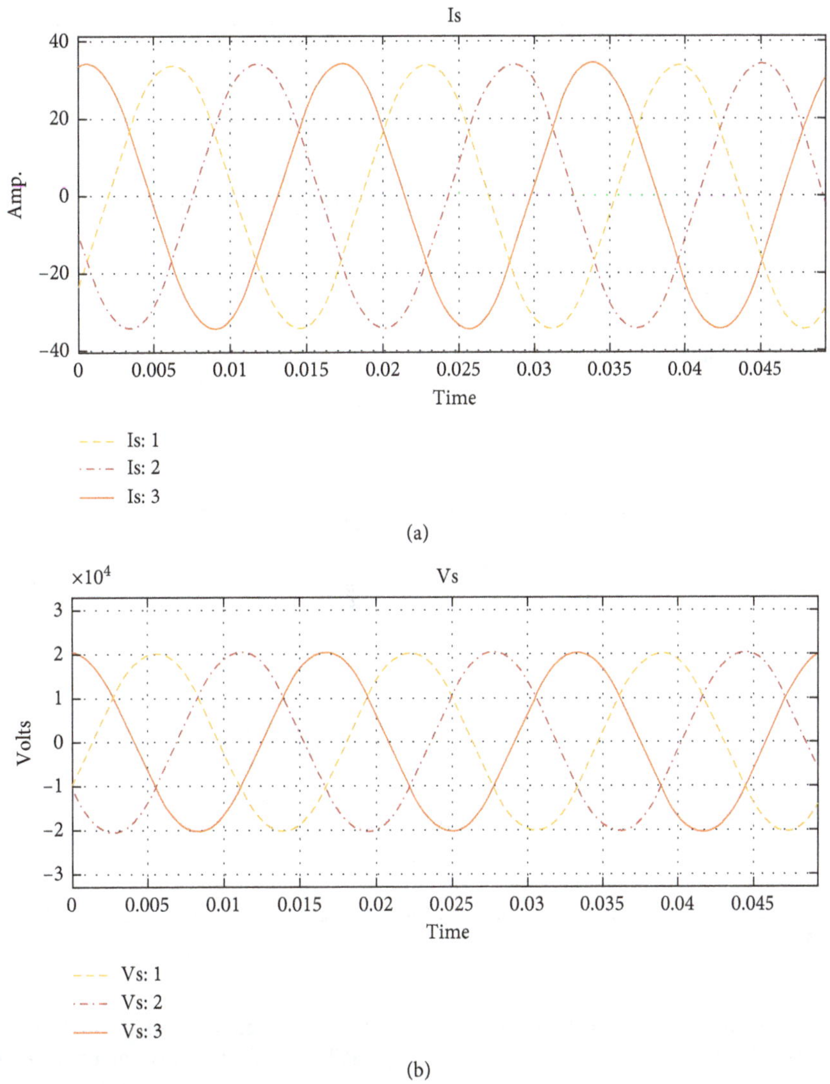

FIGURE 5: PMU current and voltage signals with noise-free.

The values of resistance, reactive inductance, and shunt admittance are changing in 60 steps. The parameters identification errors of the six loading scenarios are averaged for the three methods. The voltage and current waveforms are assumed to be noise-free. Results of the average errors for the resistance, reactive inductance, and shunt admittance are shown in Figure 6. The maximum errors for each method are stated in Table 4. The results reveal that SMT is more effective in calculating the DL parameters.

The TPCMT shows weakness in calculating the shunt admittance for short lines. This is expected as the method was developed specifically for medium transmission lines. However, it performs very well when the DL length is ranging between 10 and 30 km which is a common sort of DLs.

5.2. Positive Sequence Quantities. Both OFT and SMT have excellent performance in identifying the DL parameters using positive sequence quantities. The average and maximum errors recorded in the simulated studies are presented in Figure 7 and Table 4, separately. It is observed form the results that TPCMT fails to calculate the line parameters using positive sequence voltage and current quantities. Therefore, the results were excluded from Figure 7. The results demonstrate that SMT is superior to OFT in calculating the line parameters using positive sequence quantities.

5.3. Phase Quantities with Noise. Actual voltage and current signals of any distribution system are not pure sinusoidal. Noise is always impeded in the signals due to several factors, e.g., harmonics produced from electronic based devices. The electronic devices could be at residential areas such as televisions, computers, laptops, electronic games, and so on. There are a number of applications that produce harmonics at the industrial sector, for example capacitor bank, variable frequency drives, and other electronic based equipment.

Accordingly, all input signals to PMUs will be associated with additional harmonics beside the fundamental frequency (60 Hz) as in the Kingdome power system.

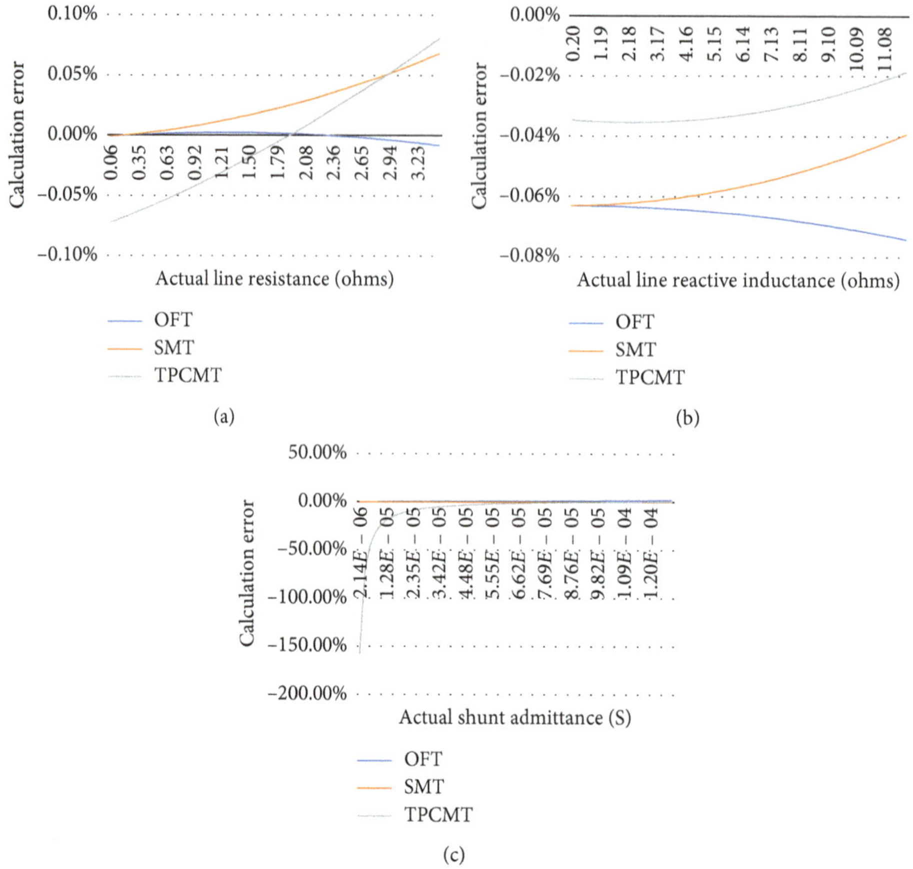

FIGURE 6: Calculation average errors of the six loading scenarios for the three methods under Category 1. (a) Resistance. (b) Reactive inductance. (c) Shunt admittance.

TABLE 4: Maximum errors of the six loading scenarios considering the variation of the line parameters.

Category	Method	R	X_L	X_C
1	1	0.04%	0.07%	3.35%
	2	0.09%	0.15%	0.07%
	3	0.12%	0.04%	High
2	1	0.11%	0.13%	3.45%
	2	0.08%	0.15%	0.11%
	3	High	High	High
3	1	0.45%	1.10%	2.22%
	2	0.38%	1.08%	**1.22%**
	3	1.87%	0.52%	High
4	1	0.34%	0.59%	1.49%
	2	0.34%	0.59%	**0.03%**
	3	0.10%	0.04816%	High

Although PMU measurements showed an improved accuracy compared to other devices, this performance is not fully materialized in the actual field due to errors from other channels such as instrumentation, CT, and potential transformer (shortly PT) and etc.

Figure 8 shows the voltage and current signals obtained from PMU devices considering noisy system.

The OFT and SMT have extraordinary performance when applying the phase values to noisy system (Figure 9). TPCMT still shows weakness in estimating the line parameters, especially for short lines capacitance. As the line length increases as TPCMT converges for identifying the X_C.

The maximum error recorded in the simulated studies is shown in Table 4. From the calculated average and maximum errors of the six loading scenarios for the three methods considering phasor quantities, it is concluded that SMT is superior to the other techniques for noisy system.

5.4. Phase Quantities with Inaccuracy Mitigation for Noisy Systems. It is observed form the simulated case studies that the error follows specific trend under different line parameters, irrespective of the loading conditions. Knowing the error trend will ease predicting the error magnitude and hence mitigating it. This category proposes to apply inaccuracy mitigation measures to improve the line parameter calculation errors. The measures are developed based on line characteristics and possible loadings. The proposed inaccuracy mitigation measure concept is illustrated in Figure 10 and given by the following formula:

$$\ddot{y} = \dot{y}(1 + \varepsilon(\dot{y})), \tag{15}$$

where $\dot{y}$ is the originally calculated value and $\ddot{y}$ is the enhanced measurement. The symbol ε is taken from the

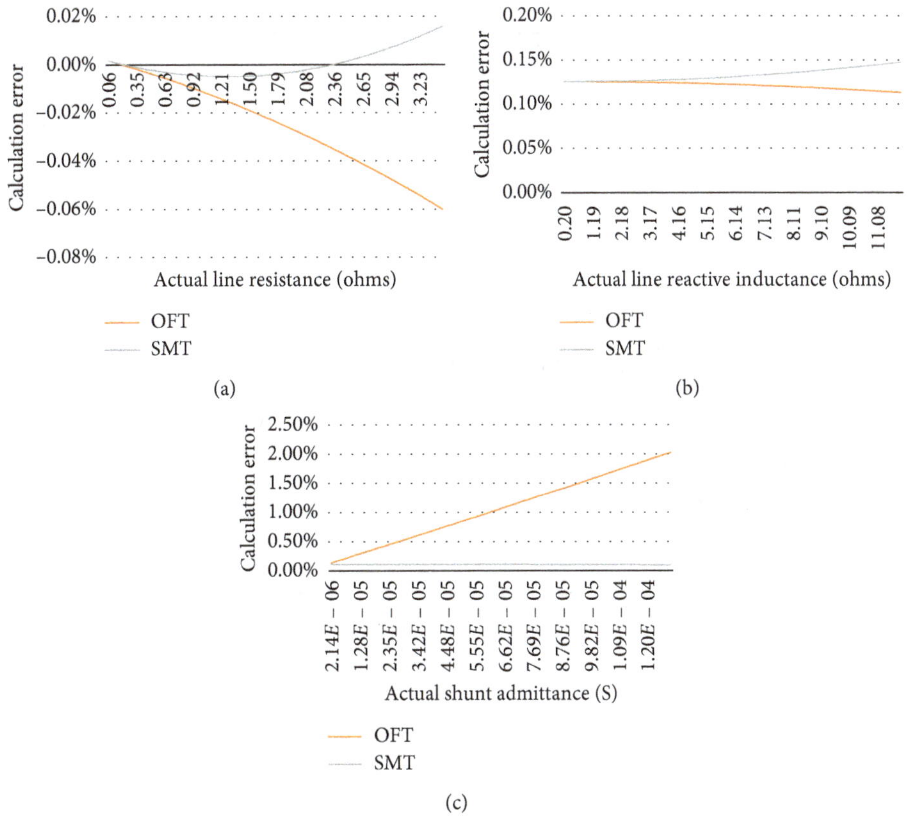

FIGURE 7: Calculation average errors of the six loading scenarios for the OFT and SMT techniques under Category 2. (a) Resistance. (b) Reactive inductance. (c) Shunt admittance.

predeveloped inaccuracy mitigation measures demonstrated in Figure 10. The inaccuracy mitigation curve could take different shapes based on line loading and characteristics.

The proposed concept has been applied to Category 3, and the simulation results are illustrated in Figure 11. The results reveal significant improvements of Category 4 compared to Category 3 in Figure 9.

The inaccuracy mitigation measures will result in accuracy improvement up to 98% of the maximum error of Category 3. The maximum errors for the four categories and six loading scenarios considering the variation of the line parameters are tabulated in Table 4.

The MAD, MSE, RMSE, MAPE, and CoD have been applied to the four categories and six loading scenarios. The results for the latter are averaged into one value for each category and parameter. The results are tabulated in Table 5 to evaluate the robustness of this paper proposals. It is noticed from the table that generally the values under the proposed inaccuracy mitigation measures category (Category 4) are improved compared to those in Category 3. This shows the strength of the proposed inaccuracy mitigation concept which could be applied for ideal and noisy systems. The use of positive sequence quantities will perform very well when using OFT and SMT. However, the phase quantities will result in more accurate line parameters estimation. Unlike OFT and SMT, TPCMT does not function when using the positive sequence values. Therefore, ABCD should not be used for any asymmetrical related studies in DLs.

MAPE is found to be the only method applicable for calculating the line shunt admittance since the values of the capacitances are very small scientific numbers.

6. Conclusions

To carry out any asymmetrical related analysis at DLs such as asymmetrical fault studies, the symmetrical components should be leveraged to identify the positive, negative, and zero sequences. Therefore, robust and accurate line parameters calculation techniques are required. Based on that, three line parameters identification techniques have been applied to different case studies and evaluated using different statistical measures. The outcomes of this analysis along with the associated recommendations are as follows:

(1) The proposed inaccuracy mitigation concept will result in accuracy improvement up to 98% of the maximum error. Therefore, it is recommended to use this concept for any online impedance and admittance calculations using PMUs.

(2) The inaccuracy of the line parameters estimation follows a specific trend over different scenarios. This will allow proper inaccuracy prediction and hence mitigation.

(3) The proposed inaccuracy prediction and mitigation have resulted in a negligible deviation between

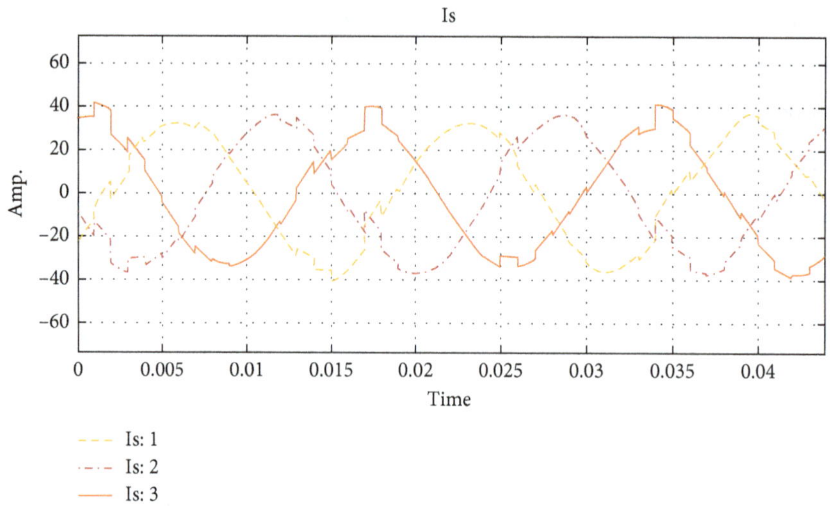

(a)

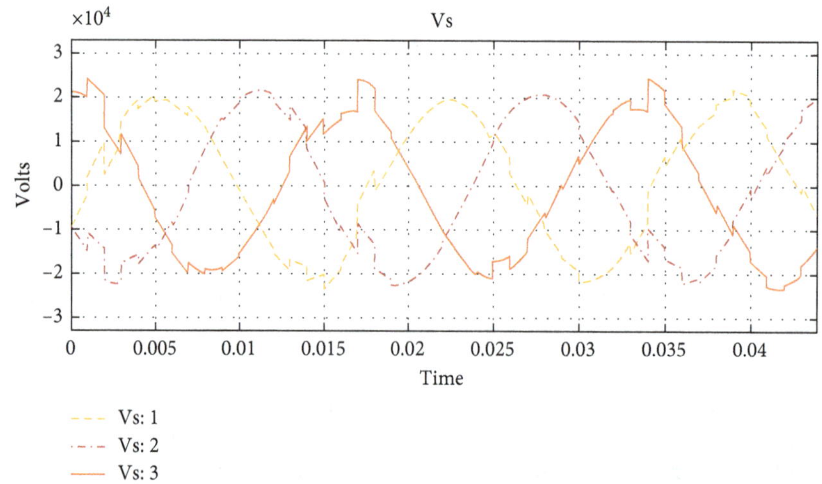

(b)

FIGURE 8: PMU current and voltage signals with noise.

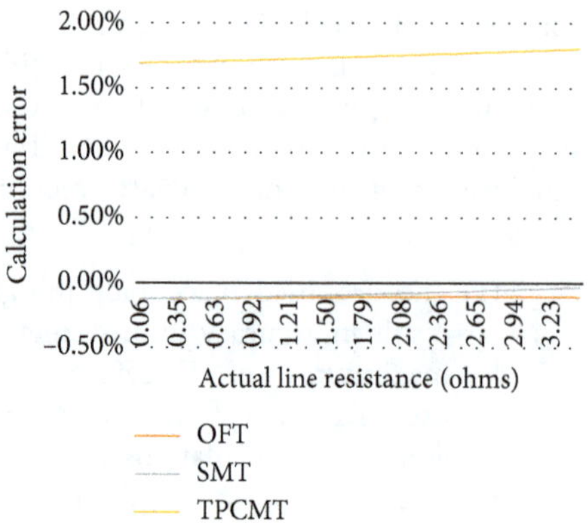

(a)

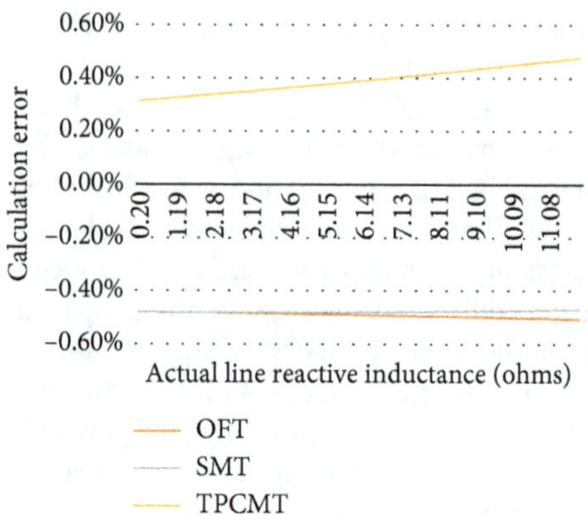

(b)

FIGURE 9: Continued.

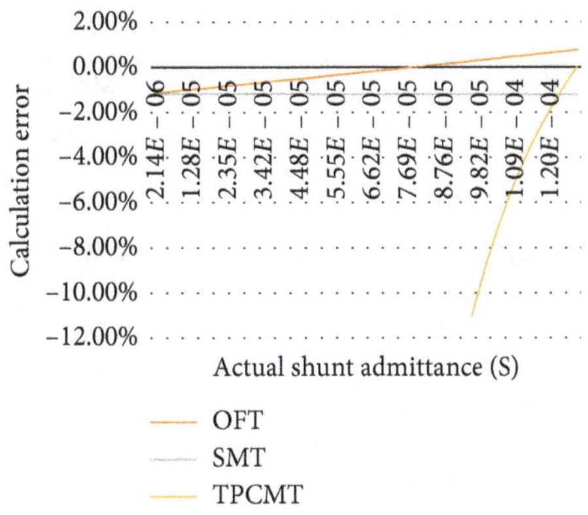

(c)

FIGURE 9: Calculation average errors of the six loading scenarios for the three methods under Category 3. (a) Resistance. (b) Reactive inductance. (c) Shunt admittance.

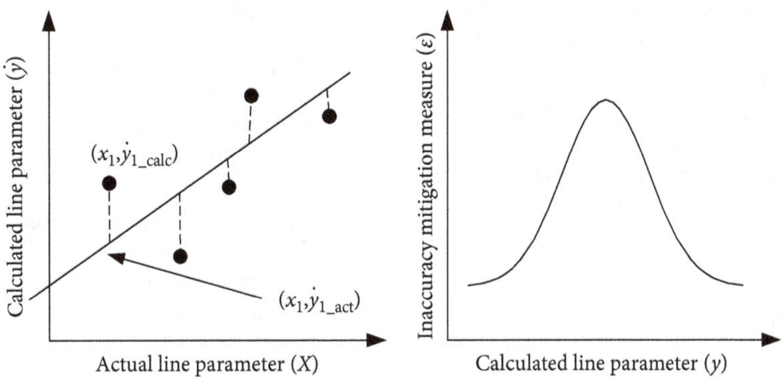

(a) (b)

FIGURE 10: Indicative graphs of the proposed inaccuracy mitigation measure concept. (a) Actual vs. calculated line parameters. (b) Proposed inaccuracy mitigation measure.

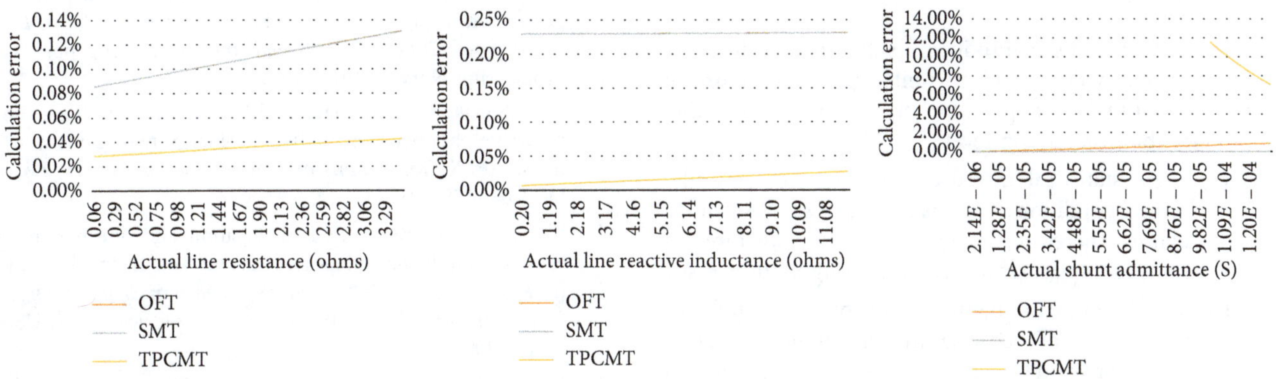

(a) (b) (c)

FIGURE 11: Calculation average errors of the six loading scenarios for the three methods under Category 4. (a) Resistance. (b) Reactive inductance. (c) Shunt admittance.

TABLE 5: Statistical measures results for all cases (%).

Parameter	Statistical measure	Method 1				Method 2				Method 3			
		Category											
		1	2	3	4	1	2	3	4	1	2	3	4
Resistance	MAD	0.05	0.14	0.41	0.41	0.13	0.09	0.38	0.41	0.14	0.00	6.20	0.14
	MSE	0.00	0.00	0.00	0.00	0.00	0.00	0.00	0.00	0.00	0.00	0.26	0.00
	RMSE	0.04	0.14	0.34	0.34	0.14	0.08	0.33	0.34	0.15	0.00	5.07	0.12
	MAPE	0.02	0.06	0.24	0.22	0.05	0.04	0.22	0.22	0.08	0.00	3.49	0.07
	R^2	99.99	99.98	99.88	99.90	99.98	99.98	99.89	99.90	99.96	0.00	98.40	99.97
Inductance	MAD	0.83	1.44	5.97	2.78	0.61	1.65	5.76	2.78	0.34	0.00	5.08	0.25
	MSE	0.00	0.01	0.32	0.08	0.00	0.02	0.30	0.08	0.00	0.00	0.18	0.00
	RMSE	0.69	1.16	4.88	2.27	0.48	1.37	4.67	2.26	0.26	0.00	4.28	0.23
	MAPE	0.13	0.24	0.98	0.46	0.11	0.27	0.96	0.46	0.06	0.00	0.79	0.03
	R^2	99.99	99.99	99.96	99.98	100.0	99.99	99.96	99.98	100.0	0.00	99.97	100.0
Capacitance	MAD	0.00	0.00	0.00	0.00	0.00	0.00	0.00	0.00	0.00	0.00	0.01	0.04
	MSE	0.00	0.00	0.00	0.00	0.00	0.00	0.00	0.00	0.00	0.00	0.00	0.00
	RMSE	0.00	0.00	0.00	0.00	0.00	0.00	0.00	0.00	0.00	0.00	0.01	0.09
	MAPE	1.83	2.15	1.42	0.88	0.12	0.21	2.36	0.03	16.68	0.00	High	High
	R^2	Zero	Zero	Zero	Zero	Zero	Zero	Zero	Zero	Zero	Zero	Zero	Zero

calculated and actual DL parameters. This proves the robustness of the proposals of this paper.

(4) Both OFT and SMT have extraordinary performance in calculating the DL parameters using positive sequence quantities. Therefore, it is recommended to use them for any asymmetrical based analysis such as unbalanced fault studies.

(5) SMT is superior to OFT in calculating the line parameters using positive sequence quantities.

(6) TPCMT does not perform when the line resistance is small (short line) and using phase quantities. This is expected as the method was developed for medium transmission lines. As the line impedance or length increase, TPCMT will boost up its resistance calculation accuracy.

(7) It is expected the TPCMT will not perform very well for capacitance identification of short DLs. Therefore, it is unrecommended to use this method for short DLs.

(8) TPCMT fails to produce result using positive sequence voltage and current signals. Therefore, it should not be applied for any asymmetrical studies at the distribution level.

(9) Some statistical measures do not function under certain conditions, such as in case of small scientific figures. Therefore, there is a need for a wide range of statistical measures to ensure covering all study cases.

(10) SMT is ranked to be the most robust technique for identifying all DL parameters under different conditions and OFT comes the second. Therefore, it is recommended to use SMT for any distribution related case studies.

Conflicts of Interest

The authors declare that they have no conflicts of interest.

References

[1] R. E. Wilson, G. A. Zevenbergen, D. L. Mah, and A. J. Murphy, "Calculation of transmission line parameters from synchronized measurements," *Electric Machines and Power Systems*, vol. 27, no. 12, pp. 1269–1278, 1999.

[2] D. Shi, D. J. Tylavsky, N. Logic, and K. M. Koellner, "Identification of short transmission-line parameters from synchrophasor measurements," in *Proceedings of 2008 40th North American Power Symposium*, pp. 1–8, Calgary, Alberta, Canada, September 2008.

[3] D. Shi, D. Tylavsky, K. Koellner, D. Wheeler, and N. Logic, "Transmission line parameter identification using PMU measurements," *European Transactions on Electrical Power*, vol. 21, no. 4, pp. 1574–1588, 2010.

[4] Y. Liao and M. Kezunovic, "Online optimal transmission line parameter estimation for relaying applications," *IEEE Transactions on Power Delivery*, vol. 24, no. 1, pp. 96–102, 2009.

[5] A. M. Dan and D. Raisz, "Estimation of transmission line parameters using wide-area measurement method," in *Proceedings of 2011 IEEE Trondheim PowerTech*, pp. 1–6, Trondheim, Norway, June 2011.

[6] I. W. Slutsker, S. Mokhtari, and K. A. Clements, "Real time recursive parameter estimation in energy management systems," *IEEE Transactions on Power Systems*, vol. 11, no. 3, pp. 1393–1399, 1996.

[7] Y. Liao, "Algorithms for fault location and line parameter estimation utilizing voltage and current data during the fault," in *Proceedings of 2008 40th Southeastern Symposium on System Theory (SSST)*, pp. 183–187, New Orleans, LA, USA, March 2008.

[8] D. Jia, W. Sheng, X. Song, and X. Meng, "A system identification method for smart distribution grid," in *Proceedings of 2014 International Conference on Power System Technology (POWERCON)*, pp. 14–19, Chengdu, China, October 2014.

[9] G. D'Antona and M. Davoudi, "Effect of phasor measurement unit on power state estimation considering parameters un-

certainty," in *Proceedings of 2012 IEEE International Workshop on Applied Measurements for Power Systems (AMPS)*, pp. 1–5, Aachen, Germany, September 2012.

[10] R. N. Mahanty and P. B. Dutta Gupta, "Application of RBF neural network to fault classification and location in transmission lines," *IEE Proceedings-Generation, Transmission and Distribution*, vol. 151, no. 2, p. 201, 2004.

[11] A. M. Prostejovsky, O. Gehrke, A. M. Kosek, T. Strasser, and H. W. Bindner, "Distribution line parameter estimation under consideration of measurement tolerances," *IEEE Transactions on Industrial Informatics*, vol. 12, no. 2, pp. 726–735, 2016.

[12] M. M. J. Al-Khabbaz, *Fault Location in Power Distribution Grids using Phasor Measurement Units*, King Fahd University of Petroleum and Minerals, Dhahran, Eastern Province, Saudi Arabia, 2018.

18

Extraction of Earth Surface Texture Features from Multispectral Remote Sensing Data

Zhenxing Zhang,[1] Feng Gao,[1] Bin Ma,[2] and Zhiqiang Zhang[3]

[1]*College of Automation, Harbin Engineering University, Harbin 150001, China*
[2]*China Ship Development and Design Center, Wuhan, China*
[3]*92730 Army, Sanya 572016, China*

Correspondence should be addressed to Feng Gao; gaofeng19@hrbeu.edu.cn

Academic Editor: Jose R. C. Piqueira

Earth surface texture features referring to as visual features of homogeneity in remote sensing images are very important to understand the relationship between surface information and surrounding environment. Remote sensing data contain rich information of earth surface texture features (image gray reflecting the spatial distribution information of texture features, for instance). Here, we propose an efficient and accurate approach to extract earth surface texture features from remote sensing data, called gray level difference frequency spatial (GLDFS). The gray level difference frequency spatial approach is designed to extract multiband remote sensing data, utilizing principle component analysis conversion to compress the multispectral information, and it establishes the gray level difference frequency spatial of principle components. In the end, the texture features are extracted using the gray level difference frequency spatial. To verify the effectiveness of this approach, several experiments are conducted and indicate that it could retain the coordination relationship among multispectral remote sensing data, and compared with the traditional single-band texture analysis method that is based on gray level co-occurrence matrix, the proposed approach has higher classification precision and efficiency.

1. Introduction

Remote sensing technology can extract high resolution regional marine environmental information in time, especially for the complex sea area. Multispectral remote sensing data reflects the interested target or regional radiation characteristics through the electromagnetic spectrum of multiband, and it has the advantages of wide range, multiphase, multiband, and high resolution. Remote sensing image could enrich the spectral characteristics of landmark and find out more detailed information, such as the structure, shape, and texture. However, in virtue of the fact that same objects possess different spectral and different objects share same spectral, the applications of remote sensing data would be serious restricted if only spectral information is taken into consideration. The earth surface texture is a good solution to the problem because of the stability characteristics [1].

The classical texture extraction and analytic approaches include gray level co-occurrence matrix method [2, 3], wavelet analysis method [4, 5], Gabor spectrum method [6], and so forth. While all these methods could only be applied to analyze the information of single band in remote sensing images, for multispectral remote sensing data, all the bands should be processed separately, which would decrease the extraction efficiency badly.

Because of the geometric characteristics of the surface object, it has a unique texture features on the remote sensing images. So, the different surface objects can be extracted through the texture features. This paper utilizes gray level difference frequency spatial to extract texture features of multiband remote sensing data. We firstly conduct principal component analysis (PCA) on the eight bands of Worldview-II multispectral images and compress these data on basis of guaranteeing against loss of spectral information. Make gray difference statistics on the compressed principle components and establish the gray level difference frequency spatial. In the experiments, the gray level difference frequency spatial is used to extract texture features, and a comparison with Gray Level

Co-occurrence Matrix (GLCM) is made. The experimental results indicate that the gray level difference frequency spatial has higher classification accuracy and efficiency.

2. Worldview-II Multispectral Remote Sensing Data

Worldview-II is one of the highest resolution remote sensing satellites, and it has the highest spatial resolution (0.46 m in the panchromatic band and 1.84 m in the multispectral bands). It provides high resolution multispectral data with eight bands, which include four conventional bands (red, green, blue, and near-infrared 1) and four characteristic bands (coastal, yellow, red edge, and near-infrared 2). The data analyzed in this paper are Worldview-II multispectral remote sensing image of the Sea Islands; the texture features of eight bands are extracted. Firstly, we calibrate the data and get the radiance data. Secondly, atmospheric correction is conducted to eliminate the influence of atmosphere and illumination, and the actual reflectance of surface objects is obtained. Finally, we make orthorectification on the data through a few control points, thus eliminating the geometric distortion.

3. Compression of Multispectral Remote Sensing Data

Principal component analysis could project the high dimensional data onto a low-dimensional space. It takes the variance in size as the evaluation standard of information quantity; the greater the variance, the more information it provides [7, 8]. On the premise of keeping useful information of multispectral remote sensing data, principal component analysis could reduce the correlation and redundant information in order to compress multispectral remote sensing data. We transform Worldview-II multispectral remote sensing data into a column vector **W** as follows:

$$\mathbf{W} = [\mathbf{w}_1, \mathbf{w}_2, \cdots, \mathbf{w}_n]^{\mathrm{T}}, \quad (n = 1, 2, \cdots, 8). \tag{1}$$

Principal component analysis makes a combination of **W** through linear transformation and guarantees that **P** has the largest variance after transformation, as shown in the following equation:

$$\mathbf{P} = \mathbf{L}^{\mathrm{T}}\mathbf{W}, \tag{2}$$

where $\mathbf{L} = (l_1, l_2, l_3, \cdots, l_n)$ is the m-dimensional space to be determined and **R** is the covariance matrix of **W**, thereby, the variance of **P** could be computed as follows:

$$D(\mathbf{P}) = \mathbf{L}^{\mathrm{T}}\mathbf{R}\mathbf{L}. \tag{3}$$

Thereby, solving the maximum value of $D(\mathbf{P})$ is equal to seeking the vector **L** that makes $D(\mathbf{P})$ the largest. The length **L** is limited to unit length, and then the question is converted to

$$\begin{aligned} \max \quad & D(\mathbf{P}) = \mathbf{L}^{\mathrm{T}}\mathbf{R}\mathbf{L}, \\ \text{s.t.} \quad & \mathbf{L}^{\mathrm{T}}\mathbf{L} = 1. \end{aligned} \tag{4}$$

In last equation, the covariance matrix **R** could be expressed as follows:

$$\mathbf{R} = \Gamma \mathrm{diag}(\lambda_1, \cdots \lambda_n)\, \Gamma^{\mathrm{T}}, \tag{5}$$

where $\lambda_1, \cdots \lambda_n$ is the characteristic value of **R** and $\lambda_1 \geq \lambda_2 \geq \cdots \geq \lambda_n$ is satisfied. $\Gamma = (\mathbf{a}_1, \mathbf{a}_2, \cdots, \mathbf{a}_n)$, where $\mathbf{a}_1, \mathbf{a}_2, \cdots, \mathbf{a}_n$ is the eigenvector corresponding to the unit orthogonal eigenvectors. Let $\mathbf{a} \in \mathbf{R}^n$ and multiplying Equation (5) with $\mathbf{a}^{\mathrm{T}}$ and **a** on the left and right side separately, we get

$$\mathbf{a}^{\mathrm{T}}\mathbf{R}\mathbf{a} = \mathbf{a}^{\mathrm{T}}\Gamma \mathrm{diag}(\lambda_1, \cdots \lambda_n)\, \Gamma^{\mathrm{T}}\mathbf{a}. \tag{6}$$

Let $k = \Gamma^{\mathrm{T}}\mathbf{a}$, then $\mathbf{a}^{\mathrm{T}}\mathbf{a} = \mathbf{k}^{\mathrm{T}}\mathbf{k}$; Equation (6) satisfies

$$\begin{aligned} \mathbf{a}^{\mathrm{T}}\mathbf{R}\mathbf{a} &= \mathbf{k}^{\mathrm{T}}\mathrm{diag}(\lambda_1, \cdots \lambda_n)\, \mathbf{k} = \lambda_1\mathbf{k}_1^2 + \lambda_2\mathbf{k}_2^2 + \cdots + \lambda_n\mathbf{k}_n^2 \\ &\leq \lambda_1\left(\mathbf{k}_1^2 + \mathbf{k}_2^2 + \cdots + \mathbf{k}_n^2\right). \end{aligned} \tag{7}$$

Equation (4) can be rewritten as follows:

$$\max_{\mathbf{L}^{\mathrm{T}}\mathbf{L}=1} \mathbf{L}^{\mathrm{T}}\mathbf{R}\mathbf{L} = \max_{\mathbf{k}^{\mathrm{T}}\mathbf{k}=1}\left(\lambda_1\mathbf{k}_1^2 + \lambda_2\mathbf{k}_2^2 + \cdots + \lambda_n\mathbf{k}_n^2\right) \leq \lambda_1. \tag{8}$$

If $\mathbf{a} = \mathbf{a}_1$, then $\mathbf{a}_1^{\mathrm{T}}\mathbf{R}\mathbf{a}_1 = \lambda_1$, which indicates that the maximum value of $\mathbf{L}^{\mathrm{T}}\mathbf{R}\mathbf{L}$ is at the point of $\mathbf{a}_1$ under the condition of $\mathbf{L}^{\mathrm{T}}\mathbf{L} = 1$, thereby the first PCA principle component could be expressed as $\mathbf{p}_1 = \mathbf{a}_1^{\mathrm{T}}\mathbf{W}$. The contribution rate reflects the information quantity contained in each principle component, and the contribution rate of the i-th principal component could be computed as follows:

$$\frac{\lambda_i}{\mathrm{tr}(\mathbf{R})} = \frac{\lambda_i}{\sum_{k=1}^{n}\lambda_k}. \tag{9}$$

The cumulative contribution rate of the first l principal components is as follows:

$$\frac{\sum_{i=1}^{l}\lambda_i}{\sum_{k=1}^{n}\lambda_k}. \tag{10}$$

The contribution rate indicates the ability that principle components reflect **W**. It determines the number of principal components after compression of multispectral remote sensing data.

4. Texture Features Extraction from Multispectral Remote Sensing Data

4.1. Gray Level Co-Occurrence Matrix. Gray Level Co-occurrence Matrix is the most direct and simplest texture analysis approach, which considers the spatial structure of remote sensing images [9]. It describes the image texture through the two-order combined conditional probability density among image pixels [10]. Assume the remote sensing image is of size $M \times N$; the gray level is L; the distance between two pixels is d; the angle is θ; the gray levels are separately i and j; the times that these two pixels appear simultaneously is $\mathbf{P}(i, j, d, \theta)$ which could be expressed as follows:

$$\begin{aligned} \mathbf{P}(i, j, d, \theta) = \{[(x_1, y_1), (x_2, y_2)] \mid f(x_1, y_1) &= i, f(x_2, y_2) \\ &= j; x_1, x_2 = 1, 2, \cdots m; y_1, y_2 = 1, 2, \cdots n;\}, \end{aligned} \tag{11}$$

Where m and n represent the number of remote sensing image pixels in a row and a column; $i, j = 1, 2, \cdots L$, x_1, y_1, x_2, y_2 are the pixel coordinates in the image. Figure 1 shows the spatial sketch map of GLCM. If the remote sensing image has L picture gray levels, the size of the gray level co-occurrence matrix is $L \times L$, d represents the distance of two pixels in the remote sensing image, θ represents the angle between the connection line of the two pixels and horizontal direction, and it is usually set as 0°, 45°, 90°, and 135°. The element at the a-th row and b-th column in $\mathbf{P}(i, j, d, \theta)$ represents the appearance times of all the pixel couples that are δ apart from each other in the θ direction, with gray values i and j, respectively. d is related to the image, the step is usually set as $\delta = 1$, and the central pixel to be operated and compared with the directly adjacent pixel.

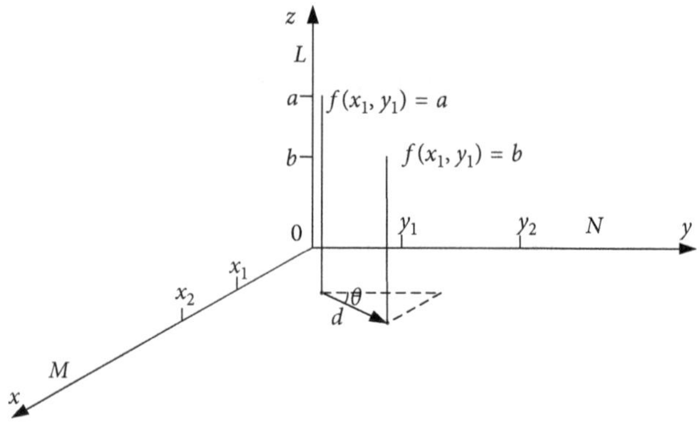

Figure 1: The space description of the GLCM.

4.2. Establishment of the Gray Level Difference Frequency Spatial. Gray level difference frequency spatial is proposed as a texture extraction approach for the multiprinciple component, which is based on the Gray Level Co-occurrence Matrix. Assume the gray level of primary gradient remote sensing image is g; firstly, iterate the principle components' remote sensing image with a rectangular window which is of size $m \times n$. Assume $L_m = \{1, 2, \cdots, m\}$ and $L_n = \{1, 2, \cdots, n\}$ are separately the horizontal and vertical spatial domains of the window and $G = \{0, 1, \cdots, g-1\}$ is the gray level. $L_n \times L_m$ is the windows' pixel set with ranking sequence in the row and column, specifying I as the conversion formula of pixels in $L_n \times L_m$ to G:

$$I : L_n \times L_m \longrightarrow G. \tag{12}$$

In different windows of principle components' image, the appearance probability of the pixel couples with distance d, direction θ, and gray level difference Δ composing the GLDFS and it is named $\mathbf{P}$, in which $(m1_{Pi}, n1_{Pi})$, $(m2_{Pi}, n2_{Pi})$ are coordinates of two pixels in the $i-th$ principal component $L_n \times L_m$; $I(m1_{Pi}, n1_{Pi})$, $I(m2_{Pi}, n2_{Pi})$ are gray levels corresponding to the two pixels; and $(m2_{Pi}, n2_{Pi})$ locates in the θ direction of $(m1_{Pi}, n1_{Pi})$ with distance d. The two pixel gray level's difference is expressed as follows:

$$\Delta_i = \left| I(m1_{Pi}, n1_{Pi}) - I(m2_{Pi}, n2_{Pi}) \right|, \tag{13}$$

where $\Delta_i \in [0, g-1]$. Make statistics of the gray level difference in each principal component and map the result to k-dimensional space according to the distance d and direction θ. The coordinates of the spatial points are $(\Delta_1, \Delta_2, \cdots, \Delta_k)$, and k is the number of the principle components' remote sensing images. Meanwhile, make statistics for each principal component at the four directions 0°, 45°, 90°, and 135°, thereby generating four n-dimensional spaces, and # indicates the number of the elements in the space; the statistical methods are as follows:

$$\mathbf{P}^k(\Delta_1, \Delta_2, \cdots \Delta_k, d, 0^0) = \#\left\{ \begin{array}{l|l} (m1_{P1}, n1_{P1}), (m2_{P1}, n2_{P1}) \in (L_n \times L_m) \times (L_n \times L_m) & m1_{P1} - m2_{P1} = 0, |n1_{P1} - n2_{P1}| = d, \ |I(m1_{P1}, n1_{P1}) - I(m2_{P1}, n2_{P1})| = \Delta_1 \\ (m1_{P2}, n1_{P2}), (m2_{P2}, n2_{P2}) \in (L_n \times L_m) \times (L_n \times L_m) & m1_{P2} - m2_{P2} = 0, |n1_{P2} - n2_{P2}| = d, \ |I(m1_{P2}, n1_{P2}) - I(m2_{P2}, n2_{P2})| = \Delta_2 \\ \quad\vdots & \quad\vdots \\ (m1_{Pk}, n1_{Pk}), (m2_{Pk}, n2_{Pk}) \in (L_n \times L_m) \times (L_n \times L_m) & m1_{Pk} - m2_{Pk} = 0, |n1_{Pk} - n2_{Pk}| = d, \ |I(m1_{Pk}, n1_{Pk}) - I(m2_{Pk}, n2_{Pk})| = \Delta_k \end{array} \right\}, \tag{14}$$

$$\mathbf{P}^n(\Delta_1, \Delta_2, \cdots \Delta_k, d, 45^0) = \#\left\{ \begin{array}{l|l} (m1_{P1}, n1_{P1}), (m2_{P1}, n2_{P1}) \in (L_n \times L_m) \times (L_n \times L_m) & (m1_{P1} - m2_{P1} = d, n1_{P1} - n2_{P1} = -d) \text{ or } (m1_{P1} - m2_{P1} = -d, n1_{P1} - n2_{P1} = d), \ |I(m1_{P1}, n1_{P1}) - I(m2_{P1}, n2_{P1})| = \Delta_1 \\ (m1_{P2}, n1_{P2}), (m2_{P2}, n2_{P2}) \in (L_n \times L_m) \times (L_n \times L_m) & (m1_{P2} - m2_{P2} = d, n1_{P2} - n2_{P2} = -d) \text{ or } (m1_{P2} - m2_{P2} = -d, n1_{P2} - n2_{P2} = d), \ |I(m1_{P2}, n1_{P2}) - I(m2_{P2}, n2_{P2})| = \Delta_2 \\ \quad\vdots & \quad\vdots \\ (m1_{Pk}, n1_{Pk}), (m2_{Pk}, n2_{Pk}) \in (L_n \times L_m) \times (L_n \times L_m) & (m1_{Pk} - m2_{Pk} = d, n1_{Pk} - n2_{Pk} = -d) \text{ or } (m1_{Pk} - m2_{Pk} = -d, n1_{Pk} - n2_{Pk} = d) \ |I(m1_{Pk}, n1_{Pk}) - I(m2_{Pk}, n2_{Pk})| = \Delta_k \end{array} \right\}, \tag{15}$$

$$\mathbf{P}^k(\Delta_1, \Delta_2, \cdots \Delta_k, d, 90^0) = \#\left\{ \begin{array}{l|l} (m1_{P1}, n1_{P1}), (m2_{P1}, n2_{P1}) \in (L_n \times L_m) \times (L_n \times L_m) & |m1_{P1} - m2_{P1}| = d, n1_{P1} - n2_{P1} = 0, \ |m1_{P2} - m2_{P2}| = d, n1_{P2} - n2_{P2} = 0 \\ (m1_{P2}, n1_{P2}), (m2_{P2}, n2_{P2}) \in (L_n \times L_m) \times (L_n \times L_m) & |m1_{P2} - m2_{P2}| = d, n1_{P2} - n2_{P2} = 0, \ |I(m1_{P2}, n1_{P2}) - I(m2_{P2}, n2_{P2})| = \Delta_2 \\ \quad\vdots & \quad\vdots \\ (m1_{Pk}, n1_{Pk}), (m2_{Pk}, n2_{Pk}) \in (L_n \times L_m) \times (L_n \times L_m) & |m1_{Pk} - m2_{Pk}| = d, n1_{Pk} - n2_{Pk} = 0, \ |I(m1_{Pk}, n1_{Pk}) - I(m2_{Pk}, n2_{Pk})| = \Delta_k \end{array} \right\}, \tag{16}$$

$$\mathbf{P}^n(\Delta_1, \Delta_2, \cdots \Delta_k, d, 135^0) = \#\left\{ \begin{array}{l|l} (m1_{P1}, n1_{P1}), (m2_{P1}, n2_{P1}) \in (L_n \times L_m) \times (L_n \times L_m) & (m1_{P1} - m2_{P1} = d, n1_{P1} - n2_{P1} = d) \text{ or } (m1_{P1} - m2_{P1} = -d, n1_{P1} - n2_{P1} = -d), \ |I(m1_{P1}, n1_{P1}) - I(m2_{P1}, n2_{P1})| = \Delta_1 \\ (m1_{P2}, n1_{P2}), (m2_{P2}, n2_{P2}) \in (L_n \times L_m) \times (L_n \times L_m) & (m1_{P2} - m2_{P2} = d, n1_{P2} - n2_{P2} = d) \text{ or } (m1_{P2} - m2_{P2} = -d, n1_{P2} - n2_{P2} = -d), \ |I(m1_{P2}, n1_{P2}) - I(m2_{P2}, n2_{P2})| = \Delta_2 \\ \quad\vdots & \quad\vdots \\ (m1_{Pk}, n1_{Pk}), (m2_{Pk}, n2_{Pk}) \in (L_n \times L_m) \times (L_n \times L_m) & (m1_{Pk} - m2_{Pk} = d, n1_{Pk} - n2_{Pk} = d) \text{ or } (m1_{Pk} - m2_{Pk} = -d, n1_{Pk} - n2_{Pk} = -d) \ |I(m1_{Pk}, n1_{Pk}) - I(m2_{Pk}, n2_{Pk})| = \Delta_k \end{array} \right\}. \tag{17}$$

4.3. Texture Features' Description of the Gray Level Difference Frequency Spatial. Haralick proposed Gray Level Co-occurrence Matrix 14 properties to describe the image's texture features, and Sor proposed Gray Level Co-occurrence Matrix 10 properties aiming at the SAR images [11]. But some properties are relevant, resulting in information redundancy; besides, part of properties is not suitable for analysis of remote sensing images' texture features [12–14]. This paper makes analysis of seven texture features with GLDFS and could be computed as follows:

(1) Energy:

$$T_1 = \sum_{\Delta_1}\sum_{\Delta_2}\cdots\sum_{\Delta_n}\{\mathbf{P}^n(\Delta_1,\Delta_2,\cdots,\Delta_n)\}^2. \quad (18)$$

(2) Entropy:

$$T_2 = -\sum_{\Delta_1}\sum_{\Delta_2}\cdots\sum_{\Delta_n}\mathbf{P}^n(\Delta_1,\Delta_2,\cdots,\Delta_n) \cdot \log(\mathbf{P}^n(\Delta_1,\Delta_2,\cdots,\Delta_n)). \quad (19)$$

(3) Autocorrelation:

$$T_3 = \sum_{\Delta_1}\sum_{\Delta_2}\cdots\sum_{\Delta_n}(\Delta_1,\Delta_2,\cdots,\Delta_n)\mathbf{P}^n(\Delta_1,\Delta_2,\cdots,\Delta_n). \quad (20)$$

(4) Correlation:

$$T_4 = \sum_{\Delta_1}\sum_{\Delta_2}\cdots\sum_{\Delta_n}(\Delta_1,\Delta_2,\cdots,\Delta_n)\mathbf{P}^n \cdot (\Delta_1,\Delta_2,\cdots,\Delta_n)\ \frac{-\mu_{\Delta_1}\mu_{\Delta_2}\cdots\mu_{\Delta_n}}{\sigma_{\Delta_1}\sigma_{\Delta_2}\cdots\sigma_{\Delta_n}}. \quad (21)$$

(5) Highlight degree of clustering:

$$T_5 = \sum_{\Delta_1}\sum_{\Delta_2}\cdots\sum_{\Delta_n}\left(\Delta_1+\Delta_2+\cdots\Delta_n-\mu_{\Delta_1}\mu_{\Delta_2}\cdots\mu_{\Delta_n}\right)^3\mathbf{P}^n \cdot (\Delta_1,\Delta_2,\cdots,\Delta_n). \quad (22)$$

(6) Dark degree of clustering:

$$T_6 = \sum_{\Delta_1}\sum_{\Delta_2}\cdots\sum_{\Delta_n}\left(\Delta_1+\Delta_2+\cdots\Delta_n-\mu_{\Delta_1}\mu_{\Delta_2}\cdots\mu_{\Delta_n}\right)^4\mathbf{P}^n \cdot (\Delta_1,\Delta_2,\cdots,\Delta_n). \quad (23)$$

(7) The maximum similarity:

$$T_7 = \underset{\Delta_1,\Delta_2,\cdots,\Delta_n}{\mathrm{MAX}}\ \mathbf{P}^n(\Delta_1,\Delta_2,\cdots,\Delta_n) \quad (24)$$

The energy, entropy, autocorrelation, and correlation are the extentions of Haralick's method in high-dimensional space, and the maximum similarity is generalization of Soh's method, the highlight degree, and the dark degree of clustering, which are the simulation calculations of human perception.

5. Analysis of Experimental Results

The experiment data are based on the multispectral remote sensing data of 8 different bands, which were recorded by Worldview-II on 8th April, 2009. In order to ensure the same landform features, the reef selected by the experiment is uncovered with plants consisting of sedimentary and metamorphic rocks. Gray value variance has a regular pattern when counted on space and texture feature shows up with strong intensify as well, so this paper tests and verifies the gray level difference frequency spatial texture feature extracting method through the classification of landform and compares it with the Gray Level Co-occurrence Matrix. The gray level difference frequency spatial and Gray Level Co-occurrence Matrix choose worldview-II remote sensing data of the same region which cohere well, and thus the comparability of the experiment data is assured. Experimental classification uses SVM, which is based on structural risk minimization principle. SVM is a better solution to small samples and nonlinear problems.

5.1. The Experimental Data Compression. PCA is applied to compression of 8 different bands of Worldview-II multispectral remote sensing data. In this paper, the first three principal components are derived as the feature of the gray level difference frequency spatial texture feature. Figure 2 shows the contribution rate as well as the cumulative contribution rate of each main component of the Worldview-II 8-band multispectral data after principal component analysis transform. Figure 3 shows the three principal component image data after principal component analysis transformation and compression. Figure 3(a) shows the first principal component, and the contribution rate is 74.12%. Figure 3(b) shows the second principal component, and the contribution rate is 22.16%. Figure 3(c) shows the third principal component, and the contribution rate is 2.73%.

5.2. The Classification and Analysis of Sample Data Experiment. Deriving the texture features of the three principal components after being compressed with the gray level difference frequency spatial and deriving the texture features of the four regular bands and four special bands which are compared separately with Gray Level Co-occurrence Matrix, the distance d of both methods is 1; merely comparing the gray-scale changes of the adjacent pixel, the window size is 13×13. Features including energy, self-correlation, correlation, cluster dark, and maximum self-similarity are selected; the sample size is 1350 among which the reef, sea, and foam samples are 350 separately. The SVM pattern is applied for the purpose of dividing the data into three. The result of landmark feature classification with the gray level difference frequency spatial and Gray Level Co-occurrence Matrix are shown in Tables 1 and 2. The recognition rate is based on the typical testing samples randomly selected from the samples including reef, sea, and foam data. Calculating the percentage of correct

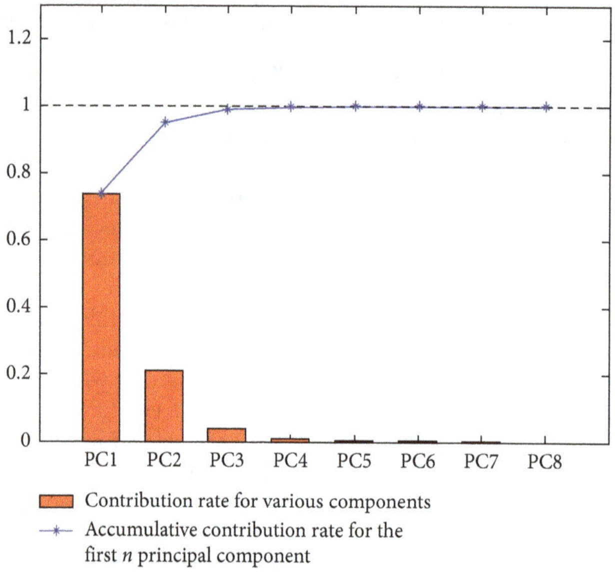

FIGURE 2: The contribution rate and the cumulative contribution rate of each principal component.

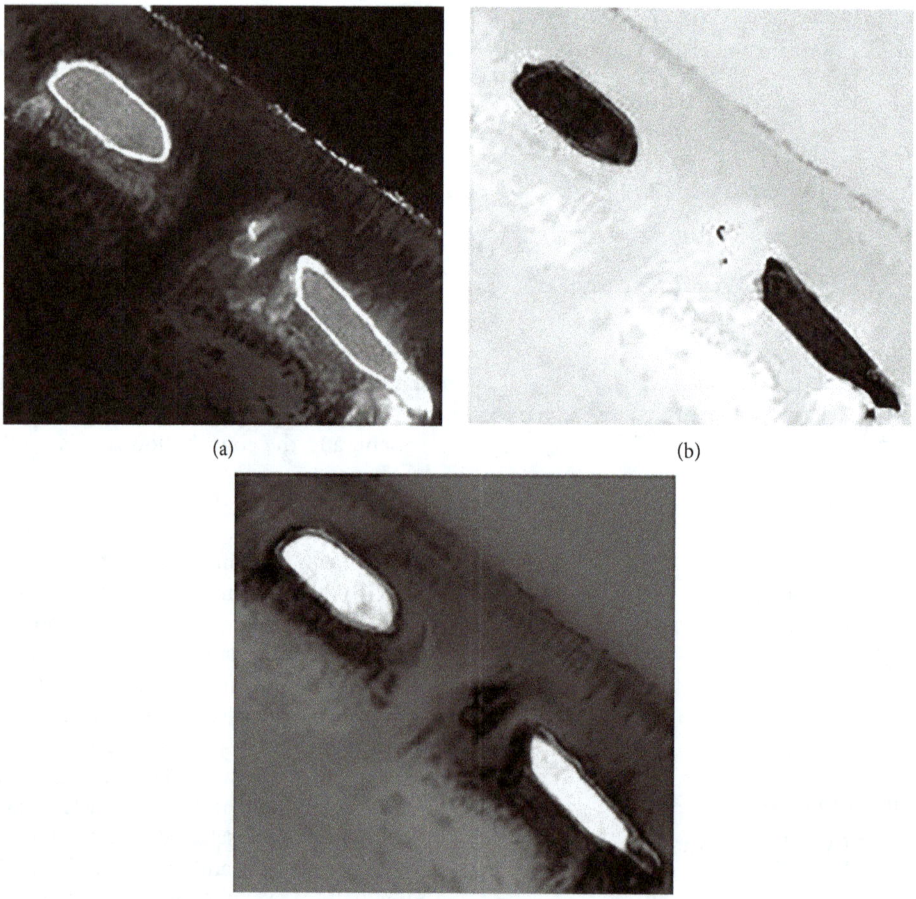

(a)

(b)

(c)

FIGURE 3: Image after PCA transformation and compression: (a) first principal component, (b) second principal component, and (c) third principal component.

identification number for each landmark class with its corresponding total samples, the average value is the right recognition number of the three types of landmarks with the gross sample number. It can be figured out that the gray level difference frequency spatial is capable of dealing with three types of landmark features at the same time and has better

TABLE 1: The recognition rate of GLDFS and GLCM of conventional bands.

Recognition rate (%)	Blue band	Green band	Red band	Near-infrared 1 band	Principal component data
	GLCM of conventional bands				GLDFS
Reef	60.2	64.6	62.9	79.8	89.3
Seawater	72.3	50.3	34.2	18.9	92.8
Foam	81.2	79.4	49.6	89.3	70.7
Average	65.4	71.3	30.9	62.4	85.9

TABLE 2: The recognition rate of GLDFS and GLCM of characteristic bands.

Recognition rate (%)	Coastal band	Yellow band	Infrared band	Near-infrared 2 band	Principal component data
	GLCM of characteristic bands				GLDFS
Reef	70.1	64.3	91.4	95.4	93.1
Sea	69.3	30.4	39.8	2.8	94.5
Foam	74.9	49.8	7.9	93.8	69.8
Average	63.2	55.2	54.6	71.2	83.6

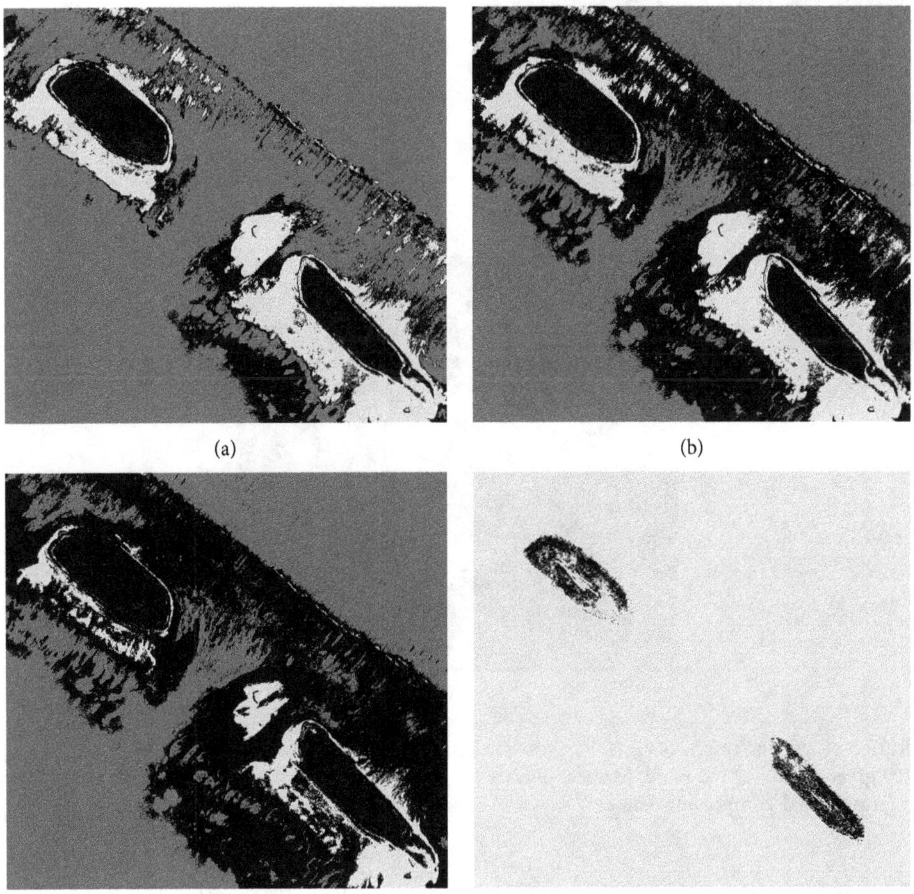

(a) (b) (c) (d)

FIGURE 4: Continued.

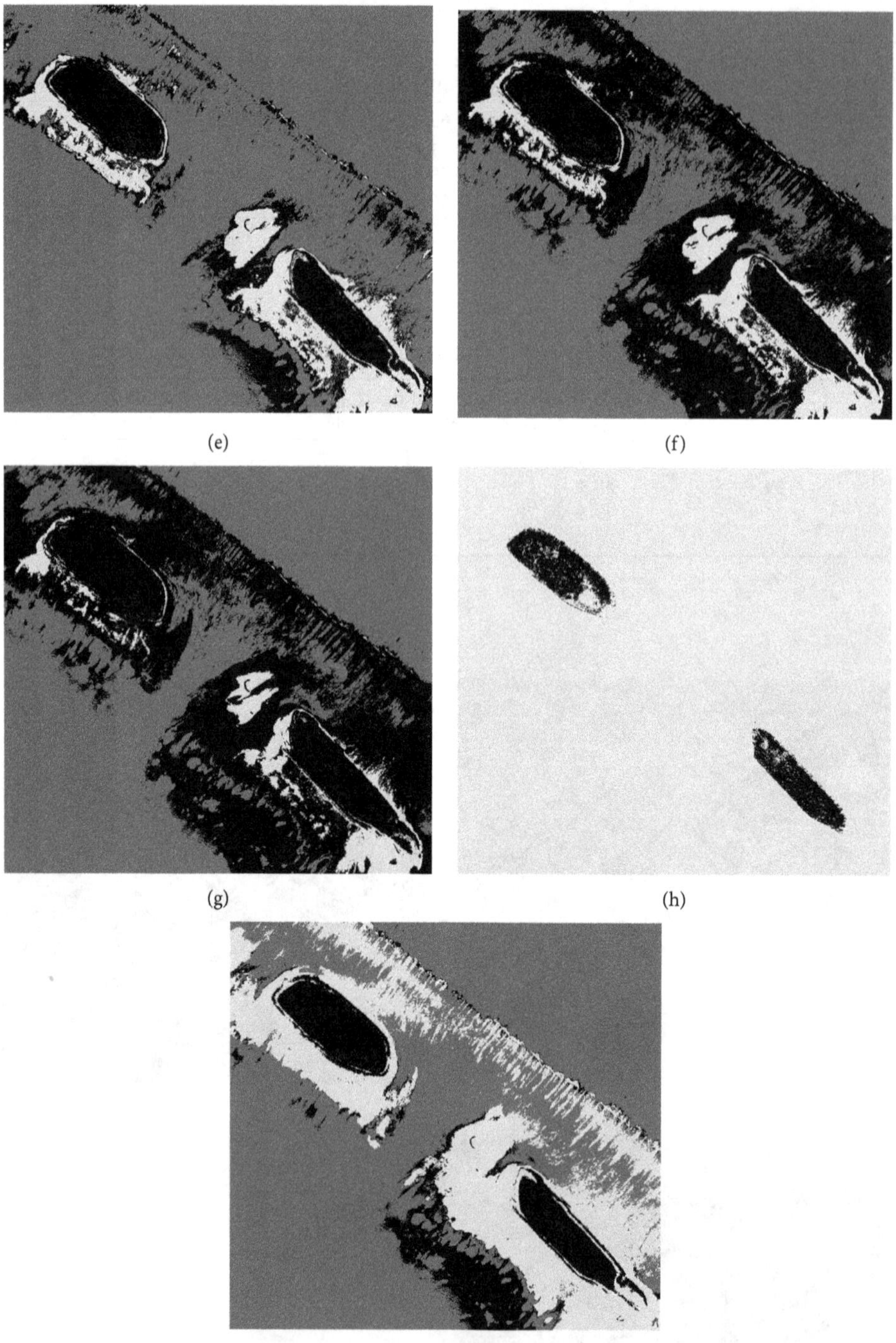

Figure 4: The classification processing results of GLDFS in conventional bands and characteristic bands and GLCM in principal component data: (a) blue band, (b) green band, (c) red band, (d) near-infrared 1 band, (e) coastal band, (f) yellow band, (g) infrared band, (h) near-infrared 2 band, and (i) principal component data.

recognition efficiency and quality than traditional Gray Level Co-occurrence Matrix.

The classification using the gray level difference frequency spatial and the classification using Gray Level Co-occurrence Matrix in conventional bands and characteristic bands of processing results about principal component data can be seen from Figure 4. In Figure 4, black indicates the reef, dark gray indicates foam, and light gray indicates sea. Figures 4(a)–4(d) are the classification result of conventional bands of the Gray Level Co-occurrence Matrix. Figures 4(e)–4(h) are the classification result of characteristic bands of the Gray Level Co-occurrence Matrix. Figure 4(i) is the classification result of the principal component data of the gray level difference frequency spatial. This paper presents a method named the gray level difference frequency spatial for the principal components data processing and has

TABLE 3: The comparision of SVM multisclassification efficiency.

Experimental data	Conventional bands	Special bands	Principal component data
Extraction method of input variables	GLCM		GLDFS
Recognition rate of reef (%)	93.2	87.3	91.6
Recognition rate of seawater (%)	93.9	89.7	96.1
Recognition rate of foam (%)	81.6	74.3	76.1
Recognition rate of average (%)	78.2	86.3	83.9
Spending of time (s)	93	89	45

a better classification results with three kinds of surface features. The method of Gray Level Co-occurrence Matrix has poor identification with reef when it is processing blue bands and coastal bands, and some island and reef are identified as foam mistakenly; when green bands and yellow bands is processing, the identification to seawater is poor, some seawater is identified as reef and island mistakenly; when red band and red edge band are processing, the identification to seawater and foam is poor, some seawater and foam is identified as reef and island mistakenly; when No. 1 and No. 2 bands of near-infrared are processing, seawater is identified as foam. The method of Gray Level Co-occurrence Matrix has high recognition to some landmark, but for the multilandmarks of multispectral remote sensing data, it cannot be guaranteed to have higher recognition rate in classification. If a variety of surface landmarks is classified, the single-band data need to be processed separately. In the experiment, texture features are extracted from the conventional band and special band data using Gray Level Co-occurrence Matrix method and from the principal component data using gray level difference frequency spatial method. The extracted results are processed and used as SVM input variables of multi-landmark classification. Finally, the classification time consumption is compared and analyzed. The efficiency of texture feature extracted by two methods using SVM classification is as shown in Table 3, and the results show that the gray level difference frequency spatial to the main component of compressed data has better efficiency in ensuring a higher recognition rate, but takes long time. The main reason is that the conventional band and special bands contain four bands of data; Gray Level Co-occurrence Matrix requires processing individual band and makes the results of 4 bands as SVM input variables to multilandmark classification. Due to the increased dimension of the input variables, the efficiency of calculation of SVM is reduced and the classification time becomes longer.

6. Conclusions and Discussions

According to the Worldview-II multispectral remote sensing data, this paper proposed a texture feature method based on gray-scale difference in space-frequency. This method compresses the multispectral remote sensing data after having it disposed, carries out gray-scale statistic on the main components after the compression process, and builds the gray-scale frequency difference space model. The method is validated by taking the data collected by Worldview-II. From the results, it can be seen that the gray level difference frequency spatial can extract texture features and recognize and classify multiband, multi-landmark, and multi-spectral remote sensing data at the same time and has an advantage over tradition Gray Level Co-occurrence Matrix on consideration of both recognition efficiency and quality.

Conflicts of Interest

The authors declare that they have no conflicts of interest.

Acknowledgments

This work was funded by the National Key Research and Development Project of China (2017YFC1404100, 2017YFC1404102).

References

[1] J. Feng and N. Shu, "A novel texture feature extraction of hyperspectral," *Journal of WuHan University of Technology*, vol. 31, no. 3, pp. 11–17, 2009.

[2] R. M. Haralick, K. Shanmugam, and I. H. Dinstein, "Textural features for image classification," *IEEE Transactions on Systems, Man, and Cybernetics*, vol. 3, no. 6, pp. 610–621, 1973.

[3] G. N. Srinivasa and G. Shobba, "Statstical texture analysis," *Proceedings of World Academy of Science, Engineering and Technology*, vol. 11, pp. 196–201, 2006.

[4] C. S. Lu, P. C. Chung, and C. F. Chen, "Unsupervised texture segmentation via wavelet transform," *Pattern recognition*, vol. 30, no. 5, pp. 729–742, 1997.

[5] M. K. Bashar, T. Matsumoto, and N. Ohnishi, "Wavelet transform-based locally orderless images for texture segmentation," *Pattern Recognition Letters*, vol. 24, no. 15, pp. 2633–2650, 2003.

[6] A. K. Jain and F. Farshid, "Unsupervised texture segmentation using Gabor Filters," *Pattern recognition*, vol. 24, no. 12, pp. 1167–1186, 1991.

[7] A. Kaarna, "Integer PCA and wavelet transforms for multispectral image compression," in *Proceedings of International Geoscience and Remote Sensing Symposium*, pp. 1853–1855, Sydney, Australia, July 2001.

[8] C. Jutten and J. Herault, "Independent component analysis versus PCA," in *Proceedings of European Signal Processing Conference*, pp. 643–646, Grenoble, France, September 1988.

[9] G. M. Xian, "An identification method of malignant and benign liver tumors from ultrasonography based on GLCM texture features and fuzzy SVM," *Expert Systems with Applications*, vol. 37, no. 10, pp. 6737–6741, 2010.

[10] R. F. Walker, P. T. Jackeay, and I. D. Longstaff, "Recent developments in the use of the Co-occurrence matrix for texture recognition," in *Proceedings of International Conference on Digital Image Processing*, pp. 463–471, Santorini, Greece, July 1997.

[11] L. K. Soh, "Texture analysis of SAR sea ice imagery using gray level Co-occurrence matrices," *IEEE Transactions on*

Geoscience and Remote Sensing, vol. 37, no. 2, pp. 780–795, 1999.

[12] R. Jobanputra and D. A. Clausi, "Preserving boundaries for image texture segmentation using gray level co-occurring probabilities," *Pattern Recognition*, vol. 39, no. 2, pp. 234–245, 2006.

[13] D. A. Clausi, "An analysis of co-occurrence statistics as a function of gray level quantization," *Canadian Journal of Remote Sensing*, vol. 28, no. 1, pp. 45–62, 2002.

[14] P. Philippe, O. Richard, and C. Dubes, "Performance evaluation for four classes of textural features," *Pattern Recognition*, vol. 25, no. 8, pp. 819–833, 1992.

19

Improved Stochastic Gradient Matching Pursuit Algorithm based on the Soft-Thresholds Selection

Zhao Liquan and Hu Yunfeng

College of Information Engineering, Northeast Electric Power University, Jilin 132012, China

Correspondence should be addressed to Zhao Liquan; zhao_liquan@163.com

Academic Editor: Panajotis Agathoklis

The preliminary atom set exits redundant atoms in the stochastic gradient matching pursuit algorithm, which affects the accuracy of the signal reconstruction and increases the computational complexity. To overcome the problem, an improved method is proposed. Firstly, a limited soft-threshold selection strategy is used to select the new atoms from the preliminary atom set, to reduce the redundancy of the preliminary atom set. Secondly, before finding the least squares solution of the residual, it is determined whether the number of columns of the measurement matrix is smaller than the number of rows. If the condition is satisfied, the least squares solution is calculated; otherwise, the loop is exited. Finally, if the length of the candidate atomic index set is less than the sparsity level, the current candidate atom index set is the support atom set. If the condition is not satisfied, the support atom index set is determined by the least squares solution. Simulation results indicate that the proposed method is better than other methods in terms of the reconstruction probability and shorter running time than the stochastic gradient matching pursuit algorithm.

1. Introduction

Compressed sensing (CS) [1, 2] theory has three core problems: sparse representation of signals, and design of the measurement matrix, and design of reconstruction algorithms. The reconstruction algorithm is directly related to the accuracy of the recovery signal and the convergence rate of the algorithm, which determines whether the theory is feasible. The recovery algorithm is described as the recovery of the high-dimensional original signal from the low-dimensional measurement vectors. At present, many reconstruction algorithms have been proposed to recover the signal, which include convex optimization methods, combinational methods, and greedy pursuit methods. Convex optimization methods approximate the signal by transforming nonconvex problems into convex ones, which include basis pursuit (BP) [3], the gradient projection for sparse reconstruction (GPSR) [4] algorithm, iterative threshold (IT) [5], interior-point method [6], Bergman iteration (BT) [7], and total variation (TV) [8]. Although convex optimization algorithms have fewer observations, their higher complexity is not suitable for practical applications.

Combinational methods include Fourier sampling [9,10], chaining pursuit (CP) [11], heavy hitters on steroids pursuit (HHSP) [12]. Combinational methods are low in complexity; however, the accuracy of the recovery signal is not as good as convex optimization algorithms. Greedy algorithms have many advantages, such as simple structure, fast convergence, and low complexity, becoming the first choice for the recovery algorithm.

To date, many greedy pursuit algorithms have been proposed, including the first, the matching pursuit (MP) [13] algorithm. Based on this algorithm, the orthogonal matching pursuit (OMP) [14] algorithm was proposed to optimize MP via orthogonalization of the atoms of the support set. However, the OMP algorithm only selects one of the atoms of the support set at each round of iteration, and the efficiency is lower. Successors include the regularized OMP (ROMP) [15], subspace pursuit (SP) [16] algorithm, compressive sampling matching pursuit (CoSaMP) [17, 18] algorithm, and stagewise OMP (StOMP) [19] algorithm. The ROMP algorithm realizes the effective selection of the atom via the regularized rule, to improve the speed of the OMP. The SP and CoSaMP algorithms are similar. Both are proposed with the backtracking strategy. The difference between the SP and CoSaMP is the number of atoms that are selected from measurement matrix to compose preliminary atomic set in each round of iteration. The SP selects s atoms, while the CoSaMP selects $2s$ atoms. The StOMP algorithm selects multiple atoms or columns of the measurement matrix in each round of iteration via a threshold parameter. The greedy algorithms mentioned above all require the sparsity information of the signal, and the choice of the type of atoms is inflexible, which affects the convergence speed, robustness, and reconstruction performance of the algorithms.

Because the traditional greedy pursuit algorithm needs to compute the inverse matrix of the sensing matrix, this process requires a significant amount of computation time and storage space, resulting in lower reconstruction probability. In recent years, some research workers have proposed the fast version of the greedy algorithm, which avoids computation of the inverse matrix [20]. The GP algorithm uses the gradient idea of the unconstrained optimization method to replace the computation of the inverse matrix and to reduce the computational complexity and storage space of the traditional greedy algorithms [21]. However, the GP algorithm has a slow convergence and lower efficiency. In addition, the GP algorithm cannot solve the problem of large-scale data recovery. To improve those problems, the conjugate gradient pursuit (CGP) [22] algorithm and approximate conjugate gradient pursuit (ACGP) [23] algorithm are proposed, respectively. Although CGP and ACGP algorithms effectively reduce the computational complexity and storage space of traditional greedy algorithms, the reconstruction performance still needs to be improved. Based on the GP algorithm, the stagewise weak selection (SWGP) [24] algorithm was proposed to improve the reconstruction performance and convergence speed of the GP algorithm, via introduction of the conjugate direction and weak selection strategy. Motivated by the stochastic gradient descent methods, the stochastic gradient matching pursuit (StoGradMP) algorithm [25] was recently proposed for the optimization problem with sparsity constraints. The atomic selection method of the fixed number in the StoGradMP algorithm (that is, selecting 2s atoms at the preliminary stage of each iteration) will lead to the redundant atoms of the preliminary atomic set. When joined with the candidate atomic set, this will reduce the accuracy of the least squares solution and the inaccuracy of the support atomic set estimation, which affects the precision of the signal reconstruction and increases the computational complexity of the algorithm. In this study, we use the limited soft-threshold selection strategy to realize the second selection of the preliminary atom set after the preliminary stage, which will improve the reconstruction accuracy. The combination with reliability verification conditions ensures the correctness and effectiveness of the proposed method.

2. Compressed Sensing Theory

The recent work in the area of the compressed sensing [26] demonstrated that it was possible to algorithmically recover sparse (and, more generally compressible) signals from incomplete observations. The simplest model is an n-dimensional signal x with a small number of nonzero entries under no noise conditions:

$$x \in R^n, \quad \|x\|_0 \le s \le n, \tag{1}$$

where $\|x\|_0$ and s are the number of nonzero entries and sparsity level of the signal, respectively. Such signals are called s-sparse. Any signal in R^n can be represented in terms of $n \times 1$ vectors $\{\psi_i\}_{i=1}^n$. For simplicity, we assume that the basis is orthonormal. Forming the $n \times n$ basis matrix $\Psi = [\psi_1, \psi_2, \dots \psi_n]$ by stacking the vectors $\{\psi_i\}$ as columns, we can express any signal x as

$$x = \sum_{i=1}^{n} \theta_i \psi_i = \Psi\theta, \tag{2}$$

where $\theta = \Psi^T x$ is the $n \times 1$ column vector of projection coefficients, $\theta_i = \langle x, \psi_i \rangle = \psi_i^T x$ is the projection coefficient. Obviously, x and θ are the equivalent representations of the same signals in the different domains. That is, x and θ are the signals in the time domain and Ψ domain, respectively.

When x is s-sparse, let $\Psi = I$, I is the unit matrix, and $x = \theta$. We use a measurement matrix $\Phi \in R^{m\times n}\,(m \le n)$ to make a linear measurement of the projection coefficient vector θ, obtaining an observation vector $y \in R^{m\times 1}$, expressed as

$$y = \Phi x, \tag{3}$$

where Equation (3) is a linear projection of the original signals x on Φ. Note that the measurement process is nonadaptive; that is, Φ does not depend in any way on the signal x. Clearly, the dimension of y is much lower than the dimension of x. That is, this problem is an underdetermined problem; Equation (3) has infinitely many solutions. It is difficult to recover the projection coefficient vector θ directly from the observation vector y. However, the original signal x is s-sparse and Φ satisfies certain conditions in Equation (3); x can be recovered by solving the l_0-minimization problem:

$$\begin{aligned} &\min_x \quad \|x\|_0, \\ &\text{s.t.} \quad \Phi x = y, \end{aligned} \tag{4}$$

where $\|.\|_0$ is the l_0-norm of the vector, representing the number of nonzero entries. Candes and Tao demonstrated that if the s-sparse signal x is to be accurately recovered, the number of measurements m (or the dimension of observation vector y) must satisfy $m = O(s\ln(n))$, and the measurement matrix must satisfy the restricted isometric property (RIP) [27, 28].

When x is not s-sparse in the time domain, the signal recovery process cannot be directly used in the Equation (4). The signal x can be sparse representations on the sparse basis matrix Ψ. Combining Equations (2) and (3), we obtain

$$y = \Phi\Psi^T x = \tilde{\Phi}\theta, \tag{5}$$

where $\tilde{\Phi} = \Phi\Psi^T \in R^{m\times n}$ is the sensing matrix [29]. According to [30], the equivalent condition of the RIP is that the measurement matrix Φ is not correlated with the sparse basis matrix Ψ. Note that if the sensing matrix $\tilde{\Phi}$ also satisfies the RIP, the recovery signal (or projection coefficient) θ can be obtained by solving an optimal l_0-norm problem similar to Equation (4):

$$\begin{aligned} &\min_{\theta} \quad \|\theta\|_0, \\ &\text{s.t.} \quad \tilde{\Phi}\theta = y. \end{aligned} \tag{6}$$

From Equation (5), we see that Ψ is fixed, so that Φ also satisfies the RIP. The measurement matrix Φ must meet certain conditions. It is shown in [31] that when the measurement matrix is a Gaussian random matrix, the sensing matrix $\tilde{\Phi}$ can satisfy the RIP condition with a larger probability.

However in most scenarios, the signal contains noise. In this case, the measurement process can be expressed as

$$y = \tilde{\Phi}\theta + e, \tag{7}$$

where the $m \times n$ matrix $\tilde{\Phi}$ is the sensing matrix and e is the m-dimensional noise vector. θ is the s-sparse signal in the Ψ domain. In this study, for simplicity, we assume that the signal x is s-sparse, that is, $x = \theta$ and $\Phi = \tilde{\Phi}$. Therefore, Equation (7) can be written as $y = \Phi x + e$, and we minimize the following formula to recover x:

$$\begin{aligned} &\min_{x\in R^n} \quad \frac{1}{2m}\|y - \Phi x\|_2^2, \\ &\text{s.t.} \quad \|x\|_0 \le s, \end{aligned} \tag{8}$$

where s controls the sparsity of the solution to Equation (9).

To analyze (8), we combine Equation (2), where θ_i are the projection coefficients of signal x. x is considered sparse with respect to Ψ if s is relatively small compared to the ambient dimension n. Therefore, we can express the optimization (9) in the form as follows:

$$\begin{aligned} &\min_{x} \quad \underbrace{1/M \sum_{i=1}^{M} f_i(x)}_{F(x)}, \\ &\text{s.t.} \quad \|x\|_{0,\Psi} \le s, \end{aligned} \tag{9}$$

where $f_i(x), x \in R^n$, is a smooth function that is nonconvex; $\|x\|_{0,\Psi}$ is defined as the norm that captures the sparsity level of x. In particular, $\|x\|_{0,\Psi}$ is the smallest number of atoms in Ψ such that x can be represented as

$$\|x\|_{0,\Psi} = \min_{x}\left\{ s : x = \sum_{i\in T} \theta_i \psi_i \text{ with } |T| = s \right\}. \tag{10}$$

For sparse signal recovery, the set Ψ consists of n basic vectors, each of size n in the Euclidean space. This problem can be seen as a special case of Equation (10), with $f_i(x) = (y_i - <\varphi_i, x>)^2$ and $M = m$. We decompose the vector $y^{m\times 1}$ into nonoverlapping vectors y_{b_i} with a size of b and denote Φ_{b_i} as $\Phi_{b_i\times n}$. $\Phi_{b_i\times n}$ is the submatrix of measurement matrix Φ. According to Equation (9), the objective function is $F(x) = (1/2m)\|y - \Phi x\|_2^2$. We then denote $F(x)$ as

$$F(x) = \frac{1}{M}\sum_{i=1}^{M}\frac{1}{2b}\left\|y_{b_i} - \Phi_{b_i}x\right\|_2^2 \triangleq \frac{1}{M}\sum_{i=1}^{M} f_i(x), \tag{11}$$

where $M = m/b$, which is a positive integer. We treat each function $f_i(x)$ as $f_i(x) = (1/2b)\|y_{b_i} - \Phi_{b_i}x\|_2^2$, and each function $f_i(x)$ represents a collection (or block) of observations of size b. Here, we spilt the function $F(x)$ into multiple subfunctions $f_i(x)$ or block the measurement matrix Φ into the submatrix Φ_{b_i}, which will be beneficial for the calculation of the gradient.

3. StoGradMP Algorithm

CoSaMP [17] is a popular algorithm for recovering a sparse signal from its linear measurements. The idea of the CoSaMP algorithm is generalized to prove the GradMP algorithm that solves a broader class of sparsity-constrained problems. In this section, we describe the stochastic version of the GradMP, the StoGradMP [25] algorithm, where at each iteration, only the evaluation of the gradient of a function f_i is required.

The StoGradMP algorithm is described in Algorithm 1, which consists of the following steps in each iteration:

Randomize. The measurement matrix Φ is subject to random block processing, to determine the region of the submatrix $\Phi_{b_i\times n}$. Then, according to Equation (11), calculates the subfunction $f_{i_k}(x_k)$.

Proxy. Compute the gradient of $f_{i_k}(x_k)$. Here, the gradient is an $n \times 1$ column vector.

Identify. Select the column indexes of the submatrix $\Phi_{b_i\times n}$ corresponding to the maximum $2s$ components in the gradient vector, forming a preliminary index set S_k.

Merge. Merge the preliminary index set S_k and the support index set F_{k-1} of the previous iteration to form a candidate index set C_k.

Estimation. The estimation of signal b_k by a sub-optimization method is determined, which is the least squares method. Generally, b_k is the transition signal.

Prune. Select the column indexes of the measurement matrix Φ corresponding to the maximum s components in the signal estimation vector b_k that forms a support index set F.

Update. Update the signal estimation $\hat{x} = x_k$.

Check. When the residual is less than the tolerance error of the proposed method iteration, stop the iteration. If the loop index k is greater than the maximum number of iterations s, the proposed method ends and the approximation of signal $\hat{x} = x_k$ is output. If the iteration ends, the condition is not satisfied. Otherwise, continue the iteration until the halting condition is met.

Input:
Sparsity level s
Sensing matrix $\Phi^{m\times n}$
Observation vector $y^{m\times 1}$
Block size b
Tolerance used to exit loop tol
Maximum number of iterations maxIter
Initialization parameter:
$\hat{x} = 0$ {initialize signal approximation}
$k = 0$ {loop index}
done = 0 {while loop flag}
$S_0 = \varnothing$ {empty preliminary index set}
$C_0 = \varnothing$ {empty candidate index set }
$F_0 = \varnothing$ {empty support index set}
$M = \text{floor}(m/b)$ {number of block}
While (~done)
$k = k + 1$
(1) Randomize
$ii = \text{ceil}(\text{rand} * M)$
$\text{block} = b * (ii - 1) + 1 : b * ii$
$f_{i_k}(x_k) = (1/2b)\|y_{b_{i_k}} - \Phi_{b_{i_k}} x\|_2^2$
(2) Computation of gradient
$r_k = \nabla f_{i_k}(x_k) = -2 * \Phi_{b_{i_k}}{}^T (y_{b_{i_k}} - \Phi_{b_{i_k}} x_{k-1})$
(3) Identify the large $2s$ **components**
$S_k = \max(|r_k|, 2s)$
(4) Merge to update candidate index set
$C_k = F_{k-1} \cup S_k$
(5) Signal estimation by the least square method
$b_k = \Phi_{C_k}{}^{+} y$
(6) Prune to obtain current support index set
$F = \max(|b_k|, s)$
(7) Update
$x_k = b_{kF}$
$F_k = F$
(8) Check the iteration condition
If ($\|y - \Phi x_k\|_2 \le \text{tol}$ or $k \ge \text{maxIter}$)
done = 1 quit iteration
end
end
Output: $\hat{x} = x_k$ (s-sparse approximation of signal x)

Algorithm 1: StoGradMP algorithm.

4. Improved StoGradMP Algorithm

The StoGradMP algorithm selects $2s$ atoms in each iteration, where s is a fixed number. The choice of atoms is inflexible and will increase the redundancy of the preliminary atom set. This affects the reconstruction and computational speed. To solve this problem, we use the limited soft selection strategy to select atoms from the preliminary atom set.

Firstly, according to Equation (11), we randomly block the measurement matrix to obtain a stochastic block matrix (or submatrix) $\Phi_{\text{block}\times n}$, which can be expressed as

$$\begin{aligned} ii &= \text{ceil}(\text{rand} * M), \\ \text{block} &= b * (ii - 1) + 1 : b * ii, \end{aligned} \tag{12}$$

where M is the number of the block measurement matrix, $M = \text{floor}(m/b)$, and $b = \min(s, m)$ is the number of rows of the block matrix. floor(.) and rand(.) generate a maximum integer smaller than m/b and uniformly distributed pseudorandom numbers, respectively. ceil(.) obtains a minimum integer greater than rand $* M$. block is the row index of the measurement matrix Φ. From Equation (12), we know that the area of the block matrix is randomly determined. For simplicity, we express $\Phi_{\text{block}\times n}$ and y_{block} as Φ_{b_i} and y_{b_i}, respectively. Next, we compute the subfunction $f_i(x)$ as follows:

$$f_{i_k}(x_k) = \frac{1}{2b}\left\| y_{b_{i_k}} - \Phi_{b_{i_k}} x_{k-1} \right\|_2^2, \tag{13}$$

where k is the number of iterations, $i \in [M]$. Φ_{b_i} is the i-th submatrix of the measurement matrix Φ, which is stochastically determined and of size $b_i \times n$. The subfunctions $f_i(x)$ are also stochastically determined, and $f_i(x) \in F(x)$. Here, $f_i(x)$ is the i-th subfunction of $F(x)$.

After determining the area of submatrix Φ_{b_i} and the subfunction $f_i(x)$, we calculate the gradient of the subfunction, which is expressed as

$$r_k = \nabla f_{i_k}(x_k) = -2 * \Phi_{b_{i_k}}{}^T \left(y_{b_{i_k}} - \Phi_{b_{i_k}} x_{k-1} \right), \tag{14}$$

where r_k is the gradient of the subfunction $f_i(x)$ at the k-th iteration and the gradient is an $n \times 1$ column vector. $\nabla(.)$ and $\Phi_{b_i}{}^T$ represent the derivative of the subfunction $f_i(x)$ and the transpose matrix of submeasurement matrix Φ_{b_i}, respectively. x_{k-1} is the approximation of sparse signal x at the $k-1$-th iteration.

In the atomic preliminary selection stage of the stochastic algorithm, we obtain the gradient vector of the function through the derivation of randomly determined subfunctions, and then select the $2s$ largest gradient values from the gradient vector r_k, thereby determining the column index of the measurement matrix corresponding to the values of the gradient vector, and form a preliminary atom index set S_k as

$$S_k = \max\left(|r_k|, 2s \right), \tag{15}$$

where $|r_k|$ is the absolute value of the gradient vector r_k, k is the number of iterations, and $\max(|r_k|, 2s)$ is the atomic (or column) index of the measurement matrix corresponding to the maximal value from r_k that forms a preliminary atom index set S_k.

After the preliminary atomic stage, the preliminary atomic set exits the redundant atoms, which will reduce the accuracy of the least square solutions of signal and the inaccuracy of the support atomic set estimation. This will eventually affect the precision of signal reconstruction, and increase the computational complexity of the algorithm. To improve the reconstruction accuracy, we use the limited soft-threshold selection strategy to realize the second selection of the preliminary atom set.

In the atomic index set S_k, the atoms corresponding to the index are expressed as Γ, $\Gamma = [\Gamma_1, \Gamma_2, \ldots \Gamma_{2s}]$. We select the atoms that are larger than the threshold in the atomic set Γ, to form a new atomic set. The index of the new atomic set forms the new atomic index set S_k^*. This process is described as

$$\begin{aligned} \text{sigma} &= \max\left(|r_k|\right), \\ S_k^* &= \text{Find}\left(\|r_k\|_{2s} \geq t * \text{sigma}\right), \end{aligned} \quad (16)$$

where sigma is the maximum gradient value at the k-th iteration. $\|r_k\|_{2s}$ represents the largest $2s$ gradient values from the absolute value of the gradient vector, which corresponds to the atomic index set S_k. t is the threshold, whose value range is generally $[0.1 \sim 1.0]$. It should be noted that if the size of the threshold t is greater than 1, the soft-thresholds selection strategy fails. That is to say that the selection of the preliminary atomic index set S_k^* cannot be completed. Meanwhile, if the value of the threshold t is too small, it will lead to the redundant atoms cannot be effectively eliminated in the preliminary atomic set. $t * \text{sigma}$ is the soft-thresholds selection strategy. Find(.) searches the corresponding index that satisfies the soft-threshold condition and forms a new atomic index set S_k^*.

After the soft-threshold selection strategy is completed, we merge the new atomic set $\Phi_{S_k^*}$ and the support set $\Phi_{F_{k-1}}$ to update the candidate atom set Φ_{C_k}, which can be expressed as

$$\begin{aligned} C_k &= F_{k-1} \cup S_k^*, \\ \Phi_{C_k} &= \Phi_{F_{k-1}} \cup \Phi_{S_k^*}, \end{aligned} \quad (17)$$

where C_k, F_{k-1}, and S_k^* represent the candidate atomic index set, the support atomic index set, and the preliminary atomic index set, respectively. Φ_{C_k} is the candidate atomic set at the k-th iteration. $\Phi_{F_{k-1}}$ is the support atomic set at the previous iteration. $\Phi_{S_k^*}$ is the new atomic set corresponding to atomic index set S_k^* at the k-th iteration.

Before solving the suboptimization method, we must ensure that the number of rows is greater than the number of columns in the candidate atomic matrix Φ_{C_k}; that is, Φ_{C_k} is a full column-rank matrix. Therefore, we provide the first reliability verification condition; that is, if

$$\text{length}\,(C_k) \leq m, \quad (18)$$

then

$$A_k = \Phi_{C_k}, \quad (19)$$

where A_k is the candidate atomic set (or matrix) at the k-th iteration. If the condition is not satisfied, namely, the number of rows is smaller than the number of columns, the matrix is not inversed. If this occurs, we exit the loop. Let $b_k = 0$. Next, solving the least squares solution of the sparse approximation signal $\hat{x}$, it can be expressed as

$$b_k = A_{C_k}^{+} y, \quad (20)$$

where b_k is the estimation vector of the sparse signal x at the k-th iteration. This is known as the transition signal. $A_{C_k}^{+}$ is the inverse matrix of the candidate atomic set (or matrix), and y is the observing vector.

Since the soft-threshold selection strategy is used to complete the second selection of the preliminary atomic set, the size of the candidate atomic index set C_k may be less than s. Therefore, to ensure the correctness and effectiveness of the proposed method, we provide the second reliability verification condition, that is, if

$$\text{length}\,(C_k) < s, \quad (21)$$

then

$$F = C_k, \quad (22)$$

where s is the size of sparsity of the signal. $\text{length}\,(C_k)$ is the size of the candidate atomic index set C_k. If

$$\text{length}\,(C_k) \geq s, \quad (23)$$

then

$$F = \max\left(|b_k|, s\right), \quad (24)$$

where $|b_k|$ is the absolute value of the transition signal at the k-th iteration. $\max\,(|b_k|, s)$ determines the atomic (or column) index of the measurement matrix corresponding to the maximum value from b_k and the former support atomic index set F.

Next, we update the approximation of sparse signal x, which is expressed as

$$x_k = b_{kF}, \quad (25)$$

where x_k is the approximation of the signal at the k-th iteration, and b_{kF} is the recovery signal corresponding to the support atomic index set F.

Finally, we check the iteration stopping condition. If

$$\left\|y - \Phi x_k\right\|_2 < \text{tol or } k \geq \text{maxIter}, \quad (26)$$

then

$$\text{done} = 1, \quad (27)$$

where $y - \Phi x_k$ represents the residual at the k-th iteration and tol is the tolerance error of the algorithm iteration. max Iter is the maximum number of iteration, with special value $500 * M$ in this study. done = 1 represents the algorithm stops and outputs the signal approximation $\hat{x} = x_k$. If the iteration stopping criteria is not satisfied, then the iteration is continued until the condition is satisfied. The entire procedure is shown in Algorithm 2.

5. Results and Discussion

In this section, we used the signal with s-sparsity as the original signal. The measurement matrix is randomly generated with a Gaussian distribution. All performance is an average after running the simulation 100 times using a computer with a quadcore, 64-bit processor, and 4G memory.

In Figure 1, we compare the reconstruction probability of different sparsity of the proposed algorithm with different measurements in different threshold conditions. We set the sparsity level set as $s \in [8, 12, 16, 20]$. The threshold parameter set is $t \in [0.1, 0.2, 0.3, 0.4, 0.5, 0.6, 0.7, 0.8, 0.9, 1.0]$. From Figures 1(a)–1(e), we can see that when the size of the threshold parameter is 0.1, 0.2, 0.3, 0.4, and 0.5, the reconstruction probability of the proposed method is very

```
Input:
  Sparsity level s
  Sensing matrix Φ^{m×n}
  Observation vector y^{m×1}
  Block size b
  Tolerance used to exit loop tol
  Maximum number of iterations maxIter
Initialize parameter:
  x̂ = 0 {initialize signal approximation}
  k = 0 {loop index}
  done = 0 {while loop flag}
  S_0 = ∅ {empty preliminary index set}
  C_0 = ∅ {empty candidate index set}
  F_0 = ∅ {empty support index set}
  M = floor(m/b) {number of block}
While (~done)
    k = k + 1
  (1) Randomize
    ii = ceil(rand * M)
    block = b * (ii − 1) + 1 : b * ii
    f_{i_k}(x_k) = (1/2b)||y_{b_{i_k}} − Φ_{b_{i_k}} x_{k−1}||_2^2
  (2) Computation of gradient
    r_k = ∇f_{i_k}(x_k) = −2 * Φ_{b_{i_k}}^T (y_{b_{i_k}} − Φ_{b_{i_k}} x_{k−1})
  (3) Identify the large 2s components
    S_k = max(|r_k|, 2s)
  (4) Soft-threshold selection strategy
    sigma = max(|r_k|)
    S_k^* = Find(||r_k||_{2s} ≥ t * sigma)
  (5) Merge to update the candidate index set
    C_k = F_{k−1} ∪ S_k^*
    Reliability verification conditions 1
    If length(C_k) ≤ m
      A_k = Φ_{C_k}
    else
      if k == 1
        b_k = 0
      end
        break;
      end
  (6) Estimation of signal by least square method
    b_k = A_{C_k}^+ y
  (7) Prune to obtain current support index set
    Reliability verification conditions 2
    If (length(C_k) < s)
      F = C_k
    else
      F = max(|b_k|, s)
    end
  (8) Update
    x_k = b_{kF}
    F_k = F
  (9) Check the iteration stopping condition
    If (||y − Φx_k||_2 < tol or k ≥ maxIter)
      done = quit iteration
    end
end
Output: x̂ = x_k(s-sparse approximation of signal x)
```

Algorithm 2: Proposed algorithm.

close, with almost no difference. In Figure 1(f), when the size of the threshold parameter is 0.6, the proposed method can complete signal reconstruction with less measurements, compared with other thresholds, when the size of sparsity is 16 and 20. In Figures 1(g)–1(j), we see that when the size of the threshold is 0.7, 0.8, 0.9, and 1.0, the proposed method requires more measurements to complete the signal reconstruction under the same sparsity level.

Figure 2 compares the reconstruction probability of different measurements of the proposed method with different thresholds under the same sparsity level. We set the sparsity set as $s = [8, 12, 16, 20]$. The threshold parameter set is consistent with the threshold parameter set in Figure 1. From Figure 2, we see that different threshold parameters have certain influences on the reconstruction probability of the signal for different sparsity levels. In Figures 2(a) and 2(b), we can see that when the sparsity levels are $s = 8$ and $s = 12$, the reconstruction probability of all thresholds conditions is very close, except for the thresholds 0.8, 0.9, and 1.0. Meanwhile, from Figures 2(c) and 2(d), we can see that when the sparsity levels are $s = 16$ or $s = 20$ and the sizes of the threshold are 0.5, 0.6, and 0.7, the reconstruction performance of the proposed method is better than the reconstruction probability of the other thresholds for different measurements. Therefore, from Figure (2), it demonstrates that the soft-threshold strategy is more advantageous for larger sparsity. In particular, when the sparsity level $s = 20$ and threshold $t = 0.6$, the reconstruction probability of the proposed method is better than the reconstruction probability of other threshold conditions. Based on the analysis of Figures 1 and 2, we can see that that when the threshold is $t = 0.6$, the proposed method has better performance. Therefore, in the following simulations, we set the threshold as 0.6.

In Figure 3, we compared the average runtime of different thresholds of the proposed algorithm with different measurements. From Figures 1 and 2, we see that the reconstruction probability of the proposed method is 100% when the number of the measurements is greater than or equal to 145. In particular, when the sparsity level is $s = 20$ and the threshold is 0.6, the reconstruction probability is better than the reconstruction probability of other thresholds. Therefore, we set the range of the number of the measurements as [145 200] in the simulation of Figure 3. From this, we see that the proposed algorithm with $t = 0.8$ has the shortest runtime, and the next shortest are the proposed algorithm with $t = 0.7$ and the proposed algorithm with $t = 0.6$. The proposed method with $t = 0.9$ has the longest runtime. This means that the selection of the threshold parameter t is important to the runtime of the proposed method.

From the aspects of reconstruction probability and runtime of the proposed method, we conclude that when the size of threshold parameter is $t = 0.6$ and the sparsity level is $s = 20$, reconstruction performance of the proposed method

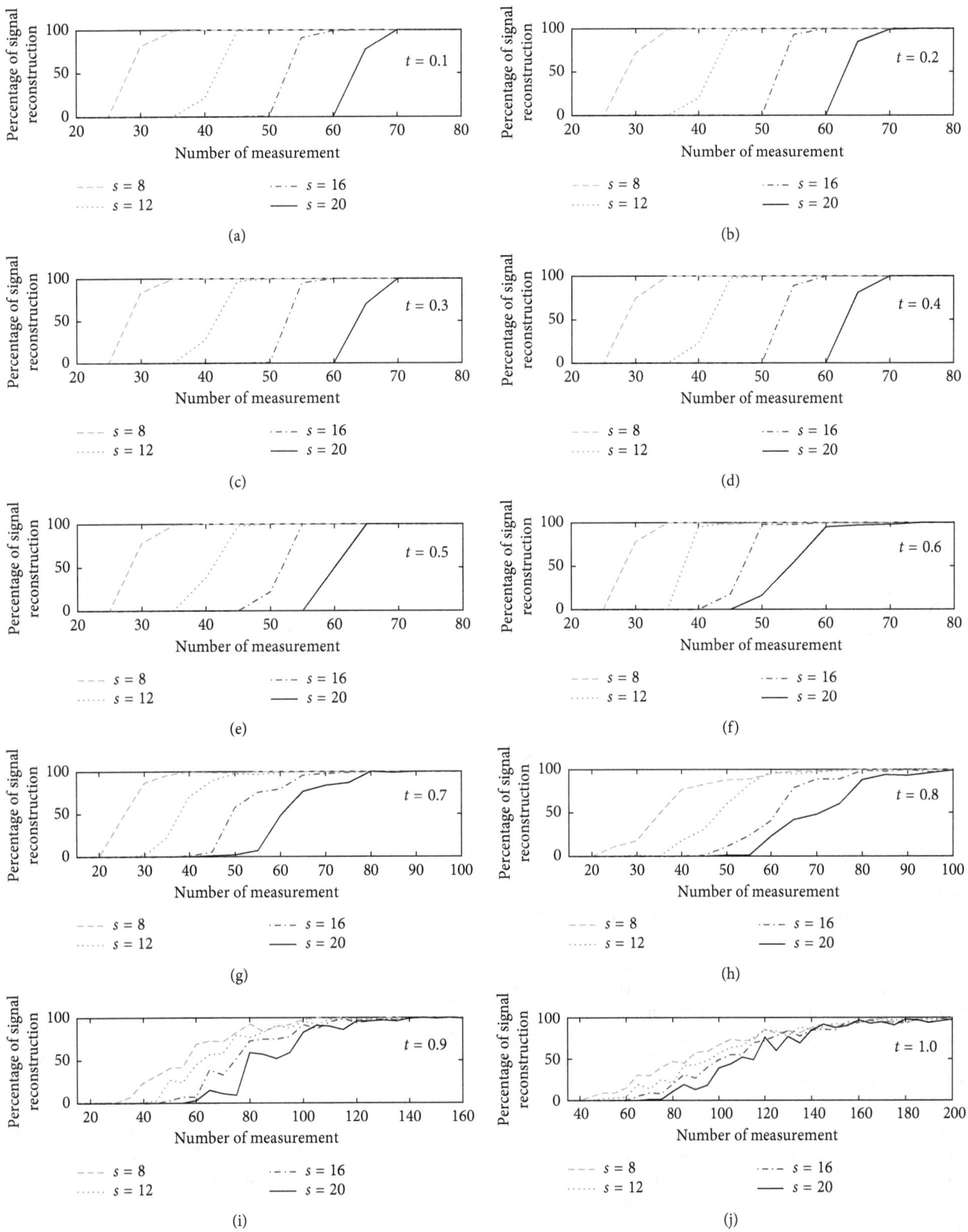

FIGURE 1: Reconstruction probability of different sparsity with different measurements in different threshold conditions ($n = 256$, $s \in [8, 12, 16, 20]$, $t \in [0.1\ 1.0]$, Gaussian signal).

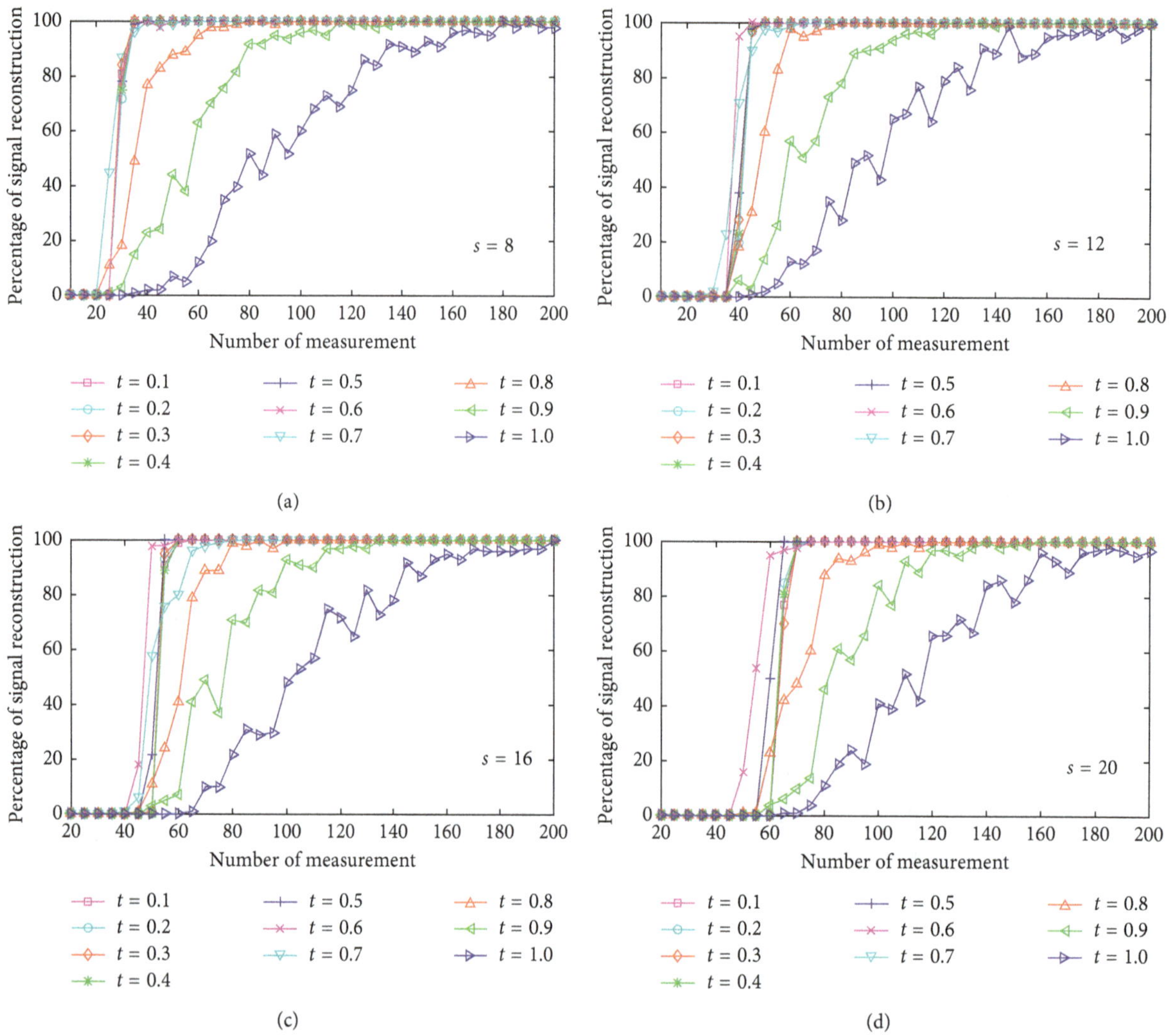

FIGURE 2: Reconstruction probability of different thresholds with different measurements in different sparsity conditions ($n = 256$, $t \in [0.1\ 1.0]$, $s \in [8, 12, 16, 20]$, Gaussian signal).

is better than the reconstruction performance of the other threshold conditions for different measurements. Therefore, in the following without special instructions, the default threshold is 0.6 and the sparsity level is $s = 20$.

In Figure 4, we compared the reconstruction performance of the proposed algorithm in the single reconstruction. From Figure 4, we see that the recovery error is less than 5×10^{-15}, which is much less than the tolerance error of the proposed method iteration. This shows that the reconstruction of the proposed method is ideal.

In Figure 5, we compare the reconstruction probability of different sparsities of the proposed method and the StoGradMP in different measurements. We set the sparsity set as $s = [8, 12, 16, 20]$. The size of the threshold is 0.6. From Figure 5, we see that when the sparsity level is $s = 8$, the reconstruction probability of the proposed algorithm and the StoGradMP algorithm is nearly identical for all measurements. When the sparsity level is $s = 8$, the reconstruction probability of the StoGradMP algorithm and

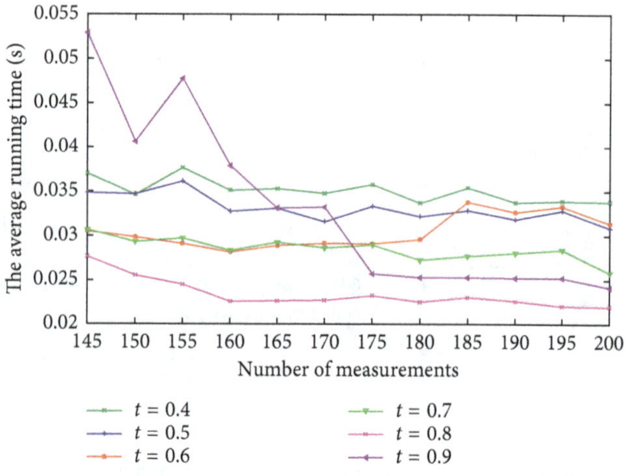

FIGURE 3: The average running time of different thresholds of the proposed algorithm with different measurements ($s = 20$, $n = 256$, $t \in [0.4\ 0.9]$, Gaussian signal).

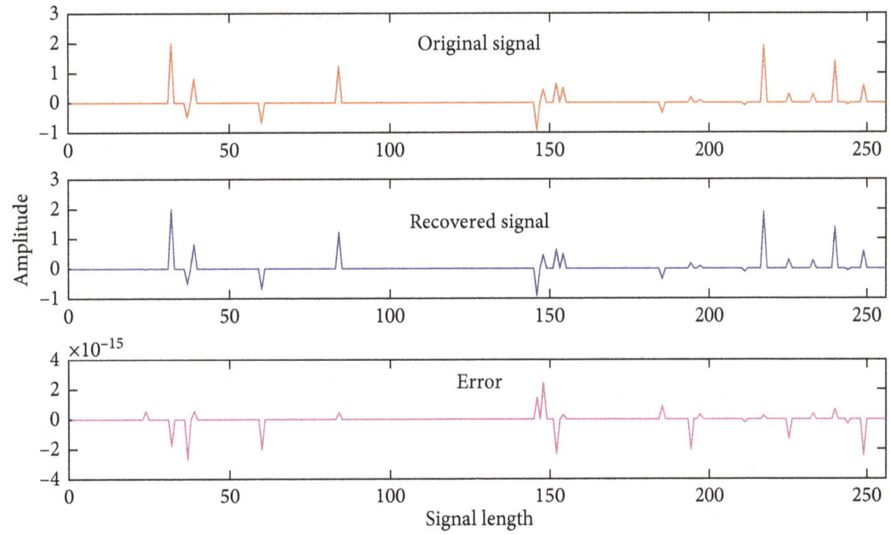

FIGURE 4: Comparison of original signal and recovery signal in the single reconstruction ($n = 256$, $s = 20$, $m = 100$, $t = 0.6$).

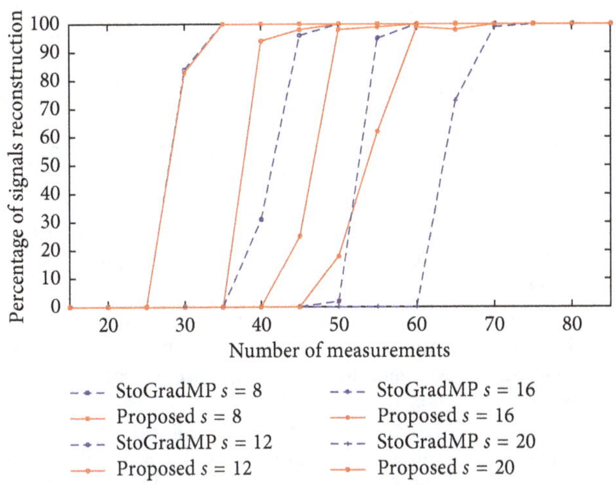

FIGURE 5: Reconstruction probability of different sparsities of the StoGradMP and proposed algorithms with different measurements (n = 256, t = 0.6, $s \in$ [8, 12, 16, 20], Gaussian signal).

the proposed algorithm is nearly identical. When $s = 12$, 16, and 20, the reconstruction probability of the proposed method is higher than the StoGradMP algorithm under the same sparsity level. Among them, when the sparsity level $s = 20$, the difference between the reconstruction probability of the two algorithms is the largest.

In Figure 6, we compared the reconstruction probability of the proposed algorithm with StoGradMP, GradMP, and StoIHT algorithms for different measurements. From Figure 5, we see that the reconstruction probability is 0% when the sparsity is equal to 20 and the number of the measurements is smaller than or equal to 40. Therefore, we set the range of the measurements as [40 130] in the simulation of Figure 6. From Figure 6, we see that when $40 \le m \le 46$, the reconstruction probability of almost all algorithms is equal to 0%. For $46 \le m \le 64$, the reconstruction probability of the proposed algorithm ranges from 0 to 97% and has a higher reconstruction probability than other methods. When

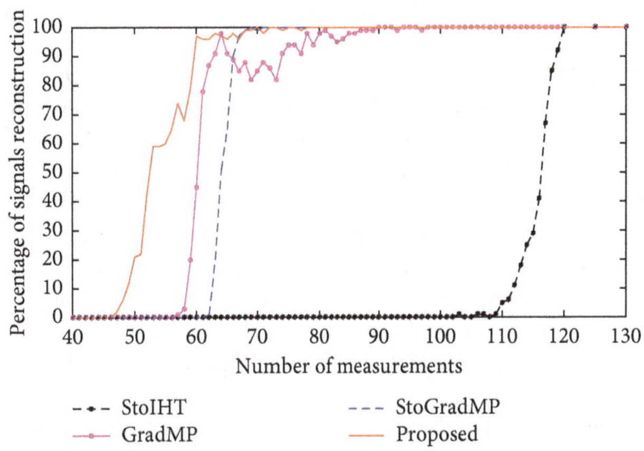

FIGURE 6: Reconstruction probability of different algorithms with different measurements ($n = 256$, $s = 20$, $t = 0.6$, Gaussian signal).

$56 \le m \le 64$, the reconstruction probability of t GradMP begins to increase, from approximately 0% to 97%, while the reconstruction probability ranges from 0 to 51%, and the reconstruction probability is 0%. When $64 \le m \le 73$, the reconstruction probability of the proposed method and StoGradMP ranges from to 97–100% and 51–100%, respectively. However, the reconstruction probability of the GradMP algorithm declines from 98% to 82%. When $73 \le m \le 108$, the reconstruction probability of the GradMP algorithm increases from 82% to 100% and the reconstruction probability of the proposed method and StoGradMP algorithm is 100%, with almost no change. However, the reconstruction probability of the StoIHT algorithm is still 0%. When $108 \le m \le 120$, the reconstruction probability of the StoIHT is increased from 0% to 100%. When $120 \le m$, all algorithms can complete the reconstruction.

In Figure 7, we compare the average runtime of different algorithms for different measurements. From Figure 6, we see that the reconstruction probability of all recovery

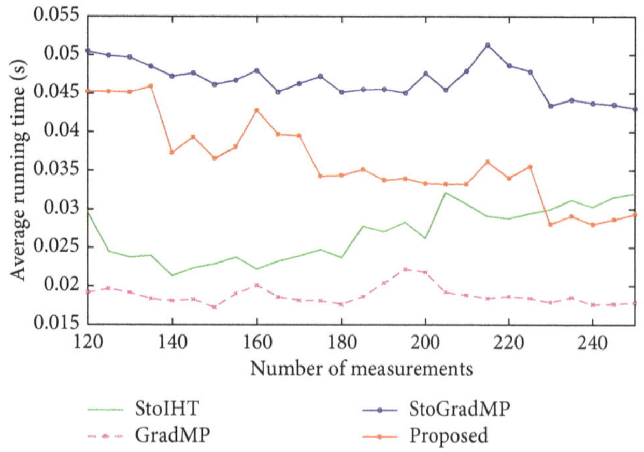

FIGURE 7: The average running time of different algorithms with different measurements ($n = 256$, $s = 20$, $t = 0.6$, Gaussian signal).

algorithms is 100% when the sparsity is equal to 20 and the number of measurements is greater than or equal to 120. Therefore, we set the range of the number of the measurements as [120 250] in the simulation of Figure 7, and the sparsity level is equal to 20. We see that the GradMP algorithm has the lowest runtime, and the next lowest are the StoIHT and proposed algorithms. The StoGradMP algorithm has the highest running time. From Figures 6 and 7, we see that although the runtime of the proposed method is much more than the GradMP and StoIHT algorithms, its reconstruction probability is much better than the other two algorithms. This means that the proposed algorithm has lower complexity than other algorithms, except for the GradMP and StoIHT algorithms in different measurements.

Based on the above analysis, the proposed method with threshold $t = 0.6$ has better reconstruction performance for different measurements than others and lower computational complexity than StoGradMP.

6. Conclusion

In this study, an improved reconstruction method was proposed. The proposed method utilizes the limited soft-threshold selection strategy to select the most relevant atoms from the preliminary atom set, which could reduce the redundancy of the preliminary atoms set and improve the accuracy of the support atom set estimation, thereby improving the reconstruction precision of the signal and reducing the computational complexity of the algorithm. In addition, the combination with reliability verification conditions ensured the correctness and effectiveness of the proposed method. The simulation results proved that the proposed method has better performance than other recovery methods.

Conflicts of Interest

The authors declare that they have no conflicts of interest.

Acknowledgments

This work was supported by the National Natural Science Foundation of China (61271115) and the Foundation of Jilin Educational Committee (JJKH20180336KJ). The authors would also like to thank Yanfei Jia for proof reading of this paper.

References

[1] E. J. Candès, J. Romberg, and T. Tao, "Robust uncertainty principles: exact signal reconstruction from highly incomplete frequency information," *IEEE Transactions on Information Theory*, vol. 52, no. 2, pp. 489–509, 2004.

[2] D. L. Donoho, "Compressed sensing," *IEEE Transactions on Information Theory*, vol. 52, no. 4, pp. 1289–1306, 2006.

[3] S. Chen, D. L. Donoho, and M. A. Saunders, "Atomic decomposition by basis pursuit," *Siam Journal on Scientific Computing*, vol. 20, no. 1, pp. 33–61, 1998.

[4] M. A. T. Figueiredo, R. D. Nowak, and S. J. Wright, "Gradient projection for sparse reconstruction: application to compressed sensing and other inverse problems," *IEEE Journal of Selected Topics in Signal Processing*, vol. 1, no. 4, pp. 586–597, 2007.

[5] I. Daubechies, M. Defrise, and C. D. Mol, "An iterative threshold algorithm for linear inverse problems with a sparsity constraint," *Communications on Pure and Applied Mathematics*, vol. 57, no. 11, pp. 1413–1457, 2003.

[6] K. Koh, S. J. Kim, and S. Boyd, "An interior-point method for large-scale l1-regularized logistic regression," *IEEE Journal of Selected Topics in Signal Processing*, vol. 1, no. 4, pp. 606–617, 2008.

[7] S. Osher, Y. Mao, B. Dong, and Y. Wotao, "Fast linearized bregman iteration for compressive sensing and sparse denoising," *Mathematics of Computation*, vol. 8, no. 1, pp. 93–111, 2008.

[8] L. I. Rudin, S. Osher, and E. Fatemi, "Nonlinear total variation based noise removal algorithms," in *Proceedings of Eleventh International Conference of the Centre for Nonlinear Studies on Experimental Mathematics*, pp. 259–268, Los Alamos, NM, USA, 1992.

[9] A. C. Gilbert, S. Guha, P. Indyk, S. Muthukrishnan, and M. Strauss, "Near-optimal sparse Fourier representations via sampling," in *Proceedings of ACM Symposium on Theory of Computing*, pp. 152–161, Montréal, Québec, Canada, May 2002.

[10] A. C. Gilbert and M. Strauss, "Improved time bounds for near-optimal sparse Fourier representations," in *Proceedings of SPIE Wavelets XI*, pp. 1–15, Bellingham, WA, USA, September 2005.

[11] A. C. Gilbert, M. J. Strauss, J. A. Tropp, and R. Vershynin, "Algorithmic linear dimension reduction in the l_1 norm for sparse vectors," in *Proceedings of 44th Annual Allerton Conference on Communication, Control, and Computing*, pp. 1–27, Allerton, IL, USA, September 2006.

[12] A. C. Gilbert, M. J. Strauss, J. A. Tropp, and R. Vershynin, "One sketch for all: fast algorithms for compressed sensing," in *Proceedings of 39th Annual ACM Symposium on Theory of Computing*, pp. 237–246, San Diego, CA, USA, June 2007.

[13] S. Mallat and Z. Zhang, "Matching pursuit with time-frequency dictionaries," *IEEE Transactions on Signal Processing*, vol. 41, no. 12, pp. 3397–3415, 1993.

[14] J. A. Tropp and A. C. Gilbert, "Signal recovery from random measurements via orthogonal matching pursuit," *IEEE Transactions on Information Theory*, vol. 53, no. 12, pp. 4655–4666, 2007.

[15] D. Needell and R. Vershynin, "Signal recovery from incomplete and inaccurate measurements via regularized orthogonal matching pursuit," *IEEE Journal of Selected Topics in Signal Processing*, vol. 4, no. 2, pp. 310–316, 2010.

[16] D. Wei and O. Milenkovic, "Subspace pursuit for compressive sensing signal reconstruction," *IEEE Transactions on Information Theory*, vol. 55, no. 5, pp. 2230–2249, 2009.

[17] D. Needell and J. A. Tropp, "CoSaMP: iterative signal recovery from incomplete and inaccurate samples," *Applied and Computational Harmonic Analysis*, vol. 26, no. 3, pp. 301–321, 2009.

[18] M. A. Davenport, D. Needell, and M. B. Wakin, "Signal space CoSaMP for sparse recovery with redundant dictionaries," *IEEE Transactions on Information Theory*, vol. 59, no. 12, pp. 6820–6829, 2012.

[19] D. L. Donoho, Y. Tsaig, I. Drori, and J. L. Starck, "Sparse solution of underdetermined systems of linear equations by stage-wise orthogonal matching pursuit," *IEEE Transactions on Information Theory*, vol. 58, no. 2, pp. 1094–1121, 2012.

[20] A. Sancetta, "Greedy algorithms for prediction," *Bernoulli*, vol. 22, no. 2, pp. 1227–1277, 2016.

[21] T. Blumensath and M. E. Davies, "Gradient pursuits," *IEEE Transactions on Signal Processing*, vol. 56, no. 6, pp. 2370–2382, 2008.

[22] T. Blumensath and M. E. Davies, "Gradient pursuit for nonlinear sparse signal modeling," in *Proceedings of 16th European Signal Processing Conference*, pp. 1–5, Lausanne, Switzerland, 2008.

[23] T. Blumensath and M. E. Davies, "Fast greedy algorithms for large sparse inverse problem," pp. 1–5, 2012, see.ed.ac.uk.

[24] T. Blumensath and M. E. Davies, "Stage-wise weak gradient pursuits," *IEEE Transactions on Signal Processing*, vol. 57, no. 11, pp. 4333–4346, 2009.

[25] N. Nguyen, D. Needell, and T. Woolf, "Linear convergence of stochastic iterative greedy algorithms with sparse constraints," *IEEE Transactions on Information Theory*, vol. 63, no. 11, pp. 6869–6895, 2017.

[26] E. J. Candès, "Compressive sampling," *Marta Sanz-Solé*, vol. 17, no. 4, pp. 1433–1452, 2006.

[27] E. J. Candes and T. Tao, "Near-optimal signal recovery from random projections: universal encoding strategies?," *IEEE Transactions on Information Theory*, vol. 52, no. 12, pp. 5406–5425, 2006.

[28] E. J. Candes and T. Tao, "Decoding by linear programming," *IEEE Transactions on Information Theory*, vol. 51, no. 12, pp. 4203–4215, 2005.

[29] R. G. Baraniuk, "A lecture on compressive sensing," *IEEE Signal Processing Magazine*, vol. 29, no. 4, pp. 1–9, 2007.

[30] R. Baraniuk, "Compressive sensing," in *Proceedings of 42nd Annual Conference on Information Sciences and Systems*, pp. 1–9, Princeton, NJ, USA, March 2008.

[31] E. J. Candès, J. K. Romberg, and T. Tao, "Stable signal recovery from incomplete and inaccurate measurements," *Communications on Pure and Applied Mathematics*, vol. 59, no. 8, pp. 1207–1223, 2005.

20

Secure and Efficient Cluster-Based Range Query Processing in Wireless Sensor Networks

Liming Zhou,[1] **Yingzi Shan**,[2] **and Lu Chen**[1]

[1]*School of Computer and Information Engineering, Henan University, Kaifeng 475004, China*
[2]*Department of Finance, Yellow River Conservancy Technical Institute, Kaifeng 475004, China*

Correspondence should be addressed to Liming Zhou; lmzhou@henu.edu.cn

Academic Editor: Stefano Vitturi

In wireless sensor networks, preserving privacy is more important and has attracted more attentions. Protecting data and sensor privacy while collecting and computing query results is a challenge. In cluster-based sensor networks, when a user queries a sensitive data, the adversaries can monitor original node or gain the data in cluster node. To deal with this problem, we propose a secure and efficient scheme for cluster-based query processing in wireless sensor networks. To preserve location privacy of sensors, we use anonymity method to confuse adversaries. To protect the sensitive data, we use prefix membership verification method to prevent adversaries from gaining sensitive messages collected by sensor nodes. And we analyze the security and communication cost. The results show that our scheme can efficiently protect privacy in query processing.

1. Introduction

Wireless sensor networks have been widely deployed in various applications, such as monitoring environment, collecting temperature data, and gaining information of battlefield. Each sensor node transmits sensed data to a base station for further processing. In some applications, clustering method has been extensively studied [1] and used to organize sensor nodes, which has been considered as a useful approach. And some nodes are grouped into clusters such that sensor node sends data to a cluster head in the same cluster. Many clustering applications aimed at enhancing the energy efficiency and extending the network lifetime in wireless sensor networks.

In wireless sensor networks, when sensor nodes collect information in our daily life, we should pay attention to protect data privacy and security. For instance, a user wants to query sensitive data from a certain sensor node according to his interests. The sensor network may leak private information about the user's interests to an adversary who can gain the content from the queried data. Meanwhile, the adversary can monitor the frequency of query to analyze the user's preferences and find the related sensor nodes. Then the adversary may attack and compromise the related nodes. And the compromised node may respond to a query and send fake data to the user, which is in conflict with the privacy requirement. So query processing brings serious security challenges.

When users monitor events or analyze sensed data, data query becomes an important operation in wireless sensor networks. Recently, many existing privacy techniques can be employed in sensor network scenarios. For example, a target region transformation technique [2], range query [3–6], and top-k query [7, 8] have been well addressed. However, these schemes are not suitable for cluster-based query processing in wireless sensor networks. And many techniques do not consider the computing power, power of sensors, and capacity, which are the limiting factors in wireless sensor networks. For limited availability resources, it is important to make the trade-off between the privacy preservation and the communication overhead.

Based on the above discussions, in this paper, we propose a novel cluster-based privacy preserving query processing in wireless sensor networks. We consider the privacy issue when processing data query in wireless sensor networks. If a user wants to gain and query information from the cluster-based sensor network, we will use the anonymity method and the prefix membership verification scheme [9, 10] to protect the sensitive data against adversaries. When a cluster head receives a query message, the cluster head will randomly choose several cluster members which include the real queried node. So, it is unlikely that the adversary can monitor the real frequency of query in a cluster. Therefore, the adversary cannot gain the user's interests from analyzing the frequency of query or find the location of the real source node. Meanwhile, cluster members encode their sensed data and send them to their cluster head. The cluster heads can correctly process data queries over encoded data without knowing their real values. And an adversary cannot know the query results in the cluster heads. In our scheme, we make a balance between data confidentiality and query efficiency.

The rest of the paper is organized as follows. Section 2 gives the related work and the previous proposed techniques for data query. In Section 3, we describe the system model and the security model. Then, we present our secure and efficient cluster-based query processing scheme in Section 4. In Section 5, we present the security analysis and performance analysis. Finally, we have the conclusion in Section 6.

2. Related Work

Protecting querying region's privacy in wireless sensor networks has been drawn attention recently [2]. In [2], a querying region transformation technique is proposed to fuzzy the target region of the query according to predefined transformation functions. The transformation function maps one region into m regions so that the target region cannot be distinguished from the other uninteresting regions. Meanwhile, multiple transformation functions include uniform, randomized, and hybrid function.

A secure and efficient range query processing scheme is proposed in [6], called SafeQ. They use the prefix membership verification and neighborhood chains to encode both data and queries such that a storage node can correctly process encoded queries over encoded data without knowing their values. The prefix membership verification converts the verifications of whether a number is in a range to several verifications of whether two numbers are equal. The neighborhood chains allow a sink to verify whether the result of a query contains exactly the data items that satisfy the query. The SafeQ scheme can preserve privacy and integrity for processing range queries in two-tiered sensor networks.

Privacy preserving range query has been widely studied in two-tiered wireless sensor networks. Many range query schemes are proposed to protect privacy of range queries. CSRQ [3] employs an encoding mechanism and encrypted constraint chain to preserve data privacy and query result integrity. In [11], Zhang et al. provided an efficient secure range query protocol. In their scheme, different sensor nodes have different hash functions to encode data items for the protection of data privacy, and the correlation among data is used for verification of result.

In [12, 13], two optimized versions which verify query result completeness to reduce the communication overhead between sensors and storage nodes based on the bucketing technique are proposed. In their scheme, a bit map is broadcasted by each sensor node to the nearby sensors, which indicates which buckets have data. In each sensor node, the collected data items and the received bit maps are encrypted together. The sink can verify the completeness of the query result for a sensor by examining the bit maps. But the compromised storage nodes can estimate the values of data items by using the bucketing technique to achieve data privacy.

Privacy preserving max/min query schemes in two-tiered sensor networks are proposed in [14–17], which use the prefix membership verification scheme to privately compute the maximum or minimum data item. But their schemes cannot be suitable for cluster-based sensor networks. The power and storage are limited in cluster heads.

3. Network and Adversary Models

3.1. Network Model. Sensor networks consist of a number of different types of sensor nodes that have been deployed to monitor environment or collect data and send information to the sink in an area. Nodes are organized into clusters. A cluster head is selected in each cluster to receive and query data from cluster members. In each cluster, every sensor sends data to its cluster head. The sink collects data with a lot of resources in storage, energy, and computation.

In this paper, we assume that sensor nodes are evenly deployed in the sensor network and do not move after being deployed. All of the sensors have roughly the same capabilities, power sources, and expected lifetimes. The users can access the sensor network by the sink. The sink translates a query from a user into multiple queries which are sent to the cluster heads. The cluster heads process the queries and return the query results to the sink. All query results are sent to the sink which changes all results into a final query result and sends the final result back to the user. When a user makes a query request, the sink will send query request to each cluster head. The cluster heads collect all results and send them back to the sink. The results are forwarded through certain routing strategies that adopted the sensor networks.

3.2. Adversary Model. For various kinds of wireless sensor networks, we assume that an adversary is a motivated and funded attacker whose objective is to learn sensitive data information. The adversary has unbounded energy resource, adequate computation capability, and sufficient memory of data storage. The adversary can use the leaked sensitive data to threaten the sensor network, such as health monitoring networks. For a user's query, the adversary tries to generate fake message and send it back to the user.

Meanwhile, the adversary wants to gain the user's interests and the frequency of query in clusters. He wants to find the location information of queried nodes. The adversary may stay nearby the cluster to monitor and eavesdrop constantly. When the adversary monitors a message in a cluster, he will know the location of a sensor node. If the frequency of transmitted messages is large, the adversary will find that certain sensor node is important for the user. When the adversary compromises the sensor node, the compromised node will send fake data to the user.

4. Secure and Efficient Cluster-Based Query Processing Scheme

In this section, we propose a scheme for preserving privacy query processing in cluster-based sensor networks. Each cluster head collects the data from sensor nodes in a cluster. To preserve privacy, sensor nodes encrypt or encode their collected data, for example, DES algorithm. So the adversary cannot gain the content of transmitted data.

4.1. The Basic Idea. In order to preserve privacy query processing, we propose a secure and efficient cluster-based query processing scheme to address this problem in wireless sensor networks. After the sensor network is deployed, the cluster heads are randomly chosen. Then the cluster heads broadcast their join messages. When a node firstly receives a join message from a cluster head, it will reply to the cluster head and join the cluster. The cluster head will record the sensor's ID. Meanwhile, the sink will record all cluster heads' ID.

However, if a cluster head has the less remaining energy, it will randomly select one of its members as the new cluster head in the cluster. And the new cluster head will record the ID of all members in the cluster. Then, the sink will replace the ID of the previous cluster head with the ID of the new cluster head.

When a user wants to gain the value of a sensor node s_i, he will make a query to the sink. The sink will send the query message to the cluster head which includes the sensor node s_i. The cluster head will randomly select several cluster members which include the real queried node s_i and gain the sensed data from them. It is aimed for preventing the adversary from monitoring the real frequency of the query in a cluster. We assume that each sensor s_i shares a secret key k_i with the sink in a network. A sensor s_i encrypts its sensed n data items $d_1, d_2, \ldots, d_n$ using key k_i in time slot t, the result of which is denoted as $(d_1)_{k_i}, \ldots, (d_n)_{k_i}$. Then, s_i encode $d_1, d_2, \ldots, d_n$ as $E(d_1, d_2, \ldots, d_n)$. Moreover, s_i sends the message that includes the encrypted data $(d_j)_{k_i}$ and the encoded data $E(d_1, d_2, \ldots, d_n)$ to its cluster head. The cluster head transmits the message to the sink. When the user wants to perform query $\{[a, b]\}$, the sink encodes the range $[a, b]$ as $G([a, b])$. Then, the sink applies a secret comparing method $C(E(d_1, d_2, \ldots, d_n), G([a, b]))$ to be used for query processing over encrypted and encoded data. A data d is in range $[a, b]$ if and only if $C(E(d_1, d_2, \ldots, d_n), G([a, b]))$ is true. Then, the sink decides whether $(d_j)_{k_i}$ should be included in the query result. Meanwhile, given $E(d_j)$ and $(d_j)_{k_i}$, it is infeasible for the sink to compute $d_j\,(1 \le j \le n)$. This condition can guarantee query privacy. Figure 1 illustrates the basic idea of cluster-based query processing scheme.

4.2. Prefix Membership Verification. We protect privacy query processing by using the prefix membership verification scheme which is first introduced in [8] and later formalized in [9]. In the prefix membership verification scheme, the key idea is to convert the verification of whether a number is in a range to several verifications of whether two numbers are equal. A $k-$prefix is in the form of $\{0, 1\}^k (*)^{w-k}$, which has k leading 0s and 1s, followed by $w-k$∗s. For instance, 101∗ is a 3-prefix and it denotes the range [1010, 1011].

A prefix family consists of w bits binary number $b_1 b_2 \cdots b_w$, which is defined as the set of $w+1$ prefixes $\{b_1 b_2 \cdots b_w, b_1 b_2 \cdots b_{w-1} *, \cdots, b_1 * \cdots *, ** \cdots *\}$, where the ith prefix is $b_1 b_2 \cdots b_{w-i+1} * \cdots *$. The prefix family of x is denoted as $F(x)$. For example, the prefix family of number 11 is $F(11) = F(1011) = \{1011, 101*, 10**, 1***, ****\}$. In prefix membership verification scheme, for any number x and prefix P, $x \in P$ if and only if $P \in F(x)$.

In order to confirm whether a number x is in a range $[d_1, d_2]$, the range $[d_1, d_2]$ can be translated into a minimum set of prefixes, denoted as $S([d_1, d_2])$, the union of which is equal to $[d_1, d_2]$. Each prefix is a subrange of $[d_1, d_2]$, which follows the binary prefix format. For $[d_1, d_2]$, the number of prefixes in $S([d_1, d_2])$ is at most $2w-2$ [18], where d_1 and d_2 are two numbers of w bits. For example, $S([9, 15]) = \{1001, 101*, 11**\}$. We compute the prefix family $F(x)$ of number x and translate the range $[d_1, d_2]$ into a minimum set of prefixes $S([d_1, d_2])$. So, $x \in [d_1, d_2]$ if and only if $F(x) \cap S([d_1, d_2]) \neq \phi$.

In order to ensure whether$F(x) \cap S([d_1, d_2]) \neq \phi$, we use the operations of verifying whether two numbers are equal. Then, we convert each prefix to a corresponding unique number using the prefix numericalization scheme defined in [19]. A prefix numericalization function Nneeds to satisfy the following properties: (1) for any p, $N(p)$ is a binary string; (2) for any two prefixes p and q, $p = q$ if and only if $N(p) = N(q)$.Given a prefix $b_1 b_2 \cdots b_k * \cdots *$ of w bits, we insert 1 after b_k, then every ∗ is replaced by 0. Given a set of prefixes S, the resulting set of numericalized prefixes is denoted as $N(S)$. For example, $N(F(11)) = \{10111, 10110, 10100, 11000, 10000\}$ and $N(S([9, 15])) = \{10011, 10110, 11100\}$. Therefore, $x \in [d_1, d_2]$ if and only if $N(F(x)) \cap N(S([d_1, d_2])) \neq \phi$. For instance, $N(F(11)) \cap N(S([9, 15])) = 10110$, the number 11 is in the range [9, 15]. Figure 2 shows the process of $11 \in [9, 15]$.

4.3. Data Collection. In order to preserve sensitive data, sensor nodes send the sensed data to cluster heads and sink by a secure way. We assume that b_1 and b_2, respectively, denote the lower bound and the upper bound, the values of which are known to both sensors and the sink. And we assume that sensor s_i collects data item $d_j\,(1 \le j \le n)$ at a time slot t, and each data d_j is in the range $[b_1, b_2]$. When each sensor node s_i collects data, $s_i$sends the sensitive data by the following steps:

(1) Sort the n data, b_1, and b_2 in an ascending order. We assume $b_1 < d_1 < d_2 < \cdots < d_n < b_2$.

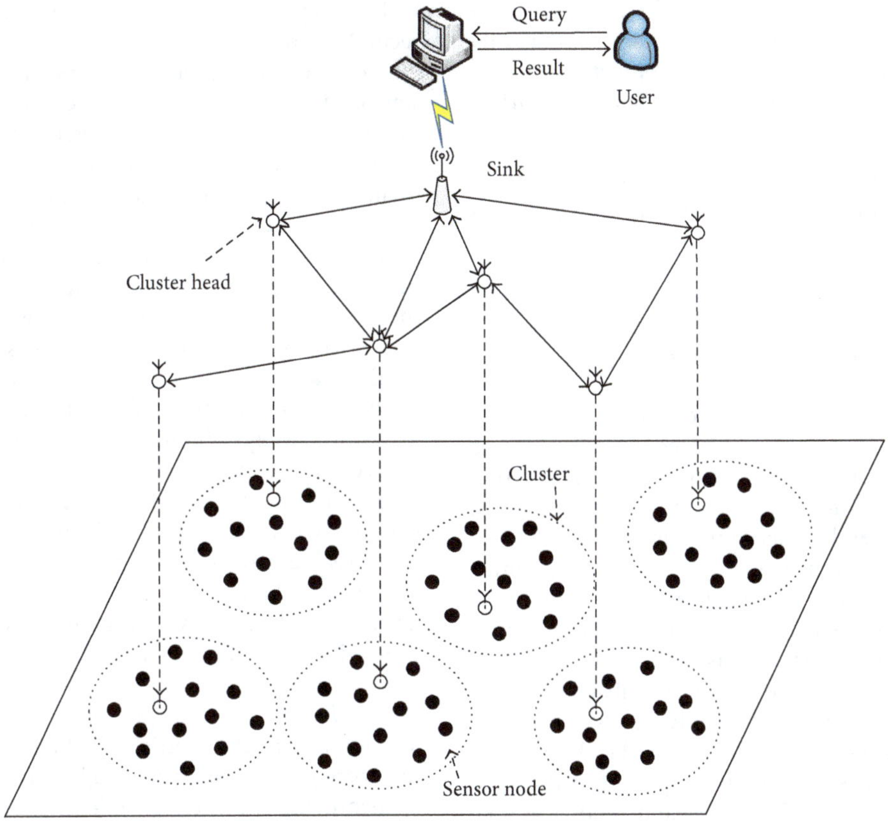

FIGURE 1: The cluster-based query processing scheme.

(2) Convert the $n+1$ ranges $[b_1, d_1]$, $[d_1, d_2], \ldots, [d_n, b_2]$ to the corresponding prefix string, that is, compute $S([b_1, d_1]), S([d_1, d_2])), \ldots, S([d_n, b_2])$.

(3) Numericalize all prefixes. For example, compute $N(S([b_1, d_1])), N(S([d_1, d_2])), \ldots, N(S([d_n, b_2]))$.

(4) Compute the keyed-hash message authentication code (HMAC) [6, 20] of each data item in numericalize prefixes using key g, which is shared by all nodes and the sink. An HMAC function using key g is denoted as HMAC_g. Compute $\mathrm{HMAC}_g(N(S([b_1, d_1])))$, $\mathrm{HMAC}_g(N(S([d_1, d_2]))), \ldots, \mathrm{HMAC}_g(N(S([d_n, b_2])))$.

(5) Encrypt every data item d_i to $(d_i)_{k_i}$ using key k_i.

(6) Sensor s_i sends the following packet to its cluster head (CH): $s_i \longrightarrow \mathrm{CH}: \langle \mathrm{id}, t, (d_i)_{k_i}, \mathrm{HMAC}_g(N(S(b_1, d_1))), \mathrm{HMAC}_g(N(S(d_1, d_2))), \ldots, \mathrm{HMAC}_g(N(S(d_n, b_2))) \rangle$.

Because the HMAC function has the one-wayness and collision resistance properties, and data items are encrypted, the cluster head cannot obtain the real values of all data items.

4.4. Filter and Query Processing. When a cluster head receives collected packets from cluster members in a cluster, the cluster head will filter the packet by sensors' id. In the query phase, the cluster head randomly selects several cluster members which include the real queried node s_i and gain the sensed data from them. It is aimed for preventing the adversary from monitoring the real frequency of query in a cluster. Therefore, in the submission phase, the cluster head needs to filter out the useless packets and obtain the real packet. Then, the cluster head transmits the real packet to the sink.

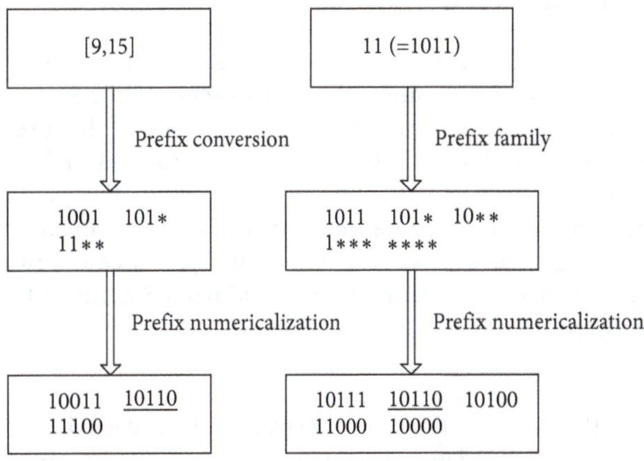

FIGURE 2: Prefix membership verification.

In the sink, it firstly converts the query range $[a, b]$ and computes prefix families $F(a)$ and $F(b)$. After the sink numericalize all prefixes as $N(F(a))$ and $N(F(b))$, it applies HMAC_g to each numericalized prefix as $\mathrm{HMAC}_g(N(F(a)))$ and $\mathrm{HMAC}_g(N(F(b)))$. When the sink receives a packet from cluster heads, it will process the packet based on the query range $[a, b]$ using the following theorem [21].

Theorem 1. *Given n numbers sorted in the ascending order* $d_1 < d_2 < \cdots < d_n$, *where* $d_j \in [b_1, b_2]\,(1 \le j \le n)$, *and a range* $[a, b]\,(b_1 < a \le b < b_2)$, $d_j \in [a, b]$ *if and only if there exist* $1 \le n_1 \le j < n_2 \le n + 1$ *such that the following two conditions hold:*

$$(1)\ \mathrm{HMAC}_g(N(F(a))) \cap \mathrm{HMAC}_g(N(S([d_{n_1-1}, d_{n_1}]))) \neq \phi$$
$$(2)\ \mathrm{HMAC}_g(N(F(a))) \cap \mathrm{HMAC}_g(N(S([d_{n_2-1}, d_{n_2}]))) \neq \phi$$

According to Theorem 1, the sink selects the smallest n_1 and the largest n_2 $(1 \le n_1, n_2 \le n + 1)$ such that $a \in [d_{n_1-1}, d_{n_1}]$ and $b \in [d_{n_2-1}, d_{n_2}]$. If $n_1 < n_2$, the data items d_{n_1}, $d_{n_1+1}, \ldots, d_{n_2-1}$ are in the range $[a, b]$; if $n_1 = n_2$, no data item is in the range $[a, b]$.

Based on the aforementioned description, Algorithm 1 shows a secure and efficient cluster-based query processing scheme. When a user wants to gain and check whether sensed data of certain node is in the range $[a, b]$ at a time slot t, the user will send a query message to the sink. And then the sink relays the message to a cluster head which include the queried node. In the cluster head, it randomly chooses several nodes which include the queried node. In the collection and submission phase, when the node collects the sensed data, it will process the data using the PMV and HMAC schemes. Then, the node sends the secure packet to the cluster head. After the cluster head filters out the useless packets, it sends the useful packet to the sink. Finally, the sink processes the packet and sends the final result to the user.

5. Performance Analysis

In order to protect privacy, we propose a secure and efficient query processing scheme to prevent an adversary from obtaining the sensitive data or finding the user's interests and location of sensor node in cluster-based sensor networks. In this section, we present the privacy analysis and communication overhead analysis. From the following analysis, we can see that our scheme brings a better network security and minimal communication overhead.

5.1. Privacy Analysis. For privacy of collected data, according to the data collection phase, sensor nodes convert the collected data by using encryption and HMAC scheme. So, the submitted information is not plaintext but encrypted and HMAC data. The HMAC function has one-wayness and collision resistance properties. And sensor nodes only share the secret key with the sink and encrypt sensitive data by the key. Therefore, it is computationally infeasible for cluster heads to obtain the value of d_i. It is difficult for the cluster head to break the privacy and gain the encryption and HMAC data. So, our scheme can efficiently protect collected data items.

For privacy of the query result, the sink obtains the query result by comparing the HMAC data items. For the HMAC data items and encrypted data, it is difficult for the adversary for computing and obtaining the values of the query result without keys. So, we can preserve the query result which is securely transmitted to the user.

For privacy of user's interests and location privacy of sensors in clusters, our scheme can efficiently preserve privacy information to prevent an adversary from monitoring user's interests and find the location information of sensors. And the adversary cannot use the content to trace the routing. We assume that an adversary monitors a local area with the intention of finding the interests of the user. We assume that each cluster has M members. The adversary wants to identify a set $D_T \subset M$ of nodes which represent the set of possible location in the local area. There is a close relationship between the analysis of query frequency of the adversary and the location privacy. When the adversary analyzes the query frequency uncertainly, it is secure to preserve the location information. In the eavesdropping area, the adversary will need to select the nodes of his analysis. We assume that the possible sensor nodes in D_{T} include queried nodes which send data to cluster head. If the size of D_{T} is very large, the adversary will find it difficult to analyze the user's interests. So, it is useful for preserving privacy information.

Let D_{P} be the set of the protected nodes. We use information-theoretic metric, called entropy [22], to measure the privacy protection provided by our scheme. The entropy of identifying the queried node in the wireless sensor network is defined as

$$c = -\sum_{i=1}^{D_{\mathrm{T}}} P_i \cdot \log_2(P_i), \tag{1}$$

where P_i is the probability that node i is the queried node, $|D_{\mathrm{T}}|$ is the number of uncertain nodes by the adversary, and $\sum_{i=1}^{|D_{\mathrm{T}}|} P_i = 1$. Therefore, the probability P_i of any sensor nodes in D_{T} being queried nodes can be estimated by $|D_{\mathrm{P}}|/|D_{\mathrm{T}}|$. Then, we denote the size of D_{T} as $R\,(|D_{\mathrm{T}}| = R)$. And let r be the size of the protected queried node's set $(|D_{\mathrm{P}}| = r)$. And we define the privacy as

$$\begin{aligned} c &= -\sum_{i=1}^{|D_{\mathrm{T}}|} \frac{|D_{\mathrm{P}}|}{|D_{\mathrm{T}}|} \cdot \log_2\left(\frac{|D_{\mathrm{P}}|}{|D_{\mathrm{T}}|}\right) \\ &= -\sum_{i=1}^{R} \frac{r}{R} \cdot \log_2\left(\frac{r}{R}\right) \\ &= r \cdot \log_2\left(\frac{R}{r}\right). \end{aligned} \tag{2}$$

The entropy represents the adversary's uncertainty about the user's interests and the location of sensors in a wireless sensor network. When the adversary believes that the nodes have the same probability to be the queried node in a cluster, the entropy is maximum value. Let D_{T}^* be the set of all nodes in a cluster and $|D_{\mathrm{T}}^*| = M$. We know that the size of D_{T} can influence the level of privacy. Therefore, the entropy is

```
initial_cluster;
sink_node;
query_node_id;
CH = cluster_head;
query_range = [a,b];
query_msg = QueryMsg(query_node_id, sink_node, query_range);
if sink receives the query_msg then
  SendQueryMsgToCH(CH, query_node_id);
  if CH receives the query_msg then
    selected_nodes = SelectNodesRandom(query_node_id);
    for node in selected_nodes then
      d_i = CollectData();
      Sort d_i, b_1 and b_2 as b_1 < d_1 < d_2 < ... < d_n < b_2
      Compute prefixes and numericalize all prefixes as
        N(S([b_1, d_1])), N(S([d_1, d_2])), ..., N(S([d_n, b_2]))
        Compute the HMAC as HMAC_g(N(S([b_1, d_1]))), HMAC_g(N(S([d_1, d_2]))),
          ..., HMAC_g(N(S([d_n, b_2])))
      Encrypt d_i by k_i as (d_i)_{k_i};
      Send packet to CH s_i ⟶ CH : ⟨id, t, (d_i)_{k_i}, HMAC_g(N(S(b_1, d_1))), HMAC_g(N(S(d_1, d_2))), ..., HMAC_g(N(S(d_n, b_2)))⟩
    end for
    packet = FilterPackets();
    SendToSink(packet);
  end if
  Compute and encode [a,b] as HMAC_g(N(F(a))), HMAC_g(N(F(b)));
  GetQueryData(packet, HMAC_g(N(F(a))), HMAC_g(N(F(b))));
end if
```

ALGORITHM 1: Secure and efficient cluster-based query processing.

$$
\begin{aligned}
c &= -\sum_{i=1}^{|D_{\mathrm{T}}|} \frac{|D_{\mathrm{P}}|}{|D_{\mathrm{T}}|} \cdot \log_2\left(\frac{|D_{\mathrm{P}}|}{|D_{\mathrm{T}}|}\right) \\
&= |D_{\mathrm{P}}| \cdot \log_2\left(\frac{|D_{\mathrm{T}}|}{|D_{\mathrm{P}}|}\right) \\
&\le |D_{\mathrm{P}}| \cdot \log_2\left(\frac{|D_{\mathrm{T}}^{*}|}{|D_{\mathrm{P}}|}\right) \\
&= r \cdot \log_2\left(\frac{M}{r}\right).
\end{aligned}
\tag{3}
$$

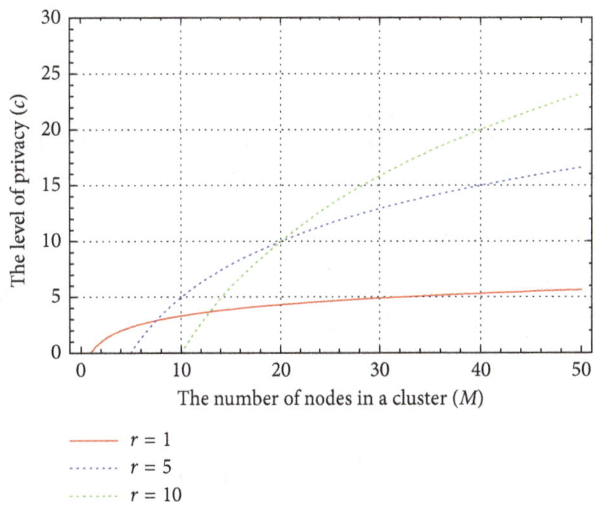

FIGURE 3: The different number of nodes in a cluster.

Figure 3 shows the relationship between the level of privacy and the different number of nodes in a cluster. When M increases, the level of privacy (c) is higher. This is due to the increased number of nodes in a cluster. The probability that the adversary finds the protected queried nodes is decreased. So the level of privacy (c) increases. For the same M, when M is greater than 20, we can see that the number of protected queried nodes (r) increases, the level of privacy (c) increases.

5.2. Energy Consumption Analysis. In cluster-based sensor networks, sensor nodes have limited energy resource. In this section, we discuss the energy consumption of sensor nodes in our scheme. In each phase, the total energy consumption includes the communication cost and computation energy consumption. We assume that the energy consumed by transmitting and receiving a data are e_{t} and e_{r}. And we assume that a cluster head randomly chooses N_{r} nodes to query.

In the data collection phase, sensor nodes will have extra computation overhead to preserve privacy of sensitive data. Given a range $[d_1, d_2]$, where d_1 and d_2 are two numbers of w bits, the number of prefixes in $S([d_1, d_2])$ is at most $2w - 2$ [18]. So a sensor computes at

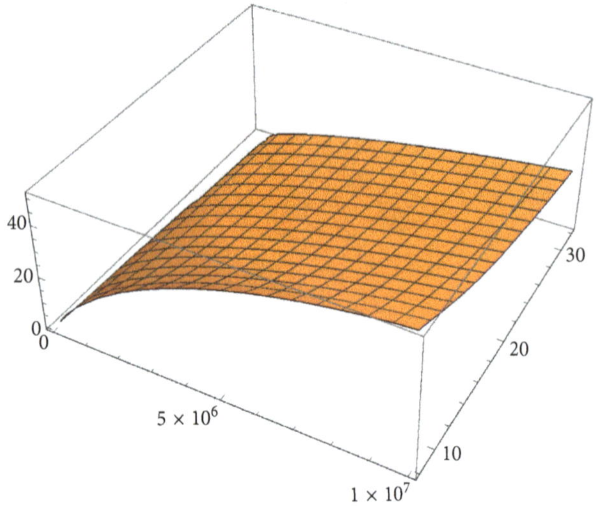

FIGURE 4: The relationship between energy consumption and the level of privacy.

most $(n+1)(2w-2)$ HMAC data. When a sensor node sends data to cluster head, it generates extra communication cost by sending encrypted data and HMAC data. We assume that each HMAC data is z_{H} bits and encrypted data is z_{D} bits. Let H_{hop} be the hop between a sensor node and a cluster head. In our scheme, $H_{\mathrm{hop}}=1$. In a cluster, the energy consumption E_{dc} is

$$\begin{aligned} E_{\mathrm{dc}} &= \sum_{i=1}^{N_r} \left[(n+1)(2w-2)\cdot z_{\mathrm{H}} + z_{\mathrm{D}} + e_{\mathrm{t}} + e_{\mathrm{r}}\right]\cdot H_{\mathrm{hop}} \\ &= N_r \cdot \left[(n+1)(2w-2)\cdot z_{\mathrm{H}} + z_{\mathrm{D}} + e_{\mathrm{t}} + e_{\mathrm{r}}\right]. \end{aligned} \tag{4}$$

In the query processing phase, the sink node computes the range $[a,b]$ and converts prefix families $F(a)$ and $F(b)$. For each value in the w bits, there are $w+1$ HMAC data items. So, the sink can perform at most $2(n+1)(2w-2)(w+1)$ comparisons. The energy consumption E_{qp} is

$$\begin{aligned} E_{\mathrm{qp}} &= \left[2(n+1)(2w-2)(w+1) + e_{\mathrm{t}} + e_{\mathrm{r}}\right]\cdot H_{\mathrm{hop}} \\ &= 2(n+1)(2w-2)(w+1) + e_{\mathrm{t}} + e_{\mathrm{r}}. \end{aligned} \tag{5}$$

Therefore, the total energy consumption E_{total} is

$$\begin{aligned} E_{\mathrm{total}} &= E_{\mathrm{dc}} + E_{\mathrm{qp}} \\ &= N_r \cdot \left[(n+1)(2w-2)\cdot z_{\mathrm{H}} + z_{\mathrm{D}} + e_{\mathrm{t}} + e_{\mathrm{r}}\right] \\ &\quad + \left[2(n+1)(2w-2)(w+1) + e_{\mathrm{t}} + e_{\mathrm{r}}\right]. \end{aligned} \tag{6}$$

According to (3) and (6), Figure 4 shows the total energy expended in the systems as the prefix number bits increase from 8 bits to 32 bits and the energy expended in the level of privacy increases from 1 to 45, for the scenario where each HMAC data and encrypted data are 256 bits. We assume that each sensor collects 100 data items at each time slot. And we assume that the energy consumed by transmitting and receiving a data are 1. Each cluster head includes 100 sensor nodes. This shows that when the prefix number bits is the same, the higher level of privacy can increase the energy of the whole sensor network.

6. Conclusions

Wireless sensor networks have been widely deployed in many applications and drawn more attentions. It is an important problem to preserve the privacy of sensitive data in cluster-based query processing in wireless sensor networks. In this paper, we propose a secure and efficient scheme to protect query processing in cluster-based sensor networks. In order to preserve privacy, sensed data items are encrypted to prevent cluster heads from obtaining the content of data. We use the prefix membership verification method to query the result without plaintext data. Meanwhile, we use anonymity method to confuse adversaries and prevent adversaries from analyzing the user's interests and finding location of the queried node. Then, we perform the privacy analysis and energy consumption analysis.

Conflicts of Interest

The authors declare that there are no conflicts of interest regarding the publication of this paper and confirm that the mentioned received funding in the acknowledgments did not lead to any conflicts of interest regarding the publication of this manuscript.

Acknowledgments

This work was supported by NSFC (Grant no. 61402015), the Science and Technology Development Plan Project of Henan Province (Grant no. 172102210189), and the Research Fund Project of Henan University (Grant no. 2016YBZR019).

References

[1] D. Gong, Y. Yang, and Z. Pan, "Energy-efficient clustering in lossy wireless sensor networks," *Journal of Parallel and Distributed Computing*, vol. 73, no. 9, pp. 1323–1336, 2013.

[2] B. Carbunar, Y. Yu, L. Shi et al., "Query privacy in wireless sensor networks," in *Proceedings of 4th Annual IEEE Communications Society Conference on Sensor Mesh and Ad Hoc Communications and Networks*, pp. 203–212, San Diego, CA, USA, 2007.

[3] H. Dai, Q. Ye, G. Yang et al., "CSRQ: communication-efficient secure range queries in two-tiered sensor networks," *Sensors*, vol. 16, no. 2, pp. 1–17, 2016.

[4] Y. T. Tsou, C. S. Lu, and S. Y. Kuo, "SER: Secure and efficient retrieval for anonymous range query in wireless sensor networks," *Computer Communications*, vol. 108, pp. 1–16, 2017.

[5] H. Dai, Q. Ye, X. Yi et al., "VP(2)RQ: efficient verifiable privacy-preserving range query processing in two-tiered wireless sensor networks," *International Journal of Distributed Sensor Networks*, vol. 12, no. 11, pp. 1–15, 2016.

[6] F. Chen and A. X. Liu, "SafeQ: secure and efficient query processing in sensor networks," in *Proceedings of the IEEE International Conference on Computer Communications*, pp. 1–9, Piscataway, NJ, USA, 2010.

[7] R. Li, A. X. Liu, S. Xiao et al., "Privacy and integrity preserving top-k query processing for two-tiered sensor networks," *IEEE/ACM Transactions on Networking*, vol. 25, no. 4, pp. 2334–2346, 2017.
[8] H. Wu and L. Wang, "Efficient and secure top-k query processing on hybrid sensed data," *Mobile Information Systems*, vol. 2016, Article ID 1685054, 10 pages, 2016.
[9] J. Cheng, H. Yang, S. H. Wong, and S. Lu, "Design and implementation of cross-domain cooperative firewall," in *Proceedings of the 15th IEEE International Conference on Network Protocols (ICNP'07)*, pp. 284–293, Piscataway, NJ, USA, October 2007.
[10] A. X. Liu and F. Chen, "Collaborative enforcement of firewall policies in virtual private networks," in *Proceedings of the 27th ACM SIGACT-SIGOPS Symposium on Principles of Distributed Computing*, pp. 95–104, New York, NY, USA, August 2008.
[11] X. Zhang, L. Dong, H. Peng et al., "Collusion-aware privacy-preserving range query in tiered wireless sensor networks," *Sensors*, vol. 14, no. 12, pp. 23905–23932, 2014.
[12] J. Shi, R. Zhang, and Y. Zhang, "Secure range queries in tiered sensor networks," in *Proceedings of the IEEE 28th Conference on Computer Communications (INFOCOM '09)*, pp. 945–953, Piscataway, NJ, USA, April 2009.
[13] J. Shi, R. Zhang, and Y. Zhang, "A spatiotemporal approach for secure range queries in tiered sensor networks," *IEEE Transactions on Wireless Communications*, vol. 10, no. 1, pp. 264–273, 2011.
[14] H. Dai, G. Yang, and X. Qin, "EMQP: an energy-efficient privacy-preserving max/min query processing in tiered wireless sensor networks," *International Journal of Distributed Sensor Networks*, vol. 2013, article 814892, 2013.
[15] Y. Yao, N. Xiong, J. H. Park, L. Ma, and J. Liu, "Privacy-preserving max/min query in two-tiered wireless sensor networks," *Computers and Mathematics with Applications*, vol. 65, no. 9, pp. 1318–1325, 2012.
[16] H. Dai, T. Wei, Y. Huang et al., "Random secure comparator selection based privacy-preserving max/min query processing in two-tiered sensor networks," *Journal of Sensors*, vol. 2016, Article ID 6301404, 13 pages, 2016.
[17] H. Dai, M. Wang, X. Yi et al., "Secure max/min queries in two-tiered wireless sensor networks," *IEEE Access*, vol. 5, pp. 14478–14489, 2017.
[18] P. Gupta and N. McKeown, "Algorithms for packet classification," *IEEE Network*, vol. 15, no. 2, pp. 24–32, 2001.
[19] Y. K. Chang, "Fast binary and multiway prefix searches for packet forwarding," *Computer Networks*, vol. 51, no. 3, pp. 588–605, 2007.
[20] H. Krawczyk, R. Canetti, and M. Bellare, "HMAC: keyed-hashing for message authentication," Tech. Rep. RFC 2104, Internet Society, Reston, VA, USA, 1997.
[21] F. Chen and A. X. Liu, "Privacy and integrity preserving range queries in sensor networks," Tech. Rep. MSU-CSE-09-26, Michigan State University, East Lansing, MI, USA, 2009.
[22] M. M. E. A. Mahmoud and X. Shen, "A cloud-based scheme for protecting source-location privacy against hotspot-locating attack in wireless sensor networks," *IEEE Transactions on Parallel and Distributed Systems*, vol. 23, no. 10, pp. 1805–1818, 2012.

21

Assessing Availability in Wireless Visual Sensor Networks based on Targets' Perimeters Coverage

Daniel G. Costa[1] **and Cristian Duran-Faundez**[2]

[1]*State University of Feira de Santana, Feira de Santana, BA, Brazil*
[2]*University of the Bío-Bío, Concepción, Chile*

Correspondence should be addressed to Daniel G. Costa; danielgcosta@uefs.br

Academic Editor: Stefano Basagni

Availability in wireless visual sensor networks is a major design issue that is directly related to applications monitoring quality. For targets monitoring, visual sensors may be deployed to cover most or all of targets, and monitoring quality may be focused on how well a set of targets are being covered. However, targets may have different dimensions and it is expected that large targets may be only partially viewed by source nodes, which may affect coverage quality and lead to a state of unavailability. In this context, this article analyzes the effect of target's size on effective coverage in wireless visual sensor networks. A new coverage metric, the Effective Target Viewing (ETV), is proposed to measure monitoring quality over a set of targets, which is exploited as a fundamental parameter for availability assessment. Results show that ETV can be used as a practical coverage metric when assessing availability in wireless visual sensor networks.

1. Introduction

An increasing demand for autonomous surveillance and control applications has fostered the development of new monitoring technologies, which has placed sensor networks into a central position. A lot of sensing applications in military, industrial, residential, health care, and smart cities scenarios may be designed exploiting the flexibility of sensor networks [1]. For those networks, when sensor nodes are equipped with a low-power camera, visual information can be retrieved from the monitored field [2, 3], opening new opportunities for monitoring in Internet of Things scenarios. In general, image snapshots, infrared images, and video streams with different coding qualities and resolutions can provide valuable information for an uncountable number of monitoring applications.

In general, visual sensors have a viewing orientation and thus a directional sensing model can be defined. In a different way of scalar sensors, designed to retrieve scalar data such as temperature, pressure, and humidity, visual sensors may view distant or close objects or scenes according to their Field of View (FoV) [4, 5]. For targets monitoring, satisfactory sensing coverage would happen when one or more targets are being viewed by deployed sensors, which means that they are partially or completely inside the area defined by the sensors' FoV.

Actually, targets may have different dimensions, potentially impacting target monitoring quality. While small targets may be sometimes more likely to be viewed, large targets may not be satisfactorily covered by deployed visual sensors. In fact, when covering a set of targets, it is usually required that every target is being viewed by at least one visual sensor, but there may be parts of targets that may not be viewed. For some applications, targets have to be viewed in all possible perspectives and monitoring quality should be accounted for all covered perspectives. As an example, visual sensors may view the front or back side of a target, providing different information for monitoring applications. For another group of applications, however, viewing perspectives may not be an issue, since enough parts of the targets are being viewed.

A system can be assumed as available when the expected services can be provided when requested. While some network environments can tolerate some states of unavailability, critical monitoring applications may be severely impaired. Therefore, a central issue in Wireless Visual Sensor Networks (WVSN) is availability assessment, since we want to say if a particular application may be assumed as available along the time. Generally, availability will be affected by hardware and coverage failures, but different availability metrics concerned with different availability issues may be defined to support the overall process of availability assessment [6].

Frequently, visual sensors may be deployed on a region of interest with many fixed or moving targets, where source nodes may view more than one target at a time. In this scenario, it is worth estimating the coverage quality for different configurations of visual sensors, potentially supporting efficient design and deployment of visual sensor networks. Evaluating the effect of different targets parameters on visual sensing coverage may then be beneficial for WVSN. Particularly, assessing availability for monitoring of small or large targets may be of paramount importance, especially for critical applications, as in automatic traffic control, industrial automation, public security, and rescue operations, just to cite a few.

This article addresses the problem of availability assessment in wireless visual sensor networks. For that, a geometrical model is defined to compute target viewing by visual sensors, for any size of targets modelled as circumferences. Based on it, a new coverage metric is defined to compute the viewed perimeter of targets, which is referred to as the Effective Target Viewing (ETV). This metric indicates the average percentage of the viewed perimeter of all considered targets. Monitoring availability can then be assessed based on ETV, along with monitoring requirements of applications, directly indicating if an application may be assumed as available or not. To the best of our knowledge, the contributions of this article have not been proposed before.

The remainder of this article is organized as follows. Section 2 presents some related works. Section 3 brings the statements and definitions of targets coverage. The proposed coverage metric and availability assessment approach are defined in Section 4. Section 5 presents numerical results, followed by conclusions and references.

2. Related Works

For wireless visual sensor networks, monitoring applications may require that a minimum number of targets are being viewed. The monitoring quality may then be associated to a percentage of coverage, which might guide deployment [7] and coverage optimization algorithms [8, 9]. In a different perspective, target viewing may be related to network availability [6], exploiting visual sensing redundancy to compensate failures in sensor nodes. Actually, sensing redundancy in WVSN is not straightforward and there are some relevant issues that should be properly considered [6, 10], as the perception of redundancy depends on applications monitoring requirements [11]. Target viewing may also be maximized when adjustable visual sensors are deployed, and the monitoring quality will be a function of visual redundancy over targets [12]. For all these cases, target viewing may be performed in different ways and with different objectives in wireless visual sensor networks.

Efficient sensing coverage will be deeply related to the way sensors are deployed. In deterministic deployment, sensors are neatly placed to achieve optimized coverage and many works have been concerned with optimization of the number of sensors required to cover a monitored field [13, 14]. On the other hand, for many monitoring scenarios, sensors are expected to be randomly deployed, bringing particular coverage problems [3, 9]. In general, nodes placement optimization is a relevant problem for scalar and visual sensor networks [4, 15, 16].

In general, visual sensors will be deployed for area, target, or barrier coverage [17]. After random deployment, camera-enabled sensors may be scattered over a monitored field, with unpredicted positions and orientations. For such sensors, coverage metrics are desired when assessing the sensing quality of wireless sensor networks. The work in [18] proposes a metric to measure the coverage quality of wireless visual sensor networks, computing the probability of a randomly deployed network to be K-Coverage, where every point is covered by at least K sensors. For higher values of K, more visual sensors will be viewing the same area of a monitored field. In a different way, a metric is proposed in [19] to compute the coverage quality for target sensing. The impact of sensor deployment for visual sensing coverage is discussed in [7]. In [4], different issues for coverage estimation and enhancement are addressed.

When sensors may adjust the viewed area, sensing coverage may be optimized [20, 21]. The work in [22] computes an optimal configuration for visual sensors with changeable orientations, where visual coverage is based on the definition of nondisjoint cover sets. The work in [12] adjusts the sensors' FoV to optimize the network coverage, achieving maximized viewing of a monitored field: sensors are reconfigured to increase sensing redundancy over defined targets. Optimal coverage is a relevant problem that has driven many research efforts in wireless visual sensor networks, but visual monitoring availability is also concerned with other relevant issues in these networks.

A core element of availability is sensing redundancy. In general, sensing redundancy is based on overlapping of sensing areas, but the way such overlapping will be considered when defining redundancy will depend on monitoring requirements of applications [6, 10]. Actually, sensing redundancy may be exploited to extend the network lifetime, when redundant nodes are deactivated, but redundancy selection is still a challenging issue in wireless visual sensor networks. In [23], algorithms for redundancy selection in WVSN were proposed. In a similar way, the work in [24] also addressed redundancy selection for availability enhancement, but it considers the targets perspectives when defining if sensors that are viewing the same target can be assumed as redundant. Sensing redundancy is also exploited in [25] when assessing availability for target coverage.

Besides redundancy, availability may be also concerned with the way targets are being viewed. Sometimes, different

Table 1: Visual sensing coverage in wireless sensor networks.

Issue	Approach	Main research challenges
Coverage metric	[18, 19, 21, 30]	Perception of coverage quality; use of quantitative or qualitative measures.
Coverage adjustment	[9, 12, 14, 16, 31]	Computing of optimal orientations; minimum set of sensors; maximal coverage.
Sensing redundancy	[10, 11, 23–25]	Viewing perspectives; sensing failures; availability enhancement.
Target viewing	[24, 26–28]	Target coverage; targets' contours.

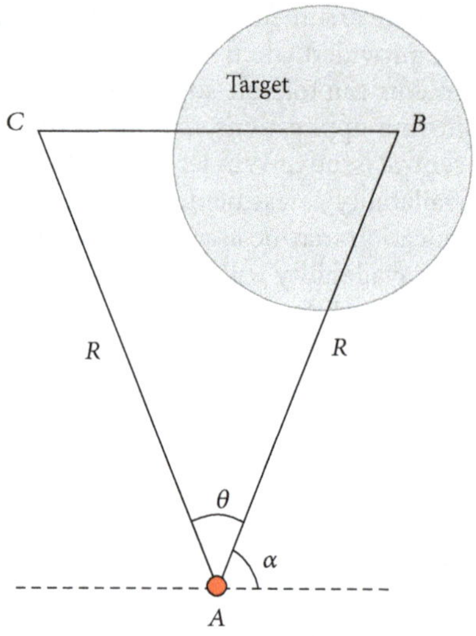

Figure 1: Field of View of a visual sensor.

parts of targets' contours may be different for applications. The work in [26] associates source priorities to cameras according to viewed parts of targets. In a different way, for large targets, it may be desired that the entire perimeter of targets is viewed by a set of cameras, as proposed in [27, 28]. In those works, scalar sensors (with circular sensing areas) are considered to cover targets, and the network is optimized to find the minimum number of sensors that cover the targets' perimeters. Table 1 summarizes the discussed papers, classifying them according to their contributions to visual coverage and availability enhancement and assessment.

Actually, previous works have addressed the problem of target coverage under different perspectives, for scalar and visual sensor networks. And some of them brought contributions for targets' perimeters coverage. However, availability assessment for target coverage is still an open issue, especially for monitoring of large targets, fostering the definition of new availability assessment metrics.

3. Targets' Perimeters Coverage

Visual sensors may be deployed for different tasks in a large set of monitoring and control applications. Such sensors may be expected to retrieve visual information of targets or scenes, with different particularities. For the case of target viewing, fundamental concepts have to be defined to allow proper modelling, as discussed in this section.

3.1. Sensors' Field of View. A typical wireless visual sensor network may be composed of scalar sensors, visual sensors, actuators, and sinks. For visual monitoring tasks, one must be concerned with visual sensors and the way they view a monitored field.

In general, it is expected that a WVSN will be composed of S visual sensors, which may be randomly or deterministically deployed over an area of interest. Each sensor s, $s = 1, \ldots, S$, has $(Ax_{(s)}, Ay_{(s)})$ location for 2D modelling. For randomly deployed sensors, their location after deployment may be discovered using some localization mechanism [22]. Whatever the case, it is assumed herein that sensors are static and their configurations do not change after deployment, but the proposed approach is also valid for dynamic networks.

Each visual sensor is expected to be equipped with a low-power camera, with a viewing angle θ and an orientation α. The embedded camera also defines a sensing radius R that is an approximation of the camera's Depth of Field (DoF) [3], which is the area between the nearest and farthest point that can be sharply sensed. For simplification, the Field of View of any visual sensor is defined as the area of an isosceles triangle composed of three vertices, A, B, and C. Vertex A is assumed as the visual sensor position [18], $(Ax_{(s)}, Ay_{(s)})$, while the other vertices are computed considering the values of θ, α, and R.

Figure 1 shows a graphical representation of a typical sensor's FoV.

One can compute the area of any sensor's FoV, as expressed in (1), whenever the sensing parameters of the camera are known.

$$\mathrm{FoV}_{(s)} = \frac{R_{(s)}^2 \cdot \sin\left(\theta_{(s)}\right)}{2}. \tag{1}$$

Basic formulations of trigonometry are used to compute vertices B and C for any sensor s, as expressed in

$$\begin{aligned} Bx_{(s)} &= Ax_{(s)} + R \cdot \cos\left(\alpha_{(s)}\right), \\ By_{(s)} &= Ay_{(s)} + R \cdot \sin\left(\alpha_{(s)}\right), \\ Cx_{(s)} &= Ax_{(s)} + R \cdot \cos\left(\left(\alpha_{(s)} + \theta_{(s)}\right) \mod 2\pi\right), \\ Cy_{(s)} &= Ay_{(s)} + R \cdot \sin\left(\left(\alpha_{(s)} + \theta_{(s)}\right) \mod 2\pi\right). \end{aligned} \tag{2}$$

3.2. Defining Targets. When wireless visual sensor networks are deployed for targets viewing, it is desired that the maximum number of targets will be visually covered by source sensors. In general, a target is any moving or static object that is expected to be viewed by visual sensors. Moreover, in real applications, targets may have different formats and sizes, but visual sensors may view just small parts of them.

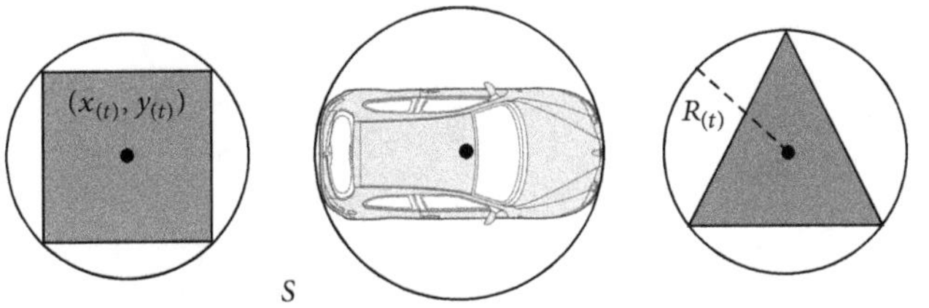

FIGURE 2: Example of targets.

A target t is defined as a generic element located at position $(x_{(t)}, y_{(t)})$, although 3D modeling may also be considered. For a total of T targets, a target t, $t = 1, \ldots, T$, has position $(x_{(t)}, y_{(t)})$ as its center and thus, for simplification, a target t is defined as a circumference with radius $R_{(t)}$ and center $(x_{(t)}, y_{(t)})$. The value $R_{(t)}$ is computed taking the greatest distance from the center of the target to its border, assuming a top-down view (observer above the monitored field). Figure 2 shows examples of generic representations of targets.

The camera's FoV will view only part of the defined circumference, which will result in a viewed perimeter lower than $\pi \cdot R_{(t)}$, which is half the perimeter defined by the circumference. Moreover, we do not consider occlusion of targets, but it could be assumed in 3D modelling.

3.3. Computing Targets Viewing. The FoV's triangle may intersect a target's circumference in different ways. The area viewed by a sensor s, defined as $V_{(s,t)}$, will be an arch of the target's circumference and thus it is defined by a pair of intersection points, $V1_{(s)} = (V1x_{(t,s)}, V1y_{(t,s)})$ and $V2_{(s)} = (V2x_{(t,s)}, V2y_{(t,s)})$. These points are computed according to the way the FoV intersects the circumference, as exemplified in Figure 3. Obviously, the basic condition for target viewing is that the Euclidean distance between the considered target's center and visual sensor position is lower than or equal to $(R_{(t)} + R_{(s)})$.

The points $V1$ and $V2$ can be computed considering the intersection of the lines defined by the vertices of FoV's triangle. More specifically, we want to compute the intersection of lines $\overline{AB}$ and $\overline{AC}$ in relation to the target's circumference. Actually, a generic line may have three different configurations concerning a circumference: it may not intersect, it may intersect in a single point (tangent line), or it may intersect in two points (secant line). Through geometry, the formulation in (3) can be considered when checking the way a line will intersect a circumference. Note that the formulation in (3) is valid for $\overline{AB}$, but $\overline{AC}$ could be considered just taking the coordinates of vertex C.

$$\begin{aligned} dx &= (Bx - Ax), \\ dy &= (By - Ay), \\ a &= \left(dx^2 + dy^2\right), \\ b &= 2 \times \left(dx \times (Ax - x_{(t)}) + dy \times (Ay - y_{(t)})\right), \\ c &= \left(Ax - x_{(t)}\right)^2 + \left(Ay - y_{(t)}\right)^2 - R_{(t)}^2, \\ \Delta &= b^2 - 4 \times a \times c, \end{aligned} \tag{3}$$

where the following conditions are found:

If $\Delta < 0$, there is no intersection.

If $\Delta = 0$, there is a tangent line.

If $\Delta > 0$, there is a secant line.

If both $\overline{AB}$ and $\overline{AC}$ are secant to a considered target's circumference, four intersection points will be computed but only the two closest to vertex A have to be taken. It is due to the fact that visual sensors are not expected to see through the targets in this work (opaque targets). If any of those two lines is tangent, the intersection vertex is the point of tangency. The formulation in (4) computes all possible vertices for tangent and secant lines; if $\overline{AB}$ (or $\overline{AC}$) is a secant line, two different values for x and y may be found, but only one value is computed for a tangent line.

$$\begin{aligned} x &= Ax + \frac{-b \pm \left(\sqrt{\Delta} \times dx\right)}{2 \times a}, \\ y &= Ay + \frac{-b \pm \left(\sqrt{\Delta} \times dy\right)}{2 \times a}. \end{aligned} \tag{4}$$

A special formulation has to be defined when $\overline{AB}$ or $\overline{AC}$ lines, or both, do not intersect the target's circumference, as depicted in Figures 3(a) and 3(d). In these cases, one or two projection lines are drawn from vertex A to line $\overline{BC}$ and these projections are tangent to the target's circumference. Actually, the tangent line is perpendicular to the radius of the target's circumference and thus a right triangle can be created, as presented in Figure 4.

There are two possibilities for the tangent line, in which length between vertex A of the considered visual sensor and the tangent point is defined as d. If d is greater than the height of the FoV's (isosceles) triangle, defined as h, the tangent point must not be considered as an intersection point. Otherwise, the tangent point is an intersection point to be considered when computing $V_{(s,t)}$. The value of d is the hypotenuse of the right triangle created also taking $R_{(t)}$ and the distance between vertex A of the visual sensor and the center of the target's circumference, defined as AT. And the value of h can be found through trigonometry when taking the other parameters of the FoV's triangle.

When $d \geq h$, we have to compute the intersection of line $\overline{BC}$ with the circumference and this can be done just adjusting (3) and (4). In this case, of course, all intersection points (one or two) must be considered. When $d \leq h$, two possible tangent points will be found. For that, we take the intersection points of target's circumference with a circumference centered at vertex A and with radius d.

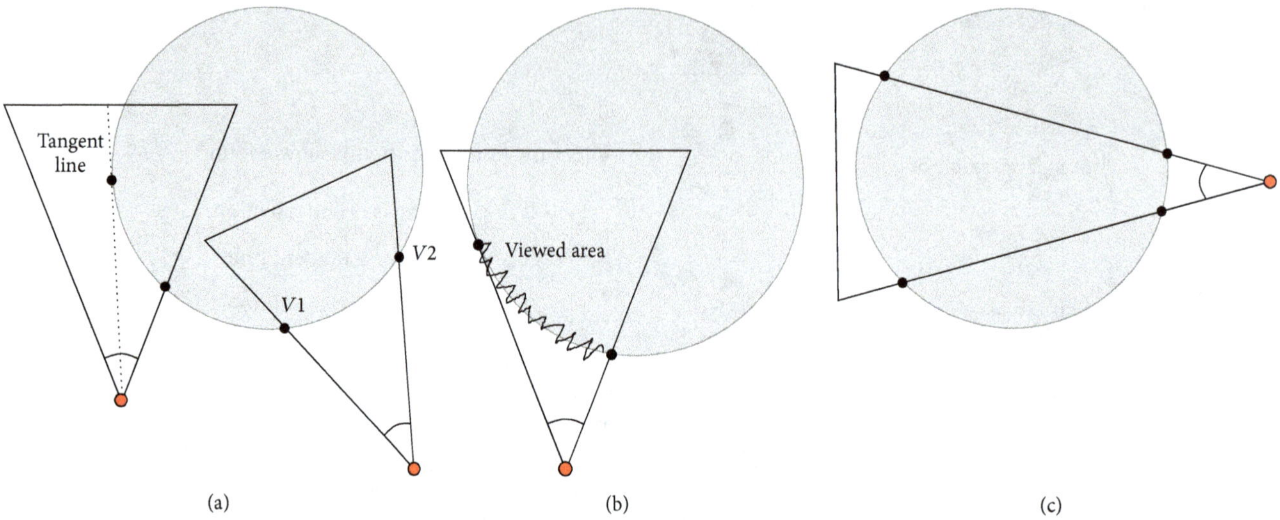

FIGURE 3: Targets coverage.

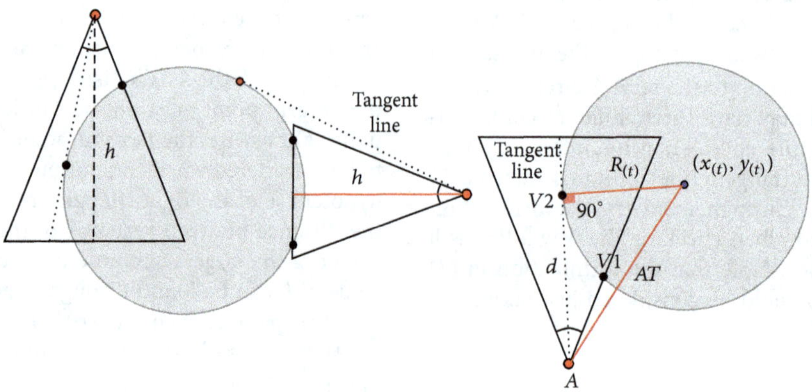

FIGURE 4: Computing tangent vertex.

Equation (5) can be used to compute those intersection points.

$$a = \frac{\left(R_t^2 - d^2 + AT^2\right)}{(2 \times AT)},$$

$$x = x_{(t)} + \frac{a \times (Ax - x_{(t)})}{AT} \pm \frac{\sqrt{R_t^2 - a^2} \times (Ay - y_{(t)})}{AT}, \quad (5)$$

$$y = y_{(t)} + \frac{a \times (Ax - x_{(t)})}{AT} \pm \frac{\sqrt{R_t^2 - a^2} \times (Ax - x_{(t)})}{AT}.$$

Two different points of tangency can be found when applying the formulation in (5). However, for two points $V1$ and $V2$, only one of them will be inside the FoV's triangle; that point will be an intersection point.

4. Proposed Availability Assessment

The availability level of monitoring applications will depend on visual and hardware characteristics of deployed sensors, as well as the network topology of the considered wireless visual sensor networks. Actually, visual monitoring applications will typically experience different levels of hardware failures and coverage failures [6]. While a hardware failure may result from energy depletion, sensors harming, connection problems, or faulty conditions [25, 29], among other factors, coverage failures happen when visual sensors cannot provide minimal acceptable information for applications functions. For example, if an application expects to view at least 70% of all targets' perimeters, it is only assumed as available whether this constraint is respected (indicating that no coverage failure happened). A practical coverage metric associated to targets viewing is then highly desired, since it can be exploited for availability assessment.

We propose the Effective Target Viewing (ETV), a metric of the coverage quality over a set of targets. ETV indicates the percentage of viewed parts of targets' perimeters. This metric is derived from the ETV(t), which indicates the percentage of the viewed perimeter of target t, while the ETV metric indicates the average values of ETV(t) for all targets t, $t = 1, \ldots, T$.

ETV is a coverage metric. However, it can be exploited to assess the availability of visual monitoring applications. In fact, ETV can be associated with an availability state, which may be "yes" (available) or "no" (unavailable). Actually, when assessing availability, monitoring applications will define the minimum acceptable ETV for the deployed visual sensors. We define M-ETV as the minimum acceptable value for the ETV of the network, while M-ETV(t) is the minimum acceptable ETV(t), for any considered target. For example, if M-ETV is 50%, it is the minimum acceptable average coverage of targets' perimeters. However, if we define M-ETV(t) as 50%, at least 50% of each target's perimeter must be viewed by visual sensors. As average results may hide the existence of targets that are not being satisfactorily viewed, M-ETV(t) may associate availability to uniform viewing over targets.

Actually, M-ETV and M-ETV(t) are parameters of applications, with no concern to deployed visual sensors and targets. In other words, as coverage failures depend on monitoring requirements [6], different applications may have different availability conditions even for the same network.

The ETV metric is computed taking the viewed parts of targets, assuming all nearby cameras. Actually, every visual sensor may view a percentage of any target's perimeter, depending on the considered parameters. It is defined that a visual sensor s may view a target t within angles interval, defined as $V_{(s,t)} = \{V1, V2\}$, which will be represented by a sector of the circumference with radius $R_{(t)}$. The viewed area is defined by the pair of intersection points, which can be used to compute an angular distance, as specified in (6). The formulation in (6) is defined by the fact that both points and the center of the circumference create an isosceles triangle with $R_{(t)}$ as one of the sides. The law of cosines is so employed to compute γ, which is the central angle of target t that determines the arc $V1V2$. This "view" will then be $(\gamma \times \pi \times R_{(t)}^2)/360$ of the considered target's circumference.

$$c^2 = a^2 + b^2 + 2ab\cos\gamma,$$

$$|V1 - V2|^2 = 2R_{(t)}^2 - 2R_{(t)}^2\cos\gamma,$$

$$\frac{2R_{(t)}^2 - |V1 - V2|^2}{2R_{(t)}^2} = \cos\gamma,$$

$$\gamma = \cos^{-1}\left(\frac{2R_{(t)}^2 - |V1 - V2|^2}{2R_{(t)}^2}\right), \quad (6)$$

$$\gamma = \cos^{-1}\left(\frac{2R_{(t)}^2 - \left((V2x - V1x)^2 + (V2y - V1y)\right)^2}{2R_{(t)}^2}\right).$$

A simple way to compute the viewed perimeters of all targets is to compute an average result for the sum of all viewed perimeters in each target. Obviously, it would compute the viewed areas assuming redundant views of the same target, which may be relevant when replacing faulty nodes [6, 10]. However, as we are computing the percentage of targets' perimeters being viewed, redundant coverage must not be accounted. In such way, the proposed ETV metric does not consider redundant views and thus its highest value for the view of any target is 100%. But if the angular distance of all values of $V_{(s,t)}$ is considered, redundant views on a target might be (erroneously) accounted, which would not correspond to the expected value of ETV. In order to avoid that problem, an algorithm was designed to avoid the accounting of redundant views, removing it from the viewed arches of the targets.

Let us define $I = [I(1), I(2), \ldots, I(2S)]$ as a vector containing all points $V1_{(s)}$ and $V2_{(s)}$, for all nodes s, sorted by their order of appearance in a counterclockwise or clockwise tour in the perimeter of the circumference defining the target t. Let us define $\Phi = [\Phi(1), \ldots, (2S)]$ as a vector containing the angle γ defined by $I(i)$, the center of t, and $I((i + 1) \mod$

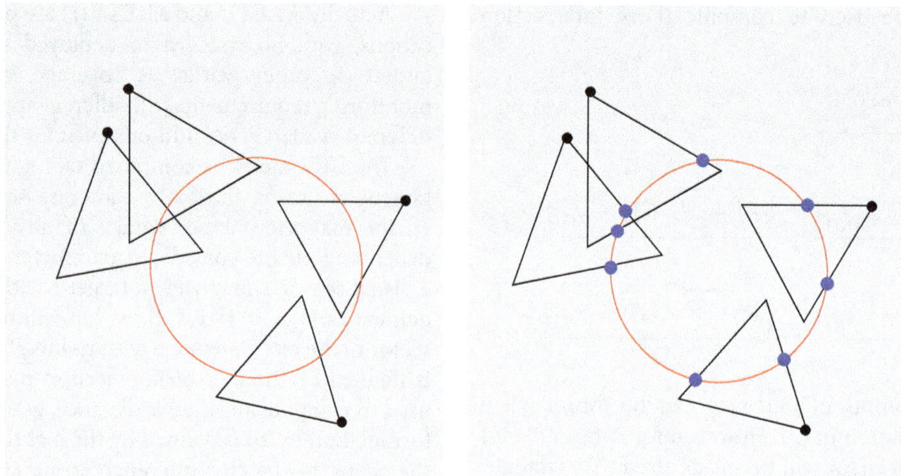

(a) (b)

FIGURE 5: Visual representation of intersection points. (a) Original target; (b) intersection points.

$2S$). We also define Υ as a numerical constant defining the maximum possible sum of the magnitudes, that is, 360°. Then, the contribution of the segment defined by a pair $(I(i), I((i+1) \mod 2S))$ to ETV is defined as presented in

$$C(\Phi, i) = \begin{cases} \Phi(i) & \text{if } s(i) = s((i+1) \mod 2S) \\ \Phi(i) & \text{if } (\widehat{s}(i) \cap \widehat{s}((i+1) \mod 2S)) \neq \emptyset \\ \Phi(i) & \text{if } \exists j; (\widehat{s}(i) \cap \widehat{s}(j) \neq \emptyset) \wedge (\widehat{s}((i+1) \mod 2S) \cap \widehat{s}(j) \neq \emptyset) \\ 0 & \text{otherwise.} \end{cases} \tag{7}$$

In (7), $s(i)$ is the visual sensor node associated to $I(i)$, and $\widehat{s}(i)$ is the arc defined by its two $V1_{(s)}$ and $V2_{(s)}$ points. The four conditions in (7) are used to decide whether the arcs confined by points in $I(i)$ and $I((i+1) \mod 2S)$ are parts of the area visualized by a sensor node. The first condition evaluates if both points were projected by the same sensor, which is possible if (1) there is an entire (nonoverlapped) area exclusively viewed by sensor $s(i)$ or (2) the target's area viewed by sensor $s(i)$ is also viewed by another sensor. The second condition evaluates if areas captured by sensors projecting points $I(i)$ and $I((i+1) \mod 2S)$ are overlapped. The third condition evaluates the case where target regions viewed by sensors $s(i)$ and $s((i+1) \mod 2S)$ are not overlapped between them but both of them are overlapped with the captured area of a common sensor $s(j)$. At last, the fourth condition will mark a nonviewed portion of the circumference.

Finally, ETV(t) and ETV can be computed as expressed in

$$\text{ETV}(t) = \frac{1}{\Upsilon} \times \sum_{i=1}^{2S} C(\Phi, i) \times 100\%,$$
$$\text{ETV} = \sum_{t=1}^{T} (\text{ETV}(t)) \times \frac{1}{T}. \tag{8}$$

A graphical example of computing the intersection points using the defined formulation is presented in Figure 5. The computed ETV(t) for this example is 47.75%.

Algorithm 1 computes ETV and ETV(t) for all targets, considering the equations previously presented. Most of computation is performed in line (6), using the proposed geometrical model.

5. Numerical Results

The proposed metrics can be used to assess availability in wireless visual sensor networks. We then defined some mathematical experiments for different parameters of visual sensors and targets, computing ETV and ETV(t). Using Matlab, Algorithm 1 was implemented, along with the defined mathematical formulations. Next subsection presents the numerical results when computing those metrics.

5.1. Computing ETV and ETV(t). Different configurations for visual sensors and targets were considered to compute ETV, assuming sensors randomly deployed and also sensors deterministically positioned in a grid-like topology. Initially, randomly visual sensors and targets were virtually positioned

```
Algorithm: ETV Computing(List[] S, List[] T)
Output: ETV, ETV(t)[]
ETV ← 0
for t ← 1; t < T.size(); t++ do
   target ← T.getVSensor (t)
   for s ← 1; s < S.size(); s++ do
      sensor ← S.getVSensor (s)
      V (target, sensor) = computePoints (target, sensor) //Angles interval
   ETV (target) ← functionC (V)
   ETV ← ETV + ETV (target)
ETV ← ETV/T.size()
```

ALGORITHM 1: ETV computing.

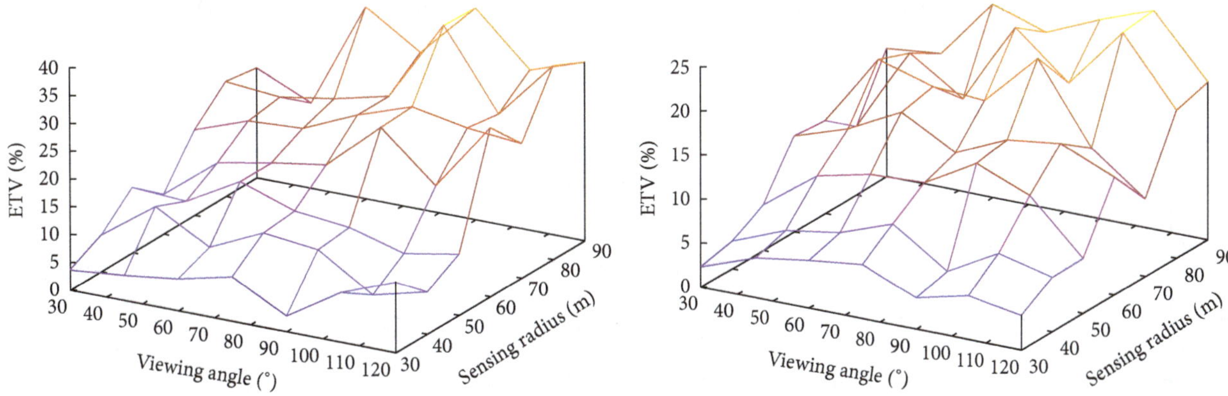

(a) (b)

FIGURE 6: ETV after random deployment: (a) 5 targets with $R_{(t)} = 20$; (b) 5 targets with $R_{(t)} = 50$.

and their parameters were considered in the defined mathematical equations. For this verification, visual sensors have $\theta_{(s)}$, $\alpha_{(s)}$, $R_{(s)}$, and $(Ax_{(s)}, Ay_{(s)})$ with random values, while targets have random values for $(x_{(t)}, y_{(t)})$ and $R_{(t)}$.

A 300 m × 600 m monitoring field is considered for computing the value of ETV for different network random configurations, as presented in Figure 6. Five different targets are randomly positioned in the monitored field for each test, taking two different fixed values for $R_{(t)}$ of all targets: 20 and 50. As random parameters are calculated, every verification is executed 10 times and only the average results are considered.

As random parameters are being considered, there is no uniform distribution for ETV in Figure 6. But, in general, ETV increases for higher sensing radius. However, as can be seen in Figure 6(b), large targets are harder to be completely viewed in average, which reduces the value of ETV.

Visual sensors were also considered in planned positions. For the next experiment, a 20 × 2-sensor network with 2 columns of sensors with 20 rows each was considered, simulating a more realistic network. In that scenario, targets are located between the columns, as it may happen when cars are being monitored on a road. Figure 7 presents a graphical example of how sensors and targets are considered for this evaluation phase, disregarding the effect of occlusion.

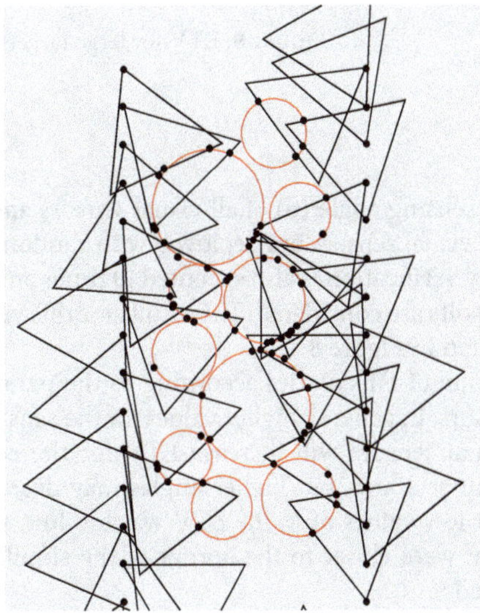

FIGURE 7: Example of sensors deployment. The ETV in this example is 51.52%.

ETV was computed when five targets are deployed in random positions (between the two columns of sensors) and with $R_{(t)} = 20$ and $R_{(t)} = 50$. We also considered different

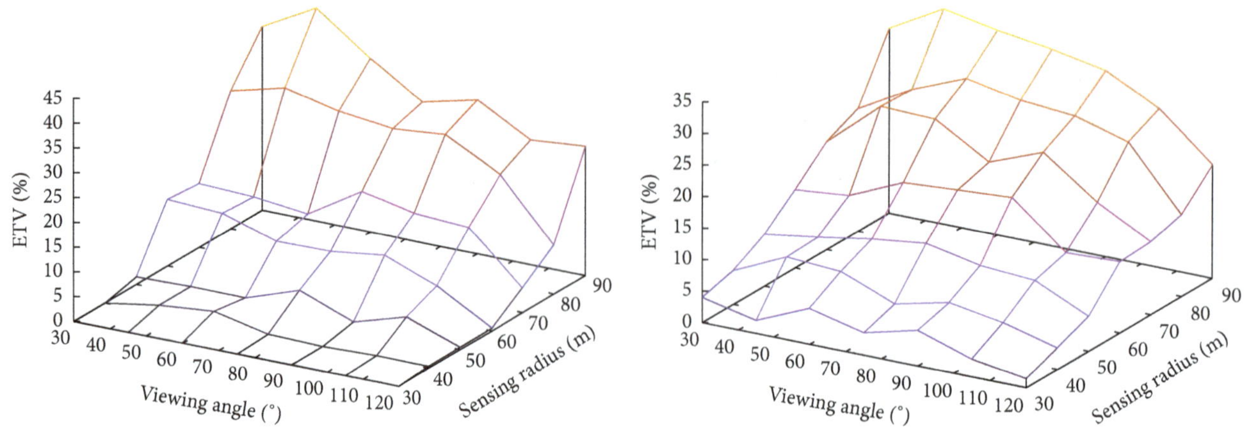

(a) (b)

FIGURE 8: ETV for a specific scenario: (a) 5 targets with $R_{(t)} = 20$; (b) 5 targets with $R_{(t)} = 50$.

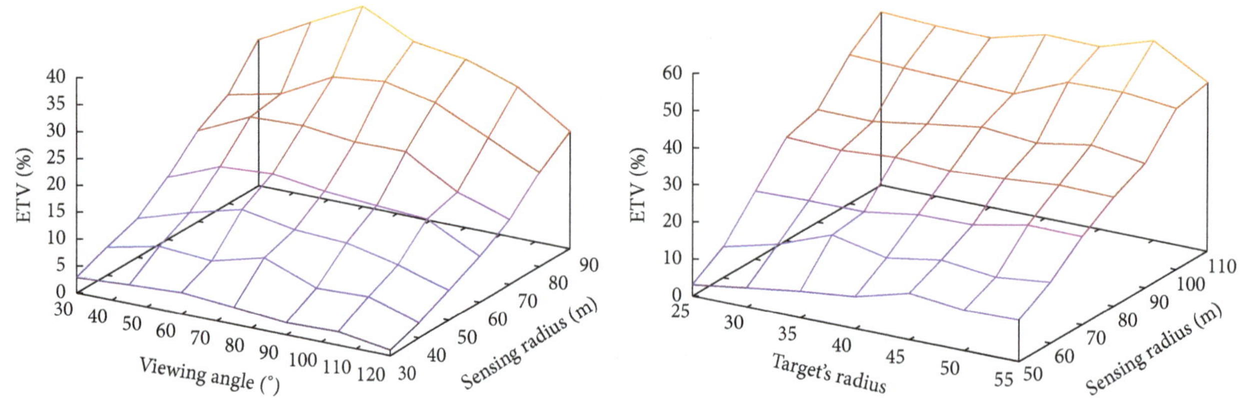

(a) (b)

FIGURE 9: ETV for large targets: (a) 20 targets with $R_{(t)} = 50$; (b) 5 targets with $R_{(t)} = 50$.

values for sensing angle (θ) of all visual sensors and sensing radius. As visual sensors are deployed with random orientations, every verification is also executed 10 times and only the average results are considered. The results for this verification are presented in Figure 8.

The value of ETV varies according to the parameters of visual sensors. In general, higher values for the sensing radius (R) of visual sensors will increase ETV for the considered deployment scenario, but higher angles may decrease ETV. In fact, for low values of R, the ETV was too low, since only targets that were closer to the border of the simulated road were viewed.

For this same scenario, more targets can be considered when assessing ETV. Figure 9(a) presents the results when 20 large targets have to be viewed. For more targets, the ETV is almost the same when also taking the same parameters, since the targets are being covered in the same way, in average. Moreover, larger targets may be harder to be completely viewed and thus the ETV may be lower.

At last, Figure 9(b) computes ETV for 20 targets with different sizes, assuming $\theta = 60^\circ$ for all visual sensors. In this verification, the value of ETV increases for higher values of R and $R_{(t)}$.

Sometimes, it may be desired to compute the lowest ETV(t) for a monitoring application, which will indicate the worst targets covering for all targets in the considered scenario. As ETV is an average value, it may hide the fact that some targets are being badly covered or even not covered at all. Figure 10 presents the computed ETV and ETV(t) for the monitoring scenario of Figure 7, with visual sensors deployed in two uniform columns and targets randomly positioned between those columns. For this evaluation, all visual sensors have $\theta = 60^\circ$ and $R = 120$ m, with random orientations (average results after 10 consecutive tests are considered).

Results in Figures 10(a) and 10(b) present ETV with similar values, indicating that in average the targets are being viewed with almost the same "quality," even for larger targets. However, when we consider the lowest achieved ETV(t),

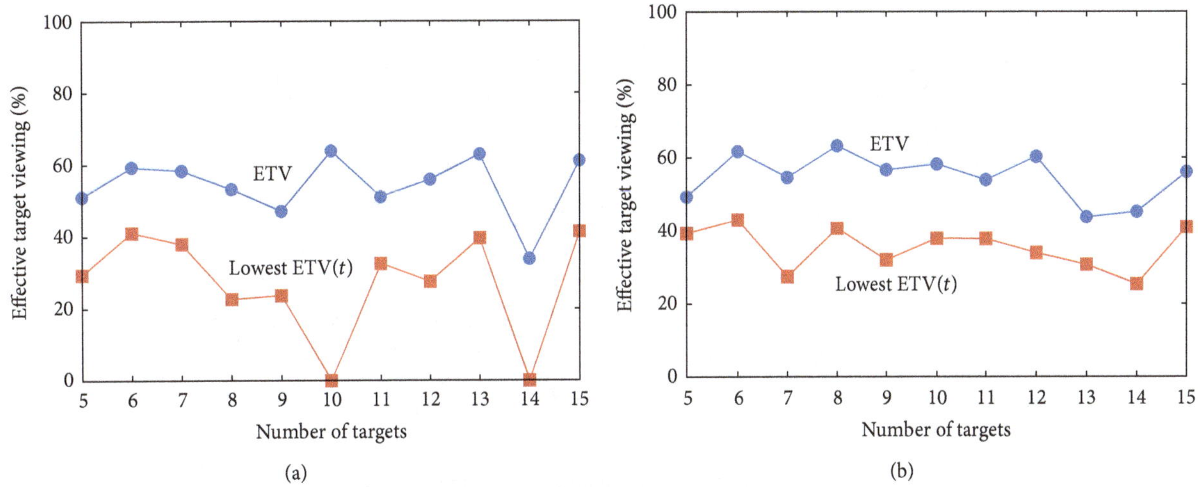

FIGURE 10: ETV and ETV(t). (a) Targets with $R_{(t)} = 20$. (b) Targets with $R_{(t)} = 50$.

TABLE 2: Availability requirements and attainable availability of some visual monitoring applications, for computed ETV and lowest ETV(t) in Figure 10(a).

Application	Targets	M-ETV	M-ETV(t)	ETV	Lowest ETV(t)	Available
1	5	40%	20%	51.10%	29.43%	Yes
2	8	40%	30%	53.26%	22.67%	No
3	10	50%	10%	63.89%	0.0%	No
4	10	50%	—	63.89%	0.0%	Yes
5	12	60%	20%	55.98%	27.55%	No
6	15	60%	30	61.14%	41.39%	Yes

results for smaller targets in Figure 10(a) show that at least one of the targets was not even covered by any of the visual sensors, which may not be acceptable for some applications.

Next subsection discusses how ETV and ETV(t) can be used when assessing availability.

5.2. Assessing Availability. In general, availability is a characteristic of the applications, instead of the networks. As different applications will have different requirements concerning visual coverage and dependability [6], any availability metric must account the characteristics of each visual monitoring application.

Considering the average results presented in Figure 10(a), availability requirements of a set of hypothetical visual monitoring applications were defined. We considered that such applications define values for M-ETV and, sometimes, for M-ETV(t) ("—" means it is not relevant for the application), directly indicating the minimal conditions for availability. The results are presented in Table 2, where an application is assumed as available when M-ETV ≤ ETV and M-ETV(t) ≤ Lowest ETV(t).

As can be seen in Table 2, network and targets configurations are not enough to determine the availability of a particular visual monitoring application, since its minimum expected level of targets coverage must be respected. And this is true even for the same network configurations, as it happens with Applications 3 and 4 in Table 2.

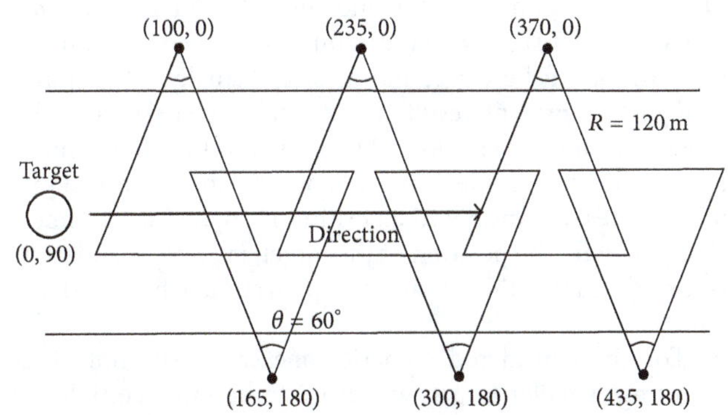

FIGURE 11: Monitoring scenario for a road with moving cars.

Availability was also assessed for a more practical application, considering targets that move through an area covered by fixed visual sensors. That scenario emulates visual monitoring over a road for moving cars, which may have different dimensions. Initially, that scenario is composed of six visual sensors deployed in two imaginary parallel lines, with three cameras positioned in each of these lines, as presented in Figure 11. For the performed verifications, all visual sensors have $R = 120$ m and $\theta = 60°$.

We consider that cars move only on one single direction, straightly from left to right in Figure 11, keeping in the center of the road. Three configurations of targets are considered for

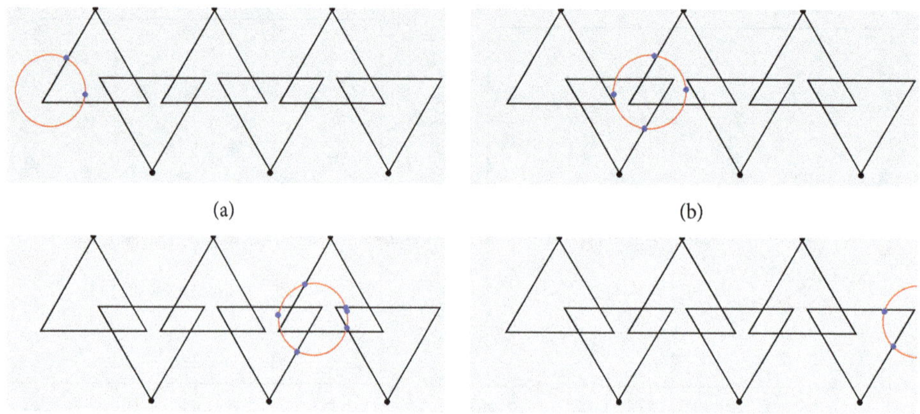

FIGURE 12: Cars moving through the considered scenario.

the tests: $R_{(t)} = 20$ m, $R_{(t)} = 40$ m, and $R_{(t)} = 70$ m. For this verification, ETV is computed for different "instants" of movement, which means that ETV is computed according to predefined positions. Figure 12 graphically presents an example of a target with $R_{(t)} = 40$ m, which is considered in fixed positions for ETV computing, in different instants. One should note that cars move only on x-axis.

The computed values for ETV are presented in Figure 13, for a single target that moves 500 m from left to right. Assuming a coordinates system where position $(0, 0)$ is at the top left corner of the road, targets move from position $(0, 90)$ to position $(500, 90)$, and the value for ETV in this scenario, with fixed cameras carefully positioned, depends on the position of the target and its size. Actually, all graphics in Figure 13 present results for the same scenario and the same movement behaviour, but only varying the number of instants of measures. In other words, for more instants of measures, the proposed algorithm is applied more times, changing only the position of the target, $(x_{(t)}, y_{(t)})$. At last, it is considered that the monitoring application defines M-ETV = 45%.

For the considered scenario, one can easily note that in average smaller targets are easier to be more completely covered by visual sensors, resulting in higher ETV. Another important conclusion is that the application will not be available when targets are in some positions, since the computed ETV will be lower than the defined M-ETV (45%). It is also interesting to note that for the smallest target it may sometimes have the lowest ETV for the experiments, because it "falls" in areas with low coverage, and that is harder to happen for larger targets.

The proposed algorithm to compute ETV is significant because it allows the identification of parts of the network with poor coverage, which may lead to states of unavailability. And this information may be exploited to change configurations of the network, for example, rotating cameras or deploying more visual sensors. In order to test this possibility, we extended the monitoring scenario in Figure 11, deploying four additional cameras, as depicted in Figure 14.

The ETV was recomputed for this new scenario, as presented in Figure 15, considering ETV computation for movement of the target after 10 m (Figure 15(a)) and 1 m (Figure 15(b)).

In general, ETV was improved for three tested sizes of the target, especially for larger targets. Actually, for $R_{(t)} = 70$ m, the application was always unavailable for the scenario in Figure 11. However, when considering the scenario with 10 visual sensors in Figure 14, application monitoring the largest target was available when it is between 125 m and 400 m.

With the performed verifications, the ETV of the defined scenarios could be assessed. Using the proposed mathematical formulations, one can estimate the way targets will be covered, which can be considered to adjust the deployed visual sensors or even trigger new deployments. We expect that this methodology can bring valuable results for wireless visual sensor network deployment, configuration, and operation.

5.3. Availability and Communication in WVSN. Availability in wireless visual sensor networks is strongly related to communication issues. Actually, the level of availability indicates how well a deployed network is retrieving data according to the monitoring requirements of the considered application, and thus states of unavailability may indicate that something is wrong or not operating as expected. And the causes of such "problems" are diverse.

A transient fault in wireless visual sensor networks will directly impact packet transmission, requiring proper mechanisms to assure some level of reliability. On the other hand, permanent faults may render part or the entire network unavailable, when the visual coverage area is reduced. In fact, if transmission paths are facing long periods of congestion, the network may become unavailable, even if enough targets are being properly viewed (packets are not being received at the sink side). High packet error rates may also impact the overall availability level of WVSN. Therefore, availability is a broader concept that comprises different levels of hardware and coverage failures [6], including communication issues.

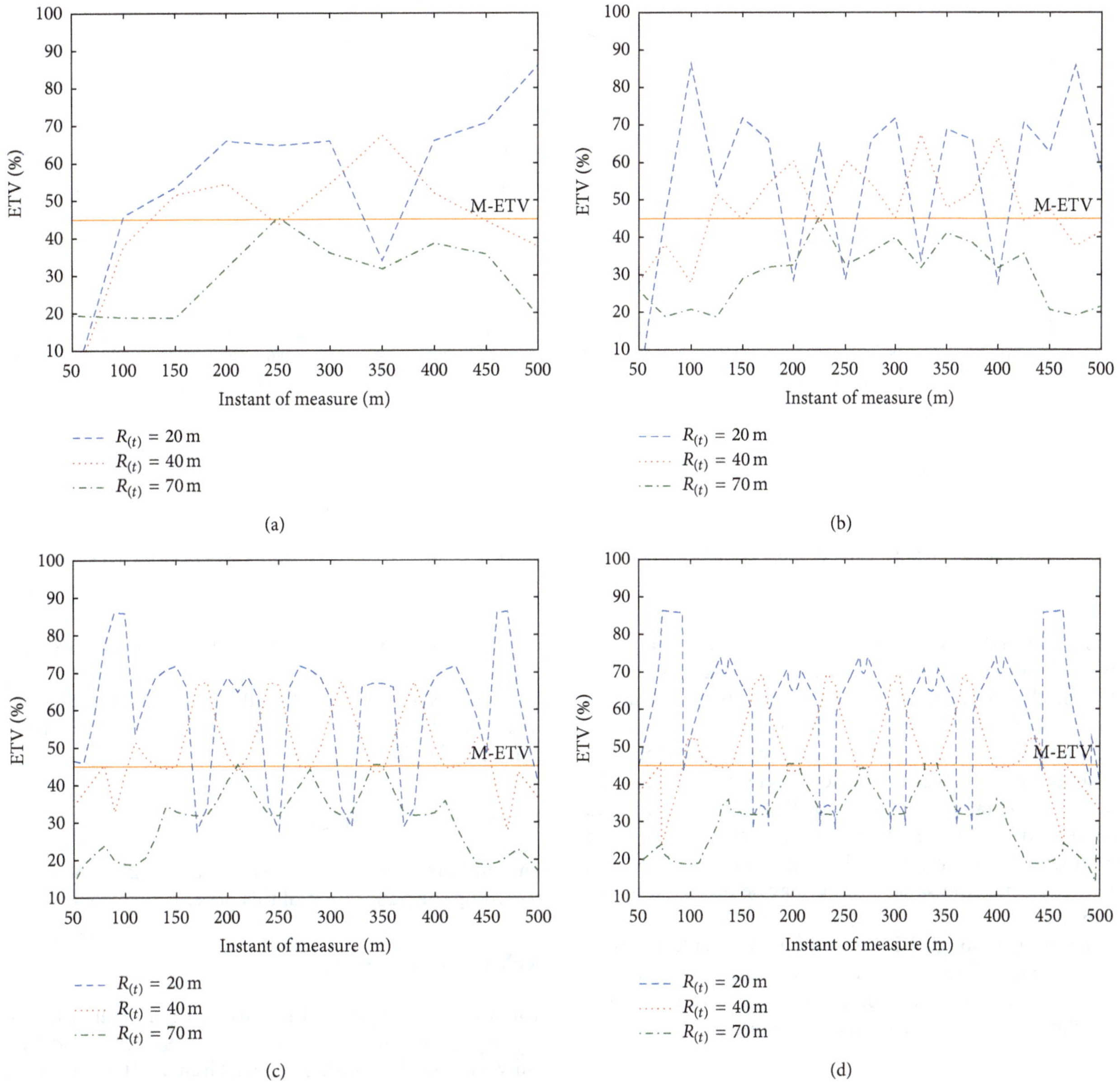

FIGURE 13: ETV when targets move through the covered area. (a) Targets move 50 m for each measure. (b) Targets move 25 m for each measure. (c) Targets move 10 m for each measure. (d) Targets move 1 m for each measure.

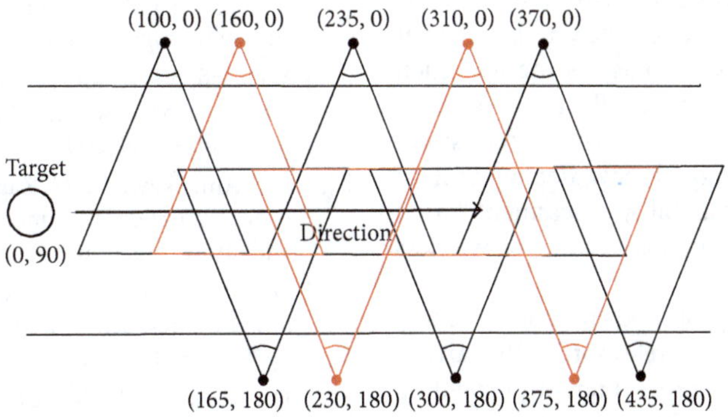

FIGURE 14: Monitoring scenario with the addition of four new visual sensors.

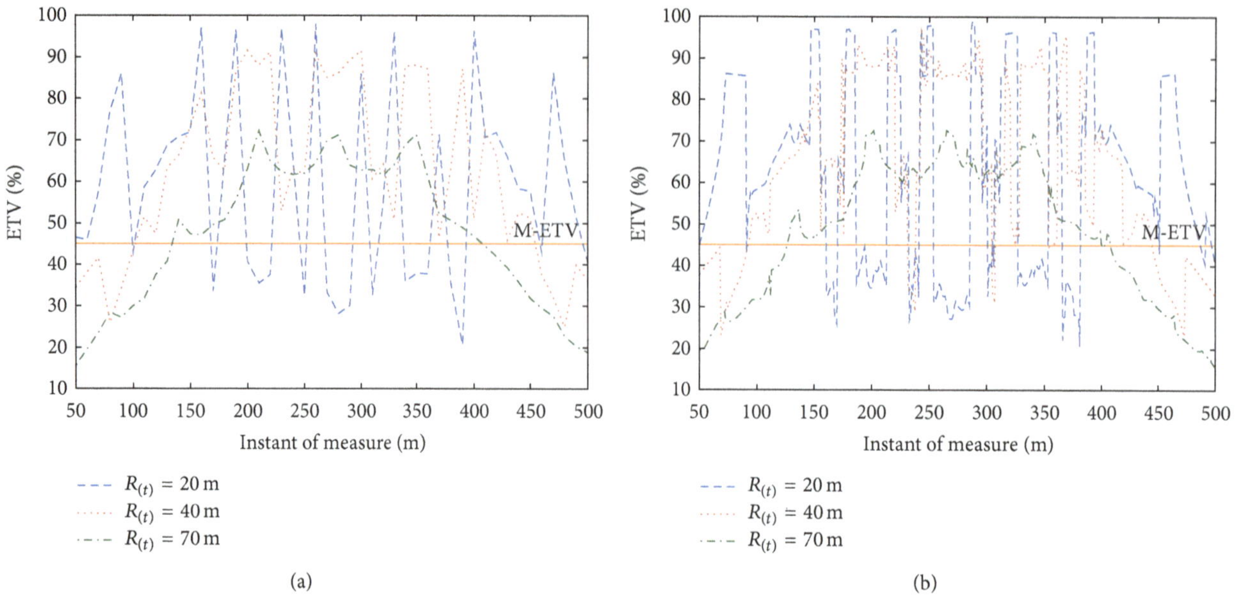

FIGURE 15: ETV after deployment of four more visual sensors. (a) Targets move 10 m for each measure. (b) Targets move 1 m for each measure.

The proposed Effective Target Viewing is a relevant metric to assess how well targets' perimeters are being viewed. But ETV should also be considered along with other parameters to more completely measure the availability of wireless visual sensor networks applications. As connectivity problems may result in hardware failures that disconnect visual sensors, the ETV may be dynamically affected by the network condition: disconnected visual sensors may be not considered when computing ETV. And thus the ETV may even be used as a QoS metric, since the value of ETV may be impacted by the network.

Therefore, although ETV is computed considering only visual sensing parameters, sensors communication may also have a relevant role when computing ETV and enhancing availability in wireless visual sensor networks.

6. Conclusions

Target monitoring in wireless visual sensor networks is a relevant research topic that still presents some relevant challenging issues, fostering investigation in this area. As targets may have different forms and sizes, it is relevant to define mathematical mechanisms to assess the way such targets will be viewed, which can then affect real WVSN. For example, a low value of ETV may trigger reposition of rotatable cameras or even suggest new deployment of visual sensors. In either way, availability assessment based on targets' perimeters can bring valuable results for wireless visual sensor networks.

As target size is central in the proposed approach, the way targets will be modelled is extremely relevant. In this article we considered circumferences to represent targets, providing a feasible and computationally viable solution. However, as future works, we will make more realistic modelling, considering convex polygons and grid of lines to represent targets, which may bring more realistic results. Moreover, real snapshots will be considered as a reference to identify the borders of the targets, allowing even more complex mathematical models. At last, 3D modelling will be also considered in future works.

Competing Interests

The authors declare that there is no conflict of interests regarding the publication of this paper.

Acknowledgments

This work was supported in part by the Brazilian Research Agency CNPq under Grant no. 441459/2014-5 and by the University of the Bío-Bío, under Grants DIUBB 161610 2/R and GI 160210/EF.

References

[1] J. Yick, B. Mukherjee, and D. Ghosal, "Wireless sensor network survey," *Computer Networks*, vol. 52, no. 12, pp. 2292–2330, 2008.

[2] Y. Charfi, N. Wakamiya, and M. Murata, "Challenging issues in visual sensor networks," *IEEE Wireless Communications*, vol. 16, no. 2, pp. 44–49, 2009.

[3] I. T. Almalkawi, M. G. Zapata, J. N. Al-Karaki, and J. Morillo-Pozo, "Wireless multimedia sensor networks: current trends and future directions," *Sensors*, vol. 10, no. 7, pp. 6662–6717, 2010.

[4] D. G. Costa and L. A. Guedes, "The coverage problem in video-based wireless sensor networks: a survey," *Sensors*, vol. 10, no. 9, pp. 8215–8247, 2010.

[5] D. G. Costa, L. A. Guedes, F. Vasques, and P. Portugal, "Adaptive monitoring relevance in camera networks for critical surveillance applications," *International Journal of Distributed Sensor Networks*, vol. 2013, Article ID 836721, 14 pages, 2013.

[6] D. G. Costa, I. Silva, L. A. Guedes, F. Vasques, and P. Portugal, "Availability issues in wireless visual sensor networks," *Sensors*, vol. 14, no. 2, pp. 2795–2821, 2014.

[7] D. Pescaru, V. Gui, C. Toma, and D. Fuiorea, "Analysis of post-deployment sensing coverage for video wireless sensor networks," in *Proceedings of the 6th Roedunet International Conference (RoEduNet '07)*, Craiova, Romania, November 2007.

[8] A. Mavrinac and X. Chen, "Modeling coverage in camera networks: a survey," *International Journal of Computer Vision*, vol. 101, no. 1, pp. 205–226, 2013.

[9] J. Ai and A. A. Abouzeid, "Coverage by directional sensors in randomly deployed wireless sensor networks," *Journal of Combinatorial Optimization*, vol. 11, no. 1, pp. 21–41, 2006.

[10] D. Costa, I. Silva, L. Guedes, F. Vasques, and P. Portugal, "Availability assessment of wireless visual sensor networks for target coverage," in *Proceedings of the IEEE International Conference on Emerging Technologies in Factory Automation (ETFA '14)*, pp. 1–8, Barcelona, Spain, 2014.

[11] D. Costa, I. Silva, L. Guedes, F. Vasques, and P. Portugal, "Selecting redundant nodes when addressing availability in wireless visual sensor networks," in *Proceedings of the IEEE International Conference on Industrial Informatics*, pp. 424–448, Porto Alegre, Brazil, July 2014.

[12] D. G. Costa, I. Silva, L. A. Guedes, P. Portugal, and F. Vasques, "Enhancing redundancy in wireless visual sensor networks for target coverage," in *Proceedings of the 20th Brazilian Symposium on Multimedia and the Web (WebMedia '14)*, pp. 31–38, November 2014.

[13] M. Younis and K. Akkaya, "Strategies and techniques for node placement in wireless sensor networks: a survey," *Ad Hoc Networks*, vol. 6, no. 4, pp. 621–655, 2008.

[14] Y. E. Osais, M. St-Hilaire, and F. R. Yu, "Directional sensor placement with optimal sensing range, field of view and orientation," *Mobile Networks and Applications*, vol. 15, no. 2, pp. 216–225, 2010.

[15] X. Sun, Y. Zhang, X. Ren, and K. Chen, "Optimization deployment of wireless sensor networks based on culture—ant colony algorithm," *Applied Mathematics and Computation*, vol. 250, pp. 58–70, 2015.

[16] H.-H. Yen, "Optimization-based visual sensor deployment algorithm in PTZ wireless visual sensor networks," in *Proceedings of the 7th International Conference on Ubiquitous and Future Networks (ICUFN '15)*, pp. 734–739, IEEE, Sapporo, Japan, July 2015.

[17] T. Sauter, "Energy-efficient coverage problems in wireless ad hoc sensor networks," *Computer Communications*, vol. 29, no. 4, pp. 413–420, 2006.

[18] L. Liu, H. Ma, and X. Zhang, "On directional K-coverage analysis of randomly deployed camera sensor networks," in *Proceedings of the IEEE International Conference on Communications (ICC '08)*, pp. 2707–2711, Beijing, China, May 2008.

[19] M. Alaei and J. M. Barcelo-Ordinas, "Node clustering based on overlapping FoVs for wireless multimedia sensor networks," in *Proceedings of the IEEE Wireless Communications and Networking Conference (WCNC '10)*, pp. 1–6, Sydney, Australia, April 2010.

[20] M. Rahimi, S. Ahmadian, D. Zats, R. Laufer, and D. Estrin, "Magic numbers in networks of wireless image sensors," in *Proceedings of the Workshop on Distributed Smart Cameras (DSC '06)*, Boulder, Colo, USA, October 2006.

[21] D. Devarajan, R. J. Radke, and H. Chung, "Distributed metric calibration of Ad hoc camera networks," *ACM Transactions on Sensor Networks*, vol. 2, no. 3, pp. 380–403, 2006.

[22] Y. Cai, W. Lou, M. Li, and X.-Y. Li, "Target-oriented scheduling in directional sensor networks," in *Proceedings of the 26th IEEE International Conference on Computer Communications (INFOCOM '07)*, pp. 1550–1558, Barcelona, Spain, May 2007.

[23] D. G. Costa, I. Silva, L. A. Guedes, F. Vasques, and P. Portugal, "Selecting redundant nodes when addressing availability in wireless visual sensor networks," in *Proceedings of the 12th IEEE International Conference on Industrial Informatics (INDIN '14)*, pp. 130–135, IEEE, Porto Alegre, Brazil, July 2014.

[24] D. G. Costa, I. Silva, L. A. Guedes, F. Vasques, and P. Portugal, "Optimal sensing redundancy for multiple perspectives of targets in wireless visual sensor networks," in *Proceedings of the 13th International Conference on Industrial Informatics (INDIN '15)*, pp. 185–190, Cambridge, UK, July 2015.

[25] D. G. Costa, I. Silva, L. A. Guedes, P. Portugal, and F. Vasques, "Availability assessment of wireless visual sensor networks for target coverage," in *Proceedings of the 19th IEEE International Conference on Emerging Technologies and Factory Automation (ETFA '14)*, pp. 1–8, Barcelona, Spain, September 2014.

[26] C. Duran-Faundez, D. G. Costa, V. Lecuire, and F. Vasques, "A geometrical approach to compute source prioritization based on target viewing in wireless visual sensor networks," in *Proceedings of the IEEE World Conference on Factory Communication Systems (WFCS '16)*, pp. 1–7, IEEE, Aveiro, Portugal, May 2016.

[27] K.-S. Hung and K.-S. Lui, "Perimeter coverage scheduling in wireless sensor networks using sensors with a single continuous cover range," *EURASIP Journal on Wireless Communications and Networking*, vol. 2010, Article ID 926075, 17 pages, 2010.

[28] K.-S. Hung and K.-S. Lui, "Perimeter coverage made practical in wireless sensor networks," in *Proceedings of the 9th International Symposium on Communications and Information Technology (ISCIT '09)*, pp. 87–92, Icheon, South Korea, September 2009.

[29] I. Silva, L. A. Guedes, P. Portugal, and F. Vasques, "Reliability and availability evaluation of wireless sensor networks for industrial applications," *Sensors*, vol. 12, no. 1, pp. 806–838, 2012.

[30] X. Bai, L. Ding, J. Teng, S. Chellappan, C. Xu, and D. Xuan, "Directed coverage in wireless sensor networks: concept and quality," in *Proceedings of the IEEE 6th International Conference on Mobile Adhoc and Sensor Systems (MASS '09)*, pp. 476–485, Macau, China, October 2009.

[31] A. Neishaboori, A. Saeed, K. A. Harras, and A. Mohamed, "On target coverage in mobile visual sensor networks," in *Proceedings of the 12th ACM International Symposium on Mobility Management and Wireless Access (MobiWac '14)*, pp. 39–46, ACM, Montreal, Canada, September 2014.

Permissions

All chapters in this book were first published in JECE, by Hindawi Publishing Corporation; hereby published with permission under the Creative Commons Attribution License or equivalent. Every chapter published in this book has been scrutinized by our experts. Their significance has been extensively debated. The topics covered herein carry significant findings which will fuel the growth of the discipline. They may even be implemented as practical applications or may be referred to as a beginning point for another development.

The contributors of this book come from diverse backgrounds, making this book a truly international effort. This book will bring forth new frontiers with its revolutionizing research information and detailed analysis of the nascent developments around the world.

We would like to thank all the contributing authors for lending their expertise to make the book truly unique. They have played a crucial role in the development of this book. Without their invaluable contributions this book wouldn't have been possible. They have made vital efforts to compile up to date information on the varied aspects of this subject to make this book a valuable addition to the collection of many professionals and students.

This book was conceptualized with the vision of imparting up-to-date information and advanced data in this field. To ensure the same, a matchless editorial board was set up. Every individual on the board went through rigorous rounds of assessment to prove their worth. After which they invested a large part of their time researching and compiling the most relevant data for our readers.

The editorial board has been involved in producing this book since its inception. They have spent rigorous hours researching and exploring the diverse topics which have resulted in the successful publishing of this book. They have passed on their knowledge of decades through this book. To expedite this challenging task, the publisher supported the team at every step. A small team of assistant editors was also appointed to further simplify the editing procedure and attain best results for the readers.

Apart from the editorial board, the designing team has also invested a significant amount of their time in understanding the subject and creating the most relevant covers. They scrutinized every image to scout for the most suitable representation of the subject and create an appropriate cover for the book.

The publishing team has been an ardent support to the editorial, designing and production team. Their endless efforts to recruit the best for this project, has resulted in the accomplishment of this book. They are a veteran in the field of academics and their pool of knowledge is as vast as their experience in printing. Their expertise and guidance has proved useful at every step. Their uncompromising quality standards have made this book an exceptional effort. Their encouragement from time to time has been an inspiration for everyone.

The publisher and the editorial board hope that this book will prove to be a valuable piece of knowledge for researchers, students, practitioners and scholars across the globe.

List of Contributors

Leandro Aureliano da Silva, Gilberto Arantes Carrijo and Eduardo Silva Vasconcelos
Department of Electrical Engineering, Universidade Federal de Uberlândia, Av. João Naves dé Avila, 2160 Bloco 3N, Campus Santa Mônica, Uberlândia, MG, Brazil

Roberto Duarte Campos and Cleiton Silvano Goulart
Department of Electrical Engineering, Faculdade de Talentos Humanos, R. Manoel Gonçalves de Rezende, 230 São Cristóvão, Uberaba, MG, Brazil

Rodrigo Pinto Lemos
Department of Electrical Engineering, Universidade Federal de Goiás, Av. Esperança, s/n. Campus Universitário, Goiânia, GO, Brazil

Rui Du, Yangyu Fan and Jianshu Wang
School of Electronics and Information, Northwestern Polytechnical University, Xi'an, Shaanxi 710072, China

Liquan Zhao and Yulong Liu
College of Information Engineering, Northeast Electric Power University, Jilin 132012, China

Markus Ulmschneider, Christian Gentner, Thomas Jost and Armin Dammann
German Aerospace Center (DLR), Institute of Communications and Navigation, Muenchner Str. 20, 82334Wessling, Germany

Yongxin Feng, Shunchao Fei, Fang Liu and Bo Qian
Communication and Network Institute, Shenyang Ligong University, Shenyang 110159, China

Bo Qian, Guolei Zheng and Yongxin Feng
School of Information Science and Engineering, Shenyang Ligong University, Shenyang 110159, China

Shexiang Ma, Jie Wang, Xin Meng and Junfeng Wang
School of Electrical and Electronic Engineering, Tianjin University of Technology, Tianjin, China

Lei Lei and She Kun
Laboratory of Cyberspace, School of Information and Software Engineering, University of Electronic Science and Technology of China, Chengdu 610054, China

Betül Yılmaz, Serhat Gökkan and Caner Özdemir
Department of Electrical and Electronics Engineering, Mersin University, Ciftlikkoy 33343, Mersin, Turkey

Weiming Cai and Qingke Qi
School of Information Science and Engineering, Ningbo Institute of Technology, Zhejiang University, Ningbo 315100, China

Xiushan Wu, Yanzhi Wang, Siguang An and Jianqiang Han
College of Mechanical and Electrical Engineering, China Jiliang University, Hangzhou 310018, China

Ling Sun
Jiangsu Key Laboratory of ASIC Design, Nantong University, Nantong, China

Meng Hou and Xiao Liu
School of Electrical Engineering and Automation, Qilu University of Technology, Jinan 250353, China

Yuan Xu
The School of Electrical Engineering, University of Jinan, Jinan 250022, China

Yong Wang, Fu Xu, Zhibo Chen, Yu Sun and Haiyan Zhang
School of Information Science and Technology, Beijing Forestry University, Beijing 100083, China

Mengdi Wang, Anrong Xue and Huanhuan Xia
School of Computer Science and Communication Engineering, Jiangsu University, Zhenjiang 212013, China

Karen Uribe-Murcia, Yuriy S. Shmaliy and Jose A. Andrade-Lucio
Department of Electronics Engineering, Universidad de Guanajuato, 36885 Salamanca, GTO, Mexico

Chen Cai and Yu-mei Zhou
Institute of Microelectronics of Chinese Academy of Sciences, Beitucheng West Road, Chaoyang District, Beijing 100029, China
University of Chinese Academy of Sciences, Yuquan Road, Shijingshan District, Beijing 100049, China

Jian-zhong Zhao
University of Chinese Academy of Sciences, Yuquan Road, Shijingshan District, Beijing 100049, China

Mustafa M. Al Khabbaz
Facilities Planning Department, Saudi Arabian Oil Company (Saudi Aramco), Dhahran 31311, Saudi Arabia

Mohamed A. Abido
Electrical Engineering Department, King Fahd University of Petroleum and Minerals, Dhahran 31261, Saudi Arabia

Zhenxing Zhang and Feng Gao
College of Automation, Harbin Engineering University, Harbin 150001, China

Bin Ma
China Ship Development and Design Center, Wuhan, China

Zhiqiang Zhang
92730 Army, Sanya 572016, China

Zhao Liquan and Hu Yunfeng
College of Information Engineering, Northeast Electric Power University, Jilin 132012, China

Liming Zhou and Lu Chen
School of Computer and Information Engineering, Henan University, Kaifeng 475004, China

Yingzi Shan
Department of Finance, Yellow River Conservancy Technical Institute, Kaifeng 475004, China

Daniel G. Costa
State University of Feira de Santana, Feira de Santana, BA, Brazil

Cristian Duran-Faundez
University of the Bío-Bío, Concepción, Chile

Index